B/N#

ADVANCES IN
ELECTRON TRANSFER
CHEMISTRY

Volume 3 • 1993

ADVANCES IN ELECTRON TRANSFER CHEMISTRY

Editor: PATRICK S. MARIANO
Department of Chemistry and Biochemistry
University of Maryland-College Park

VOLUME 3 • 1993

 JAI PRESS INC.

Greenwich, Connecticut   *London, England*

NO ANAL

QL
173, 25
A38
V. 3
CHEM

CONTENTS

LIST OF CONTRIBUTORS

Mohammad S. Farahat

Department of Chemistry
Boston University
Boston, Massachusetts

F. Peter Guengerich

Department of Biochemistry and
  Center in Molecular Toxicology
Vanderbilt University School of Medicine
Nashville, Tennessee

Jean-Marie Herrmann

Unite de Recherche Associée au CNRS
Ecole Centrale de Lyon
Lyon, France

Guilford Jones II

Department of Chemistry
Boston University
Boston, Massachusetts

Darren Lawless

Department of Chemistry and Biochemistry
Concordia University
Montreal, Quebec, Canada

Timothy L. Macdonald

Department of Chemistry
University of Virginia
Charlottesville, Virginia

Stephen F. Nelsen

Department of Chemistry
University of Wisconsin
Madison, Wisconsin

Nick Serpone

Department of Chemistry and Biochemistry
Concordia University
Montreal, Quebec, Canada

Rita Terzian

Department of Chemistry and Biochemistry
Concordia University
Montreal, Quebec, Canada

PREFACE

The consideration of reaction mechanisms involving the movement of single electrons is now becoming quite common in the fields of chemistry and biochemistry. Studies conducted in recent years have uncovered a large number of chemical and enzymatic processes that proceed via single electron transfer pathways. Still numerous investigations are underway probing the operation of electron transfer reactions in organic, organometallic, biochemical, and excited state systems. In addition, theoretical and experimental studies are being conducted to gain information about the factors that govern the rates of single electron transfer. It is clear that electron transfer chemistry is now one of the most active areas of chemical study.

The series, *Advances in Electron Transfer Chemistry*, has been designed to allow scientists who are developing new knowledge in this rapidly expanding area to describe their most recent research findings. Each contribution is in a minireview format focusing on the individual author's own work as well as the studies of others that address related problems. Hopefully, *Advances in Electron Transfer Chemistry* will serve as a useful series for those interested in learning about current breakthroughs in this rapidly expanding area of chemical research.

Patrick S. Mariano
Series Editor

PHOTOINDUCED ELECTRON TRANSFER IN FLEXIBLE BIARYL DONOR-ACCEPTOR MOLECULES

Guilford Jones II and Mohammad S. Farahat

Advances in Electron Transfer Chemistry
Volume 3, pages 1–32.
Copyright © 1993 by JAI Press Inc.
All rights of reproduction in any form reserved.
ISBN: 1-55938-320-8

1. INTRODUCTION

Electron transfer (ET) initiated by the absorption of light plays an important role in many chemical processes. Photoinduced electron transfer (PET) is the starting point in the utilization of light by nature to harvest the energy of the sun and to convert it into a form essential to the maintenance of photosynthetic biochemical pathways.[1] Our understanding of the early events associated with light-induced ET has expanded dramatically over the last decade. The original theories of ET developed by Marcus,[2] Hush,[3] and others[4,5] which were introduced in the 1950s and 1960s, have since been expanded and modified,[6] in part due to concurrent progress on experimental techniques. Spectroscopic techniques with time resolution ranging from nanosecond (nsec) to femtosecond (fsec) regimes are used routinely now in many laboratories for the study of the kinetics of fast processes.[7,8]

In recent investigations, special focus has been placed on ET that occurs within a single molecule. Synthetic efforts have provided an exceptionally broad range of architectures for testing concepts associated with predicting the rates and yields of intramolecular electron transfer (IET). Linked electron donor-acceptor (DA) systems provide structural motifs for revealing the dependences of ET on distance, geometry, symmetry, and charge. Intramolecular photochemical events are also particularly relevant as models of behavior in biological systems (e.g., a protein matrix) and important to the development of devices that "process" electrons or charge at the molecular level.

The present review will focus on a single subset of DA structures, those in which electroactive groups, D and A, are directly linked through a single bond. Charge transport across such a bond can be surprisingly fast or deceptively slow. The dynamics will turn out to depend dramatically on the extent to which linked aromatic rings associated with D and A are fused with twisted geometries. The linkage of biaryls is now appreciated as a classic example of steric control of torsional barriers that result from nonbonded interactions between aromatic substituent groups. The interactions among hydrogens in *ortho* positions for the prototype, biphenyl, are illustrated in Scheme 1. A variety of theoretical and experimental probes have established the barrier for rotation about the central bond in biphenyl (for the ground electronic state) to lie at approximately 4.0 kcal/mol.[9]

As aromatic rings are embellished in a DA structure, barriers to internal rotation are altered sharply. For binaphthyl, conformational isomerism is possible due to a barrier sufficiently high to inhibit rotation completely at room temperature.[10,11] The resolution of mirror-image rotamer forms, and the observation of optical activity for binaphthyls, is well known. Relevant to discussions below is the 9–phenylanthracene (9-PA) prototype (Scheme 1) for which a ground state equilibrium geometry having a twist angle of 83° has been determined.[12] A barrier to rotation of one ring of 11 kcal/mol for this system in the electronic ground state has also been computed, a value that is to be compared with 6.6 kcal/mole estimated for the first singlet excited state of 9-PA.[12] The central feature of these systems in the

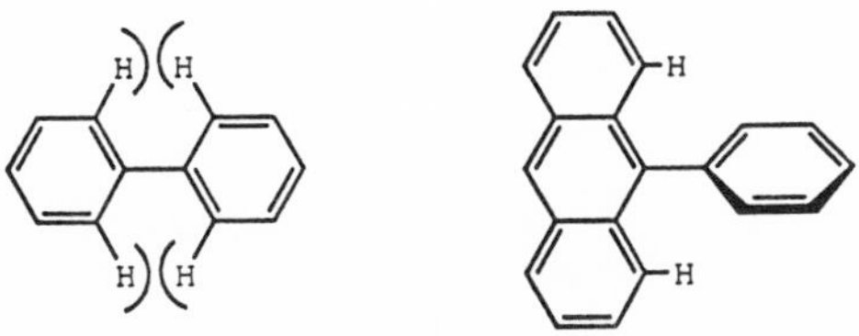

Scheme 1.

present context is the role that steric repulsion to rotation plays in "tuning" the amount of orbital overlap that can take place for linked aryl systems. The enforcement of large twist angles during the lifetime of an excited species involved in a PET is pivotal to the dynamics of charge separation and recombination. The present discussion will focus on the structural constraints that tend to favor perpendicular arrangements of linked aromatic rings and the complex role that the medium plays in stabilization of photoinduced charge.

2. INTRAMOLECULAR PHOTOINDUCED ELECTRON TRANSFER

2.1. Artificial and Bridged Systems

An area of great interest in recent years has involved the study of artificial photosynthetic model systems.[13,14] Such systems typically consist of one or more porphyrin moieties covalently attached to various other electron donors and acceptors (D and A), selected on the basis of their redox properties as well as structure. These assemblies allow systematic study of various aspects of ET that may occur between segments of the molecule (distance and orientation of D and A, and the number and types of linkage bonds). One of the main goals for studying these systems is to produce two radical sites at the opposite ends of the long molecule through several intramolecular ET steps. This feature reduces interactions between the two radical sites and results in a long-lived radical pair state, which is equivalent to storing part of the initial light energy in the form of chemical potential. Such a radical pair state in a natural system (Photosystem II) is estimated to have a lifetime on the order of 1 msec.[13]

Other studies have involved bridged D and A systems [Donor-Spacer-Acceptor, (DSA)] using various types of flexible and rigid, aromatic, saturated, and unsaturated aliphatic linkages.[15] Molecules of this type have been used to study the effect of energetics,[16] distance ,[17] medium, and the involvement of the linker groups[18] on PET. It has become evident, based on such studies, that the relative orientation of the D and A groups also plays an important role in controlling the ET process.[19,20]

In addition, symmetry considerations may be used to separate more clearly the factors influencing ET rates.[21]

Studies of elaborate artificial photosynthetic models, as well as the simpler DSA systems, have followed many years of research on inter- and intramolecular ET reactions. Before the widespread availability of instrumentation for fast kinetics measurements, indirect methods such as fluorescence quenching[22] or the determination of photoreaction quantum yields[23] were employed to obtain kinetic parameters for bimolecular processes.

Despite many advances in the area of inter- and intramolecular PET, there is still a need to better understand the crucial first step in electron transfer. Ultrafast (picosecond, femtosecond) spectroscopy allows one to study this process in the presence or absence of other secondary factors which are often employed to facilitate kinetic measurements in a slower time regime (e.g., low temperature, viscous media, hydrogen bonding media, etc.). Two areas of interest which utilize the potential of ultrafast spectroscopy involve: (1) the mechanistic details of "adiabatic" ET in solution; and (2) the dynamics of solvent motion accompanying charge transfer (CT). These two areas are in fact closely related as shown by recent theoretical[24] and experimental[25] results.

2.2 Theory of Electron Transfer : A Brief Review

Marcus[2,24a] treated electron transfer in solution classically and provided an expression for the rate constant for ET based on the transition state theory. In this theory, outer-sphere bimolecular ET is described, although the rate constant calculated in this manner is a unimolecular one. This feature is based on the assumption that the D and A involved in ET form a precursor complex with some equilibrium geometry Q_A as represented by potential energy curve R (Figure 1). Electron transfer takes place at the transition state, Q_B, before the system reaches a new equilibrium geometry for the successor complex, Q_C. The parabolic curves, R and P, in Figure 1 represent the energies of the ensemble of the donor, the acceptor, and the surrounding medium before and after electron transfer. The rate constant for ET is given by Eq. 1.[24a]

$$k_{et} = \kappa \nu \exp\left[-\frac{(\lambda + \Delta G^\circ)^2}{4\lambda kT}\right] \tag{1}$$

In this expression, κ is the transmission coefficient (electronic factor), ν is the collision frequency, and ΔG° is the free energy of reaction which includes work terms for bringing the reactants and products to their mean separation distance for ET. The reorganization energy λ introduces the effect of molecular vibrations (change in internal nuclear coordinates; λ_i) and solvent interactions (λ_o) based on the dielectric continuum model.

$$\lambda = \lambda_i + \lambda_o \tag{2}$$

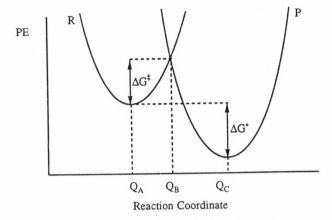

Figure 1. Potential energy curves representing the reactants (R) and products (P) in an electron transfer process.

$$\lambda_i = \sum_j \left[\frac{f_j^r f_j^p}{f_j^r + f_j^p} \right] (\Delta q_j)^2 \tag{3}$$

$$\lambda_o = (\Delta e)^2 \left[\frac{1}{2a_1} + \frac{1}{2a_2} - \frac{1}{r} \right] \left[\frac{1}{\varepsilon_{op}} - \frac{1}{\varepsilon_s} \right] \tag{4}$$

In Eq. 3, f_j^r and f_j^p are the jth normal mode force constants in the reactant and product, respectively, and Δq is the change in the jth equilibrium normal coordinate. The factor Δe is the unit of charge transferred from the donor to the acceptor, a_1 and a_2 are the radii of spheres representing them, and r is the center-to-center distance for separation of D and A. The parameters, ε_{op} and ε_s, are the optical (ε_{op} = n^2) and static dielectric constants of the solvent. An important consequence of this formalism is that a maximum value of k_{et} is reached when $\lambda = -\Delta G°$. At values of $-\Delta G° < \lambda$, k_{et} increases with an increase in driving force (increase in $-\Delta G°$); whereas, when $-\Delta G° > \lambda$, k_{et} decreases with increasing $-\Delta G°$. These two regions have been termed the normal and the inverted region, respectively.[24a]

In a quantum mechanical treatment of ET, the rate constant is related to the extent of electronic coupling of the reactants via H_{AB} (electronic coupling matrix element) and the degree of overlap of the reactant and product vibrational wavefunctions via FC (Franck Condon factor).[24a]

$$k_{et} = \frac{2\pi}{\hbar} H_{AB}^2 \, (FC) \tag{5}$$

The FC term in Eq. 5 takes into consideration both the solute and the solvent wavefunctions and is weighted according to the thermal populations of the vibrational levels involved. This formalism, although in general more difficult to apply, is more flexible in terms of introduction of various factors which may affect ET in a non-straightforward way, such as distance, orientation, involvement of spacer groups (if present), and the role of solvent.[26]

The solvent effect on ET reactions has been considered theoretically by a number of researchers.[6,24,27] In an adiabatic electron transfer reaction, the rate constant may be described by Eq. 1 where $\kappa = 1$ and the frequency factor ν may be related to the longitudinal relaxation time τ_L.[6a]

$$\nu = \left[\frac{E_r}{16\pi kT} \right]^{1/2} \frac{1}{\tau_L} \tag{6}$$

$$\tau_L = \frac{\varepsilon_\infty}{\varepsilon_0} \tau_D \tag{7}$$

$$\tau_D = \frac{V\eta}{kT} \tag{8}$$

where E_r is the solvent reorganization energy (λ_o in equation 4), ε_∞ and ε_0 are the optical and static dielectric constants of the solvent, and τ_D is the Debye relaxation time determined from the viscosity (η) and molecular volume (V) of the solvent at the absolute temperature T. Equation 1 may be rewritten for an adiabatic ET process using Eqs. 6–8, when the solvent is treated as a dielectric continuum.

$$k_{et} = \frac{C}{\tau_L} \exp \left[\frac{-(\lambda + \Delta G^\circ)^2}{4\lambda kT} \right] \tag{9}$$

and for barrierless ET ($\lambda = -\Delta G^\circ$)

$$k_{et} = \frac{C}{\tau_L} \tag{10}$$

where

$$C = \left[\frac{E_r}{16\pi kT} \right]^{1/2} \tag{11}$$

The constant C is approximately equal to 1.0 at room temperature if the solvent reorganization energy (associated with reorientation of solvent dipoles about nascent charge sites) is assumed to take on a relatively large value (about 1.3 eV, or 30 kcal/mol). Equation 10 indicates that the kinetics of barrierless adiabatic ET are mainly, and simply, governed by solvation dynamics (E_r, τ_L). Experimental tests of this result have included studies of excited state ET kinetics of directly linked DA systems[25,28,29] which will be discussed in the following sections.

3. INTER- AND INTRAMOLECULAR DONOR-ACCEPTOR SYSTEMS

When a molecule having a low ionization energy (Donor, D) and one with high electron affinity (Acceptor, A) are brought together, an interaction between the two molecules in the ground state may lead to the formation of a new species called an electron-donor-acceptor (EDA) [or charge transfer (CT) complex] (Figure 2a). The presence of such a complex is usually detected by the appearance of a new band in the absorption spectrum of a mixture of D and A (a "CT band"). Absorption of light at the wavelength of the CT band promotes an electron from an orbital associated with D to one identified with A. The theory of EDA complexes developed by Mulliken[30] has been thoroughly reviewed in a book by Foster.[31] Because ET is achieved directly by absorption of light by a ground state EDA complex, the photochemical consequences of CT excitation have received much attention as a reference point for DA systems.[32]

Recently, the kinetics of formation and subsequent relaxation of some EDA systems were studied by subpicosecond absorption spectroscopy.[33] It was shown in these studies that absorption of light by the tetracyanobenzene (TCNB)/toluene complex in acetonitrile produced a partially charge-separated state with a subsequent fast step (<1 psec) and a slower step (20 psec) before the fully charge separated (radical-ion pair) state was reached. The first step was attributed mainly to solvent reorientation, and the second step was assigned to the intracomplex structural rearrangement. Other fast kinetic investigations of excited EDA states are consistent with the model in which the excited states, CT* (at least after some initial ultrafast relaxation), consist of *contact* radical-ion pairs that appear in the 10–100 psec time domain for room temperature fluid media,[34] in keeping with the Mulliken model.[30]

Linked systems of interest in the present review can behave as EDA complexes, or may resemble systems in which D–A ground state interaction is weak. If there is no interaction between D and A in the ground state, but an excited state complex is formed, the term *exciplex* is used to describe this association (or excimer if D and A are identical).[35] Exciplexes are usually detected by a new structureless emission band that appears at a lower energy than those of either A or D. The exciplex formed between pyrene (Py) and dimethylaniline (DMA) in polar solvents has been shown to consist of the oxidized Py and reduced DMA (nearly distinct radical ions) by transient absorption spectroscopy.[36] Although the exciplex and the excited state of an EDA complex appear to be similar as represented in Figure 2, exciplexes are more likely to display varying degrees of charge separation. In fact it has been shown for the exciplex and the EDA complex between TCNB and diethylether, and for similar bimolecular complexes of reactive small ring hydrocarbons and electron–acceptors, that the two are not identical.[37] The differences involve not only the degree to which the exciplex exhibits only partial CT, but also the variable role of solvent (contact vs. solvent separated D,A species).

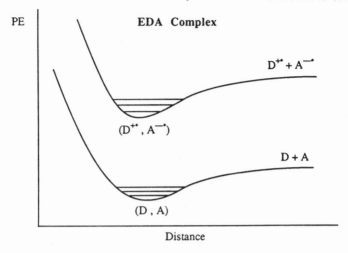

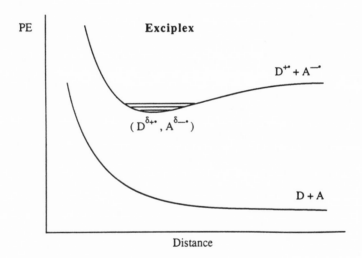

Figure 2. Ground and excited-state potential energy curves as a function of the intermolecular distance for an EDA complex and an exciplex.

For systems containing π electrons, the most stable geometry is one that maximizes the π-overlap between D and A in EDA complexes[38] as well as exciplexes.[39] In systems which have D and A linked by an aliphatic bridge, relative geometries of the two groups are restricted by the linker, although the entropic factor which disfavors formation of a bimolecular exciplex is decreased. For the A_n molecule, **1**, the optimum chain length is $n = 3$, which allows relative reorientation of D and A to achieve maximum overlap of the π-systems.[40] Transient absorption spectra of

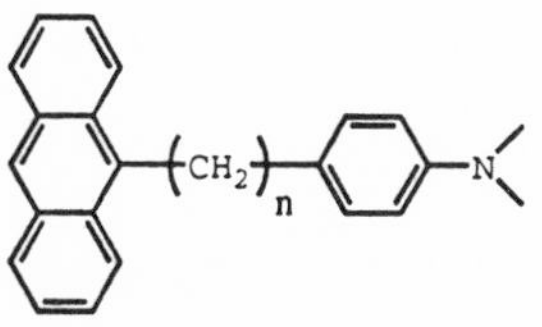

1   A_n , n = 0,1,2,3

the A_n molecules in polar solvents have confirmed the formation of a CT state upon excitation of the anthracene moiety.[28] The time constants for charge separation (τ_{CS}) increased (e.g., 0.65 psec for A_1 to 2.7 psec for A_3) and those for charge recombination (τ_{CR}) decreased (e.g., 4.0 psec for A_1 to 0.7 psec for A_3) in acetonitrile and related solvents. This trend was explained in terms of the FC factor and electronic coupling in the ET rate expression (Eq. 5).[28]

DSA molecules with rigid spacer groups (S) also show CT excited state or exciplex-like behavior. Compound 2 which has 12 sigma bonds separating the donor dimethoxynaphthalene and acceptor dicyanoethylene groups, as well as several others with shorter linkages, were found to give rise to exciplex emission in polar solvents.[41] The CT nature of the new emitting state was shown by the solvent polarity sensitivity of the emission band (generally, a red shift on increase of solvent dielectric constant) as well as time resolved microwave conductivity measurements.[41b,c] Another system which shows exciplex properties is the linked anthracene-DMA molecule 3. Transient absorption spectroscopy was used to provide evidence for the formation of the reduced anthracene moiety in this molecule.[42] Excitation of a local anthracene band of 3 in diethyl ether gave rise to an exciplex-like emission at 460 nm and a charge separated species which lived for 22 nsec.

The common features in recent studies of molecules 2 and 3 and other similar DSA systems are sensitivity of the CT emission to solvent polarity and direct observation of reduced A and oxidized D (radical-ion sites) within the molecule.

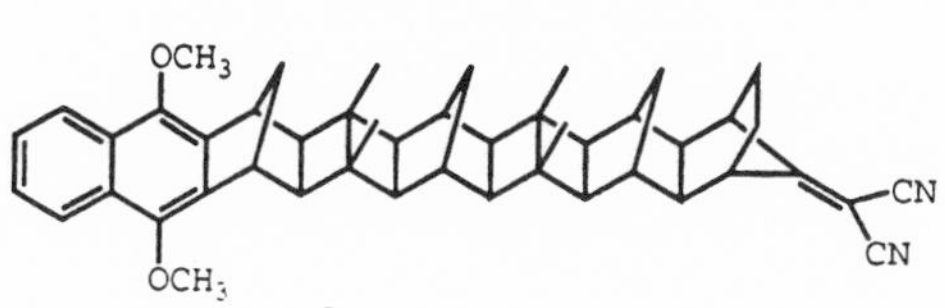

2

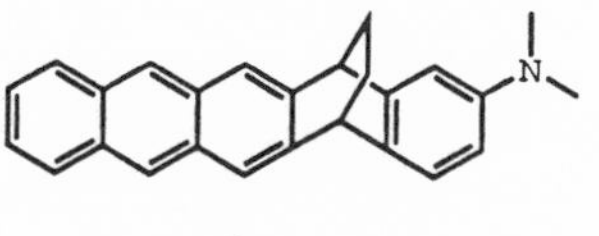

3

Because of the rigid geometry of such systems that insures minimum orbital overlap, there is little or no apparent electronic interaction between D and A groups in the ground state. However, when either D or A is excited, IET may take place to form the CT state. The feasibility of excited state ET may be determined from energetic considerations. A simple expression that allows the calculation of the free energy for excited state ET between a donor and an acceptor was introduced by Rehm and Weller.[43]

$$\Delta G^\circ = 23.06 \ (E^\circ_D - E^\circ_A) - \Delta E_{oo} - \frac{331.2}{(\varepsilon \ r_{AD})} \tag{12}$$

The free energy for ET is ΔG° in Eq.12, and E°_D and E°_A are the voltammetric reduction potentials for D ($D^{+\bullet} + e^- \rightarrow D$) and A ($A + e^- \rightarrow A^{-\bullet}$), respectively. ΔE_{oo} denotes the energy of the excited state (D or A) as the energy difference between the lowest vibrational levels of the ground state and the excited electronic state involved in ET. The parameter ε is the static dielectric constant of the medium and r_{AD} is the distance in Å separating D and A. This expression is commonly used to determine if ET is energetically feasible ($-\Delta G^\circ$ value) for a donor-acceptor system; the ΔG° value obtained in this manner may be used for rate constant calculations using the Marcus expression (Eq. 1).

4. DIRECTLY LINKED DONOR-ACCEPTOR SYSTEMS

Most systems designed for the study of IET include various types of spacer groups separating the donor and the acceptor. An important reason for the use of spacers is to minimize the interaction between D and A in the ground state and to preserve their individual characteristics as much as possible. This criterion may be met for a DA system without any spacer groups if other structural factors are important. In a system consisting of two aromatic groups as donor and acceptor which are directly connected via a single bond, the two groups may preserve their individual properties if the linked biaryl remains sufficiently twisted (recall Scheme 1). For biaryl DA systems steric effects then are mainly responsible for the absence of large interaction (reduced conjugation) between D and A.[44,45]

A clarification of the terminology used to describe intramolecular electron transfer (ICT) seems appropriate at this point in accord with the distinctions made for exciplex and CT systems above. If a new low-energy band is present in the ground state absorption spectrum of DA, the system is understood to be capable of optical electron transfer; that is, on appropriate photoexcitation an ICT state is reached that is fully comparable to the excited state of a conventional bimolecular CT (or EDA) complex.[31] In the absence of any new low-energy absorption band associated with attachment of D and A, it is understood that a locally excited chromophore (an LE state) is involved in an excited state ICT, a process that follows excitation and is presumably activated. For DA systems involving either local or

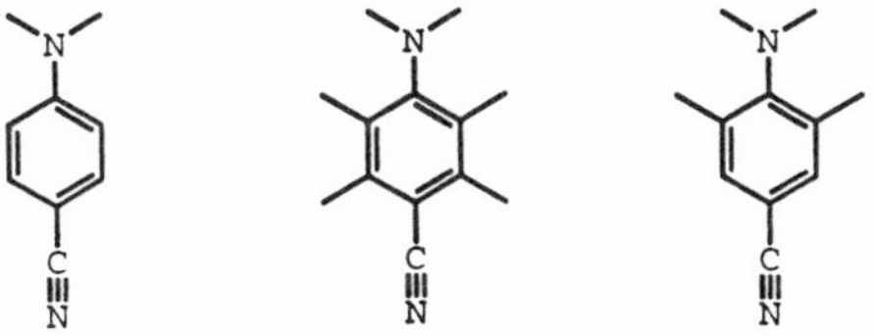

4 , DMABN   **5**   **6**

CT excitation, the resultant state is a radical-ion pair species that consists of the oxidized donor ($D^{+\bullet}$) and reduced acceptor ($A^{-\bullet}$) groups within the same molecule.

A structural variant ushers in another classification of ICT species. Studies of several prototype "direct link" DA systems have indicated that precursor excited states (either LE or ICT) may evolve into states of the twisted ICT-type (TICT).[47] Distinction of a TICT state from a locally excited (LE) state was first suggested by Grabowski et al.[45] to describe the dual fluorescence of dimethylaminobenzonitrile (DMABN, **4**)[46] and similar compounds. The motif for this family of structures consists of a nominally planar aromatic chromophore that is capable of single-bond rotation at a site that connects D and A electroactive segments. Compound **4** and its analogues are the most studied compounds which show excited state TICT behavior.[44b] For this reason it is instructive to review some important results which have been obtained for this series.

The geometry of the TICT state of **4** was shown to be one in which the plane containing the dimethylamino group is twisted to 90° with respect to the phenyl ring in the excited state (the ground state has a planar geometry).[47] The emission spectrum of **4** in solvents of medium to high polarity appears as two bands (from B^* and A^* states; Scheme 2), with the lower energy band (A^*) moving to higher wavelengths as the polarity of the solvent is increased. This result is indicative that a highly polar state is responsible for the A^* emission. Comparison between the emission properties of several analogues of **4** with more restricted geometries provided the evidence for the TICT geometry.[45,47b] Recent theoretical studies of **4** have shown that the 90° twisted geometry corresponds to a minimum in the excited

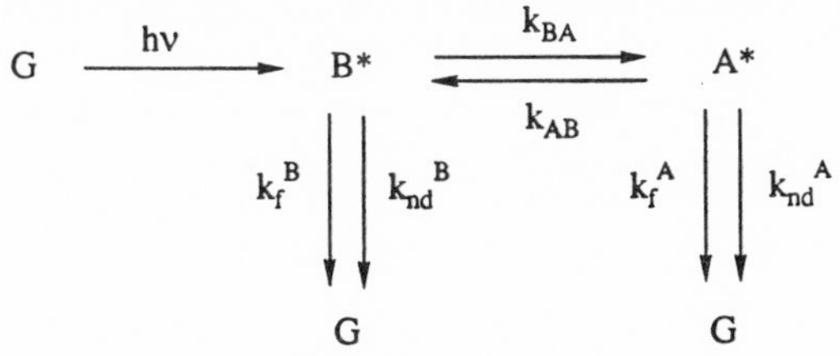

$$G \xrightarrow{h\nu} B^* \underset{k_{AB}}{\overset{k_{BA}}{\rightleftarrows}} A^*$$

$$k_f^B \quad k_{nd}^B \qquad k_f^A \quad k_{nd}^A$$

$$G \qquad\qquad G$$

Scheme 2.

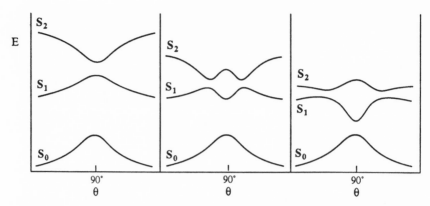

Figure 3. A schematic representation of the ground and two excited singlet state potential surfaces of a linked DA system as a function of the twist angle between the D and the A segments. Mixing of the CT state [S_2 in (a)] with the LE state [S_1 in (a)] increases as the CT state is stabilized either by solvation or by a different choice of D or A group. This mixing results in a minimum in the S_1 surface at 90° [Based on Fig. 6.35 of Ref. 48b].

CT energy surface. The twisted form is lower in energy than the locally excited state at the 90° geometry for molecules in a sufficiently polar medium and/or with a large driving force ($\Delta G°$) for ET.[44b,48] The "minimum overlap rule"[44b,45] is useful in predicting structures for which maximum charge separation should be important, and in identifying torsional modes that will control rates of excited state nonradiative decay for DA systems. The interaction of configurations (S_1 or LE, and S_2 or CT) that are typical are shown in Figure 3, along with illustrations of the geometric (structural) changes involved for DMABN and for a 7-aminocoumarin system that has been reported (Scheme 3).[49]

Recent studies on jet-cooled molecules of **4** have shown that no TICT emission is observed unless solvation by several small polar molecules is present.[50] How-

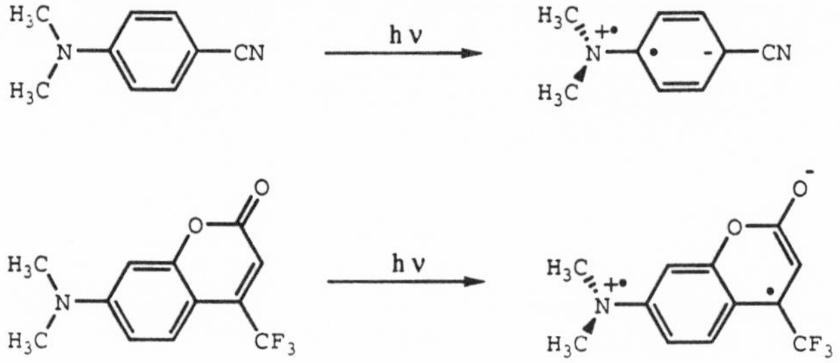

Scheme 3.

ever, unsolvated compound **5** shows a red shifted emission which is attributed to the TICT state. This compound is considerably twisted in the ground state due to steric factors associated with the repulsion of methyl groups *ortho* to the dimethylamine moiety. These results parallel those of the solution phase studies for this compound which also show TICT emission in nonpolar solvents.[45]

An alternative mechanism, in which the formation of the CT excited state is understood to result from complexation with lone-pair electrons of the solvent,[51] has been recently disputed by observing dual fluorescence for **4** in toluene which has no nonbonding electrons.[52] Global analysis of the biexponential fluorescence decays of **4** in toluene and comparison with a model compound **6**, which shows a single decay in toluene, supports the LE→TICT mechanism which was originally introduced by Grabowski, et al.[45] According to the proposed mechanism, excitation of the ground state molecule G (Scheme 2) populates the B^* state which does not have CT character (locally excited, LE); its geometry is similar to that of the ground state. When CT is feasible, either because of solvation by polar solvent, or because of a highly favorable ET (negative $\Delta G°$) in the absence of polar solvent, an equilibrium is established involving the A^* or TICT state. Both B^* and A^* may decay to the ground state via fluorescence (k_f^B and k_f^A, respectively) or via other nonradiative processes represented by k_{nd}^B and k_{nd}^A.

In many cases, excited state CT in TICT-forming molecules may represent the adiabatic limit of the ET theory. For adiabatic ET, $\kappa = 1$ in Eq. 1 and H_{AB} is large in Eq. 5. In the nonadiabatic case, mixing of the zeroth order states (represented by R and P) is small; therefore, ET is inhibited by a poor electronic factor and involves jumping between two potential energy surfaces (Figure 4). However, for strong interactions (large H_{AB}, usually associated with significant overlap of orbitals associated with D and A), mixing of the R and P states results in a modified excited state surface, and ET involves (adiabatic) movement on a single potential energy surface.[2a,24a] Depending on the relative energies of the reactants (R) and

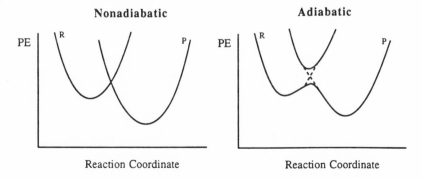

Figure 4. Adiabatic and nonadiabatic electron transfer represented by potential energy curves for the reactants and the products.

products (P), an activation energy, ΔE_a, may be present for adiabatic electron transfer.

A general treatment of LE/TICT kinetics has been reported for adiabatic ET incorporating a model with and without an activation barrier.[44a] A stochastic description of the ET process as controlled by the solvent and the generalized Langevin equation was used. It was shown that in the absence of an activation barrier, a time-dependent rate constant for ET would be expected. Analysis of DMABN (4) fluorescence in n-pentanol at –50 °C verified this point. The intensity of the TICT emission increased much faster than was expected based on the decay of the LE state.

Because of relatively large interaction between D and A groups in directly linked DA systems, the excited state ET for LE→TICT is believed to be adiabatic.[25,29,44a,b,53] Studies of the kinetics of such ET processes have allowed testing of the solvation dynamics associated with IET, features that will be discussed for specific biaryl DA systems in the following sections.

4.1. 4-(9-Anthryl)-N,N-dimethylaniline (ADMA) and its Analogues

Fluorescence studies of ADMA (7) in different solvents have shown strong solvent influences,[54] similar to the trends observed for systems such as 4. The maximum of the fluorescence band (v_{max}) decreases from 23.3 kK to 17.2 kK (430- to 580-nm shift) when solvent polarity (indicated by the dielectric constant ε) increases from n-hexane to acetonitrile (Figure 5) For the same solvent series, τ_f,

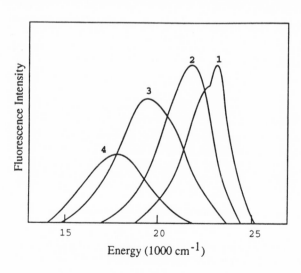

Figure 5. Fluorescence spectra of 7 (ADMA) in (1) decalin, (2) ethyl ether, (3) isobutyl alcohol, and (4) acetonitrile [Fig. 2 of Ref. 54a].

the fluorescence lifetime, increased from 2.1 to 31.4 ns at room temperature. This behavior was attributed to formation of a TICT state (Scheme 4) from the local anthracene (LE) state. The dipole moment of the TICT state was estimated to be 18 D (compared to about 2 D for the ground state) from the shift in the fluorescence band position with solvent polarity.[54a]

Ground-state interaction is evident in the lowest energy absorption band of **7** in different solvents; the bands near 430 nm are broadened and extended to lower energies as solvent polarity increases, although no distinct CT band is detected. It has been shown by electro-optical emission measurements (EOEM),[55] as well as transient absorption spectroscopy,[56] that a true zwitterion or TICT state resembling a superposition of radical–ions (Scheme 4) is achieved in the psec–nsec time domain. The observation was further supported by comparison of the spectroscopic behavior of a number of selectively hindered analogues of **7**, such as compounds **8–11**. Comparison of the fluorescence of these compounds confirmed that forma-

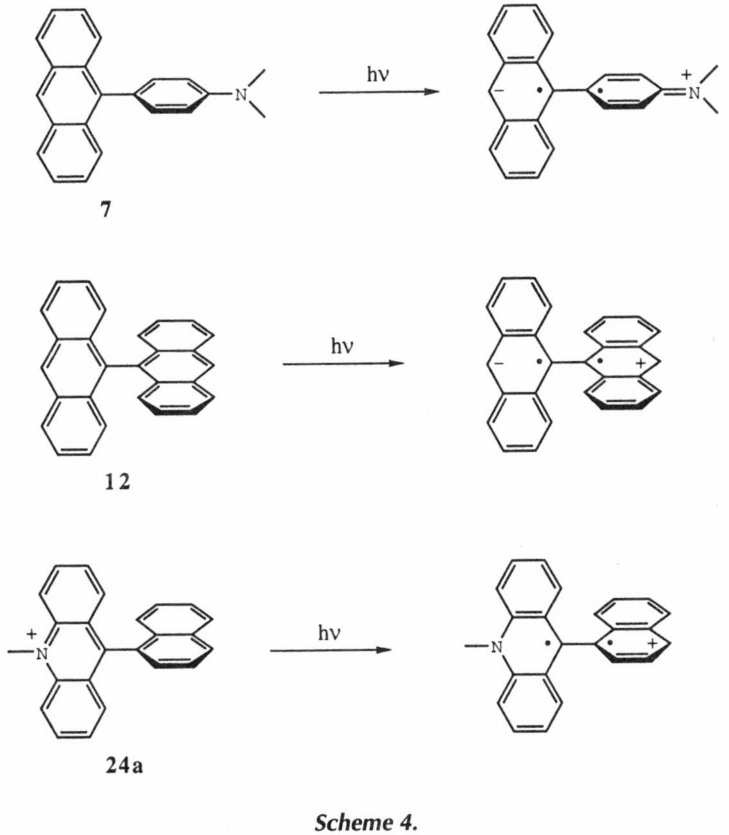

7

12

24a

Scheme 4.

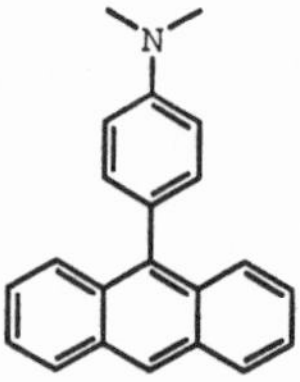

7 , ADMA

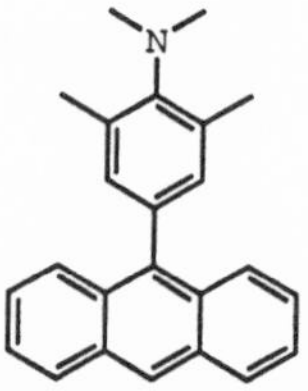

8 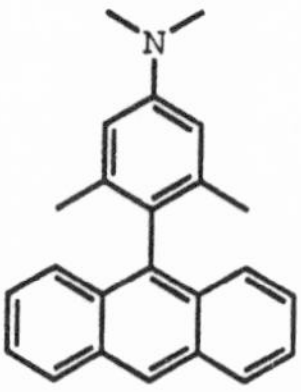   9

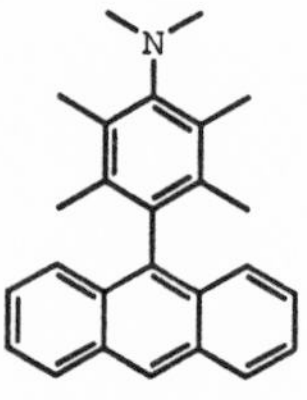

10 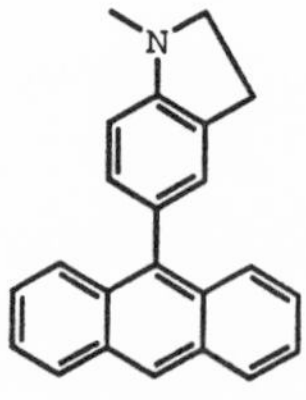   11

tion of the more polar state in these molecules involves twisting around the single bond connecting the two aromatic groups and not the dimethylamino–phenyl bond.[57]

Picosecond laser spectroscopy of ADMA (7) and a number of its analogues has revealed that multiple excited states with different degrees of CT may in fact be involved.[56,58] For compounds **7, 8**, and **11**, which have more flexibility for twisting around the biaryl bond, transient absorption spectra in solvents with intermediate polarity could not be reconstructed by a least square superposition of spectra in hexane and acetonitrile in the 540–900-nm region. The same procedure for **9** and **10**, however, resulted in good agreement between the reconstructed spectra and the actual spectra in solvents such as ethyl ether, ethyl acetate, and tetrahydrofuran.[56] Based on these results, it was suggested that a two-state model, such as presented in Scheme 2, may be suitable for the excited state behavior of the more hindered compounds such as **9** and **10**, but not for **7, 8**, and **11**. Concluding the series of studies on CT systems that encompass **7** (ADMA), **1** (A1), and intermolecular D,A complexes, Mataga and his co-workers[28] argue strongly that the degree to which solvation dynamics controls ET rates will be highly variable and depend on additional structural requirements (e.g., torsion or twisting) for charge separation. In addition, the degree of relative interaction of D and A components (strength of electronic coupling or overlap) will influence the level of relative solvation of ground, LE excited state, and CT state species.

The time constant for charge separation ($\tau_{cs} = 1/k_{et}$) for ADMA has been measured by femtosecond transient absorption spectroscopy.[28] Values of 5.0 psec for hexanonitrile and 2.7 psec for butyronitrile solvent were obtained at room temperature from the decay of the 600-nm transient associated with the LE (anthracene-like) state (accompanied by rise of a 460-nm species identified with the dimethylaniline radical cation structure). These times were much longer than those predicted by Eq. 10 (1.4 and 1.0 psec for the two solvents, respectively), and different from, but somewhat closer to, average solvation times $<\tau_s>$ determined empirically by measurement of relaxation times following excitation of polar (e.g., coumarin) fluorescence probes.[59] The authors attributed the differences to strong solvent interactions with the ground and TICT states which hinder charge separation (CS) in the excited state.

Following the most recent study of ADMA (**7**), Barbara and co-workers[29d,60b] report a new model for LE→TICT dynamics based on successful simulation of static absorption/emission data and excited state decay in the 0.3–10 psec regime. Their model emphasizes the "nested" nature of LE and CT states for this system and the importance of population of the upper state, S_2, (actually the configuration with most LE character) on initial excitation. The fast events are thus ascribed to an ultrafast S_2 (LE)→S_1(TICT) internal conversion ($\tau \cong 150$ fs) followed by a S_1 (TICT) relaxation that is controlled by solvent dynamics ($\tau_s = 1.0$–1.5 ps for DMF solvent at room temperature). The role of polar solvent then is viewed as one of stabilization of the TICT over the LE state (in the limit, a state reversal that is revealed by the dramatic Stokes shift of the fluorescence band) that can be simulated with appropriate state and solvent parametrization.

According to this model,[29d,60b] photoexcitation (396 nm) principally populates the LE (S_2) state and ET occurs, as first a nonadiabatic (state-to-state) transition, $S_2 \rightarrow S_1$ (TICT) (Figure 8). The surface jump occurs at a crossing point for $S_1 - S_2$ corresponding to the Marcus "inverted region" for intersecting parabolas[2,17] with a relatively large energy gap separating the two equilibrated states (the $S_1 - S_2$ gap was estimated at $\Delta G° = 10.5$ kcal/mole from simulations of spectral data for **7**). The latter implies that a small energy barrier is traversed, followed by adiabatic relaxation along the S_1 surface toward the TICT minimum.

In separating out the radiative and nonradiative components of fluorescence decays for D-A systems, it is often important to make simplifying assumptions (e.g., such as assuming that radiative rate constants, k_f, of organic molecules are temperature independent and only moderately sensitive to changes in medium).[61]

However, a recent survey of k_f values for TICT emission of a number of compounds, including ADMA, has indicated that the fluorescence in these compounds is thermally activated.[62] This dependence is explained[44b,63] in terms of the forbidden nature of TICT emission, a restriction which is relaxed due to a larger overlap of higher TICT vibrational levels (which are more populated by higher temperatures) with those of the ground state.

4.2. Bianthryl and Other Symmetrical DA Systems

A special case arises when the donor and the acceptor groups in a directly linked DA molecule are identical such as found in bianthryl (BA, **12**). This widely studied compound has been shown to have the main properties common to other excited state compounds that display TICT behavior.[44b,c,45,64] The fluorescence spectrum in nonpolar solvents resembles that of anthracene, but as the polarity of the solvent is increased the fluorescence band shifts to lower energies (Figure 6). This change is attributed to an increase in the contribution of TICT state emission to the total fluorescence. The mechanism presented in Scheme 2 has also been proposed for bianthryl.[63]

Quantum chemical calculations have indicated that the excited states of BA are comprised of two degenerate charge resonance states which, in the absence of strong environmental perturbation, lie higher in energy than the lowest locally excited (LE) state of anthracene.[63] However, interaction with a polar solvent is sufficient to polarize the excited state and to stabilize one CT state with respect to the other with a minimum energy at a twist angle of 90° between the two anthracene planes. In other molecules, such as **13**, the CT state is already lower in energy than the corresponding state in BA, even in the absence of a polar solvent;[63] the ground- and excited-state TICT structures for **12** are shown in Scheme 4.

The dynamics of the LE→TICT transition for BA are believed to be governed by solvation, in a fashion similar to, but not identical with, the findings for ADMA (**7**). A number of studies have made comparisons between experimental k_{et} and $1/\tau_L$ (Eq. 6) for BA in different solvents.[29,53,65] The ET rate constant determined in these studies corresponds to the forward reaction from B* to A* in Scheme 2. Due to the complexities involved in determining the individual rate constants in this scheme,

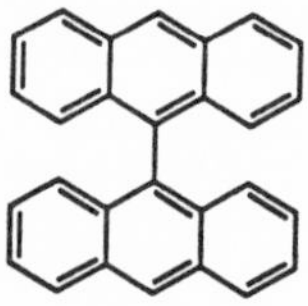

12

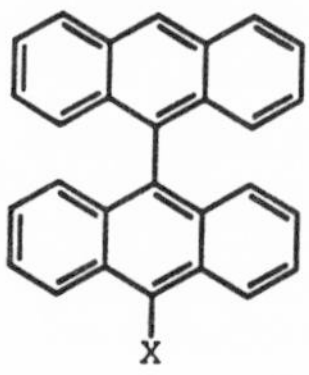

13 $\begin{cases} \text{a: } X = Cl \\ \text{b: } X = CHO \end{cases}$

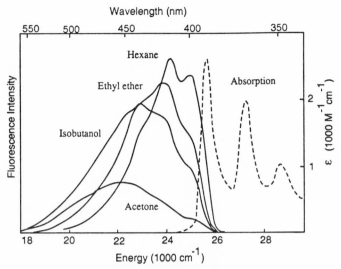

Figure 6. Fluorescence and absorption spectra of **12** (BA) in different solvents [Fig. 2 of Ref. 67].

the experimental parameters determined by different techniques are sometimes different. In the case of BA, τ_{et} ($1/k_{et}$) has been reported to be 0.7 psec and 2.0 psec in acetonitrile and butyronitrile, respectively, based on time-resolved emission data.[29a] The same parameter obtained from transient absorption spectra is reported to be 1.8 and 3.2 psec for the same solvents.[66] These differences may reflect the difficulty in spectral resolution at ultrafast times of the relatively broad featureless absorption and emission bands that are undergoing subtle changes. The possibility that multiple intermediate states take part in the ET process has been suggested again for BA (as with ADMA).[66]

The excited state ET of BA has been shown to be affected by pressure. The equilibrium constant for the LE/CT interconversion (ratio of the fluorescence quantum yields, ϕ_{CT}/ϕ_{LE}) decreases as the pressure is raised (0–10 kbars) in solvents of moderate viscosity (alcohols).[67] Using pressure to control solvent properties, Windsor, Rettig, and their co-workers[29b] have measured τ_{et} values for BA by transient absorption spectroscopy. The equilibrated transient absorption spectra (50 psec after laser excitation) in cyclohexane and acetonitrile were assumed to correspond to the LE and the TICT states, respectively (Figure 7) (no spectral shifts observed at longer times). In the viscous solvent, glycerol triacetate (GTA), nonexponential kinetics for the LE→TICT transition were observed in the range of 0.1–300 MPa pressure.

The data were fitted to an extended exponential function with the general form of Eq. 13. The highly nonexponential nature of the decay became more evident as viscosity increased with pressure. The polarity increase associated with the pres-

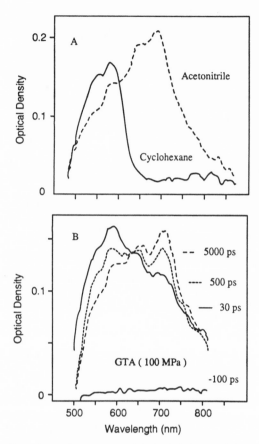

Figure 7. Transient absorption spectra of **12** (BA): (**A**) in cyclohexane and acetonitrile at 50 ps; (**B**) in glycerol triacetate (GTA) at different delay times [Fig. 1 of Ref. 29b].

sure rise was shown to only affect the equilibrium between LE and TICT states. About half of the TICT state concentration was formed in a time shorter than 25 psec (instrument limit) independent of the pressure. The authors suggested that inhomogeneous ground state solvent orientations (some favorable for ET) are responsible. The important conclusion is that for ultrafast processes in which solvent motion is largely rate-limiting, adiabatic ET will be dominated by an inhomogeneous distribution of relaxation times. The importance of inhomogeneous populations of rotamers (having an ensemble of LE electron transfer rates) has been proposed for BA in solid polymers for which photoselection of different twisted forms leads to highly nonexponential LE decay.[68]

$$F(t) = A \exp[-(t/\tau)^\beta] + C \qquad (13)$$

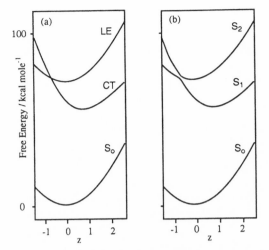

Figure 8. Theoretical estimates for the (**a**) nonadiabatic and (**b**) adiabatic free energies of ADMA in dimethylformamide as a function of the solvent coordinate [Fig. 2 of Ref. 60b].

It remains to be seen whether ultrafast kinetic probes will continue to reveal multi-exponential decays for the earliest stages of ICT, indicative of an inhomogeneous distribution of relaxation times. The latter complexity is apparent on considering a typical TICT system (e.g., BA and ADMA) for which slightly different angles of twist result in different levels of electronic interaction between D and A segments; solvent then orients in a way that differentially solvates these angular geometries that lie close in energy in the ground state.

The "state mixing" (configuration interaction) model of Barbara, Fonseca, and their co-workers[60b,65] has also been used to describe the excited state ET of bianthryl (Figure 8). In this model the theory utilizes the generalized Langevin equation (GLE) to treat solvent dynamics (friction) based on empirical data for relaxation of the polar coumarin fluorescence probes. In this model the inner sphere vibrational components (λ_i in Eq. 2) are ignored based on available experimental data regarding vibronic features in neutral versus radical-ion spectra for aromatics.[29a] This assumption was also confirmed by successful simulation of steady-state absorption and emission spectra in several solvents employing a empirical solvatochromic/vibronic description. In addition, qualitative agreement was obtained between simulated and experimental time-dependent emission spectra for BA. From the analysis of static absorption/emission spectra for BA, an LE/CT state separation of 5.0 kcal/mole is estimated. Photoinduced ET then proceeds over a larger barrier ("normal" Marcus region) than that for ADMA, consistent with the slower rates observed in fast kinetics experiments.

14

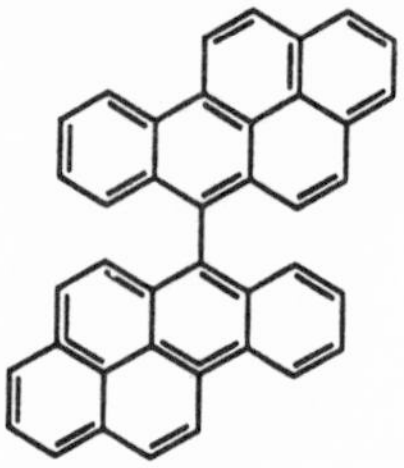

15

There have been reports of other symmetric biaryl systems which exhibit LE and TICT emission.[44,45] Compounds **14** and **15** show dual fluorescence even in the nonpolar solvent n-hexane. The energies of the TICT states relative to the ground state have been estimated for a number of symmetric biaryls from their ionization potentials (IP) and electron affinities (EA). The energies were used to predict whether excited state ET from LE to a CT state was thermodynamically feasible ($\Delta G° < 0$). The predictions were validated by detection of TICT emission for a number of symmetrical biaryl systems.[44c] Additional work is required in order to more fully describe the dynamics of these systems since only steady-state and simple lifetime data are available for all symmetrical biaryls, other than the well studied bianthryl.

4.3. Other Biaryl DA Systems

There are scattered reports of other biaryl donor-acceptor molecules which are also believed to undergo excited state ET. Most are related structurally to ADMA (**7**) and some similar trends are apparent, although thorough fast kinetic investigations of these alternate systems are so far unreported. The relative energies corresponding to the emission from the excited CT state of four diethylanilino-substituted aromatic compounds including **16** and **17** have been estimated from the oxidation and reduction potentials of the donor and acceptor groups, respectively.[69] These values were shown to correlate well with the experimentally observed emission energies for **16** and **17** in polar solvents. The fluorescence spectra were solvent polarity dependent, in a fashion similar to the findings for ADMA. However, no kinetic measurements were conducted for the excited state ET reaction.

Excited-state CT has also been observed for a number of N-phenylpyrroles and their methyl-substituted analogues represented by compounds **18** and **19**.[70] These compounds show TICT emission as characterized by solvent-polarity-dependent spectra. Some of the compounds studied, such as **19b**, showed this property even in relatively nonpolar solvents. Photoelectron spectra were used to obtain ground-

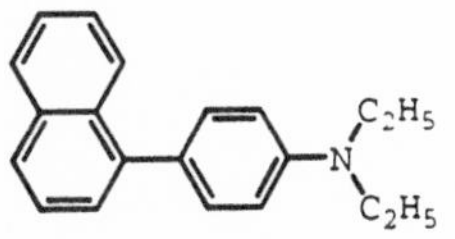

16

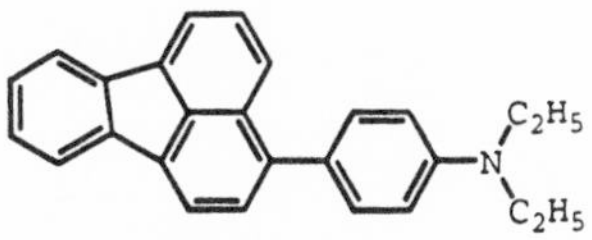

17

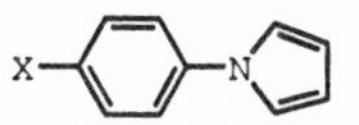

18 { a : X = H
     b : X = CN

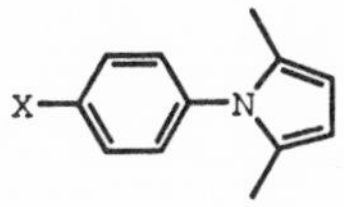

19 { a : X = H
     b : X = CN

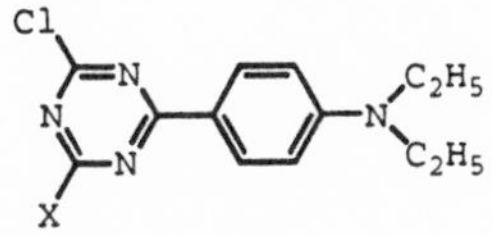

20 { a : X = Cl
     b : X = N(CH₃)(alkyl)

state twist angles which indicated a broader rotational distribution for the less hindered molecules and/or presence of poorer acceptor groups (e. g., H instead of CN in **18** and **19**). Analysis of CNDO/S-type quantum chemical calculations for **18b** revealed[70] that the perpendicular geometry does not necessarily correspond to the lowest excited CT state (TICT) as is usually expected. The minimum overlap condition[44b,45,48] which is thought to be necessary to achieve minimum energy in the excited CT state may be satisfied for this molecule at twist angles other than 90°. A number of energy minima for the excited CT state may even be present. This notion is similar to the one presented earlier to explain the picosecond transient absorption spectra of **7** in solvents of intermediate polarity.[56,58]

Another series of DA compounds reported recently are the triazinylaniline derivatives (**20a,b** and others) which were studied from the standpoint of their absorption behavior and their interactions with hydrogen-bonding solvents.[71] These compounds, designed to be used as fluorescence probes, display lowest energy absorption bands assigned to an intramolecular charge transfer (ICT) transition. These transitions are unusual in their intensity (as CT bands), indicative of a high oscillator strength. The fluorescence bands were found to be solvent

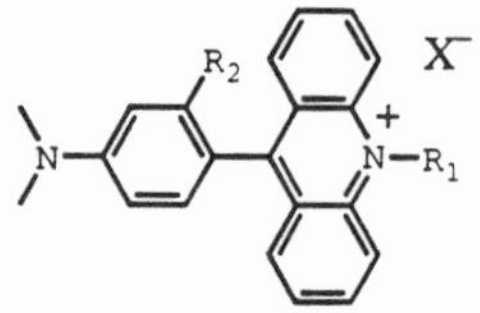

$$21 \begin{cases} a : R_1 = R_2 = H \\ b : R_1 = CH_3 , R_2 = H \\ c : R_1 = R_2 = CH_3 \end{cases}$$

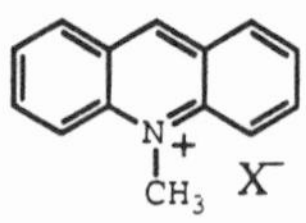

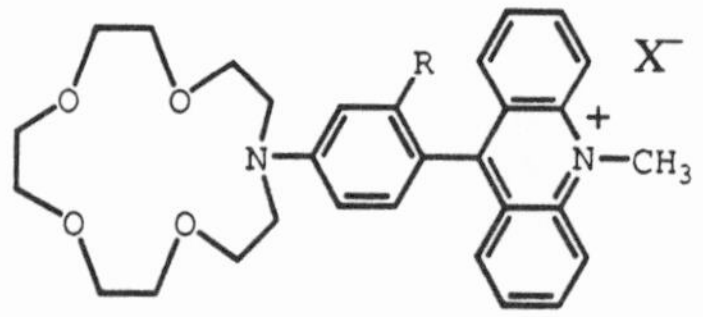

22

$$23 \begin{cases} a : R = H \\ b : R = CH_3 \end{cases}$$

polarity-dependent, but they were not assigned to TICT emission (full charge separation between triazinyl and aniline moieties at a twisted geometry). The fluorescence quantum yields and lifetimes decreased with an increase in solvent polarity, and the maximum of the ICT absorption band decreased on lowering the pH below 2.5. The latter result is consistent with a protonation of the amine functionality at lower pH which alters the nature of the CT chromophore. These properties were considered to make the triazinylaniline compounds useful as microenvironmental probes.[71] It will be of interest in future studies to determine the importance of twisting about single bonds (two types possible) in controlling excited-state dynamics for **20**.

A number of acridinium DA systems have also been designed as molecular probes; specifically, as chromoionophoric and fluoroionophoric probes.[72] Dialkyl-anilino-substituted acridinium **21a** has been shown to display ground-state ICT interaction which is identified by the appearance of a new lower energy band in its visible absorption spectrum.[73] Protonation at the dialkylanilino group reduces its donor ability to the point that the ICT transition is no longer the lowest in energy. Compounds **21a–c** do not show any detectable fluorescence when the low energy ICT band is present in their absorption spectra. However, when the aniline moiety is protonated, fluorescence is observed at about 500 nm which is similar to the fluorescence of 10-methylacridinium salt **22** (the LE state).[72c]

Compounds **23a,b** behave similarly to **21b,c** in terms of their absorption and emission spectra. In addition, complexation of the aza-crown ether part of **23a,b** with ions such as Na$^+$, Ba^{2+}, and Ag$^+$ causes the disappearance of the low-energy ICT absorption band in the visible region, as with the protonation of the dialky-

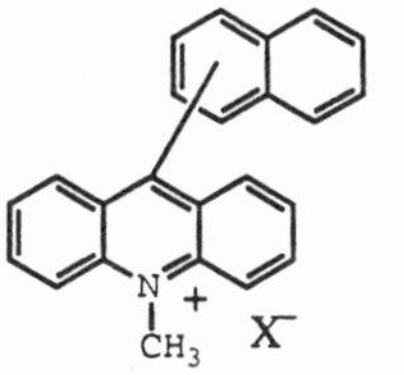

$$24 \begin{cases} a : \text{1-naphthyl} \\ b : \text{2-naphthyl} \end{cases}$$

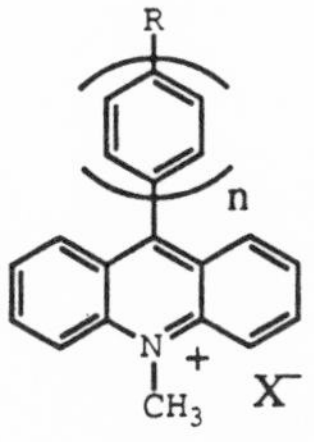

$$25 \begin{cases} a : R = H, n = 2 \\ b : R = OH, n=1 \\ c : R = OH, n=2 \end{cases}$$

lanilino group.[72b] The lack of fluorescence, again for these compounds in their uncomplexed form, was suggested to be due to the existence of a low-lying TICT state which deactivates to the ground state via fast nonradiative processes. Protonation of **23a,b** as well as their complexes with Ag^+ caused the acridinium LE fluorescence to be observed. This result is ascribed to the raising of the energy of the nonfluorescent TICT state relative to that of the LE state so that LE emission is observed.[72c] The spectral changes induced by complexation make compounds **23a,b** suitable to probe the presence of complexing ions in various environments.[72] The X-ray crystallographic data for complexes of **23** reveal that metal ion complexation in the crown portion results in deformation of the aniline nitrogen out of plane (out of conjugation with the aromatic ring).[72b] This feature is central to the elimination of the electron donor (fluorescence quenching) property of the aniline ring.

A new series of acridinium DA compounds has recently been studied which shows TICT behavior and can be "tuned" with variations in medium and structure.[74] Compounds **24a,b** and **25a** show TICT-type emission in solvents of medium to high polarity, and the wavelength of emission is solvent-dependent. The corrected emission maximum for **25a** shifts from 628 nm in acetonitrile to 560 nm in dichloromethane and is clearly well resolved from local emission associated with the acridinium chromophore (ca. 500 nm). The excited states for the **24,25** series must be of the TICT type in which *charge shift* is important. Picosecond transient absorption spectra for **24a,b** and **25a–c** appear as the superposition of the spectra of oxidized D and reduced A groups. Rapid decay (< 60 psec) of the phototransient for **25b,c** was observed in acetonitrile which indicates very efficient nonradiative decay from the CT state due to the small energy gap which separates the ground and (twisted) ICT states with the phenol groups acting as electron donors. This result is in agreement with the lack of detectable fluorescence in that solvent for **25b,c**.

Fluorescence lifetimes for the emitting compounds (e.g., **24**, 1–7 nsec range) were in agreement with the decay times of the transient absorption spectra indicat-

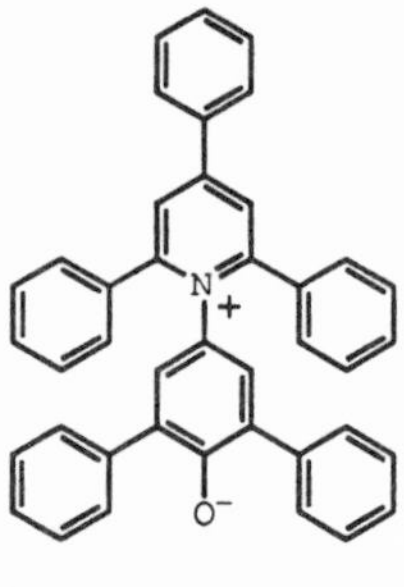

26 , Betaine–30

ing that both corresponded to the same state.[74] The special feature of these compounds is that they allow the study of electron transfer reactions which can be categorized as charge shift and charge annihilation (for the ionized acridinium phenolates) as compared to the usual case of charge separation. Different energy gap laws are expected to govern these different types of ET reactions.[75] For the acridinium derivatives, forward and back ET is controlled by the facility in reaching a partially twisted geometry; subtle dependences on the substitution of the twisting group have been observed [τ_f (ICT) for **24a** and **24b** in methylene chloride = 5.1 and 7.3 nsec].

The dynamic solvent effect on IET of the N-pyridinium phenolate **26** ("betaine-30") has been investigated.[76] This compound exhibits strong ground state ICT interaction characterized by a broad absorption band in the visible and near-infrared region. The phenolate group acts as electron donor and the pyridinium group as electron acceptor in **26**. Due to its high polarity, the ground state of **26** is stabilized in more polar solvents, and its ICT band is shifted to higher energies (855 nm in toluene to 621 nm in propylene carbonate). The high sensitivity of the position of this band has been utilized to develop an empirical solvent polarity scale.[77] The energies (in kcal/mol) corresponding to the frequency maximum for the ICT absorption band of **26** in different solvents [the $E_T(30)$ parameter] are taken as a measure of the polarity of the solvent used.

In a pleasing reversal of charge type, compared with most of the DA systems reported here, the zwitterionic *ground state* of **26** is a species displaying full charge separation, whereas the lowest excited state behaves as a neutral intermediate. This radical-pair excited state is reached directly by absorption of light in the ICT band. The rate constant (k_{et}) was measured for the charge separation (repopulation of the ground state) after laser excitation with 70 fsec pulses at 792 nm using the pump–probe technique.[76] Comparison of the experimental k_{et} values in different polar solvents with those obtained from theoretical treatments by Sumi and Marcus[27a] and by Jortner and Bixon[78] indicated that these treatments were not suitable for predicting the experimental results. The latter model emphasizes the role of high-frequency vibrational modes in ET, while the former treats ET classically. By considering the high-frequency modes, predicted values for k_{et} were closer than the classical values to the experimental ones in low-viscosity solvents.[76] Electron transfer time constants ($1/k_{et}$) correlated roughly with the average solvation times $<\tau_s>$ (determined from solvation dynamics of probe molecules) in low-viscosity solvents.

5. SUMMARY

Donor–acceptor (DA) molecules that display a direct single–bond link between D and A segments (e.g., biaryls) have been used extensively in the study of excited state electron transfer (ET) in polar solvents. Although thorough kinetics studies have been conducted only on a limited number of "short-link" DA systems, the information gained from them has clarified some aspects of the ET mechanism. There can be relatively strong interaction between donor and acceptor groups due to the direct linkage between them if geometries are reached which permit some overlap of D and A orbitals. On the other hand, the individual ground-state electronic properties of D and A are largely preserved for systems that are highly hindered for rotation about the D–A bond. Excited-state ET is considered to be adiabatic in these molecules [appreciable D–A electronic interaction and a mixing of "pure" locally excited (LE) and intramolecular charge-transfer excited (ICT) states] and may require a coupling of some twisting molecular motion and the motion of solvent molecules in reorientation about nascent dipoles.

The results for biaryls provide a contrast with findings for bridged donor-acceptor systems (DSA molecules that are transposed by having an "inert" linking group, S). It is commonly believed that these structures undergo nonadiabatic excited-state ET as a consequence of weak interaction between D and A groups, the result of relatively large distances of DA separation enforced by a rigid linkage. The experimentally observed rate constants for ET are in general faster for DA systems than DSA systems. The "forward" k_{et} values for ET in DA systems correspond to subpicosecond to tens of picosecond times, generally, for room-temperature, low-viscosity media. Back reaction for the resultant charge-separated radical-pair state [e.g., a twisted charge-transfer (CT), or twisted intramolecular charge transfer (TICT) state] remains relatively robust in simple DA systems, with decay times in the nanosecond range (a larger energy gap is spanned for return to the ground state). The adiabatic nature of ET in these directly linked DA molecules has rendered them useful for examination of ET theories and the role of solvation dynamics in ET processes.

Various formalisms used to describe adiabatic ET have provided relationships between k_{et} and $1/\tau'$, where τ' may be the longitudinal dielectric relaxation time (τ_L) (Eq. 7) which can be calculated based on a dielectric continuum model for the solvent. It is increasingly common, however, to find that electron transfer times for charge separation in DA systems do not correspond to the longitudinal relaxation times. The experimentally determined average solvation times $<\tau_s>$ associated with processes of solvent reorientation around polar fluorescent probes may serve as better predictors of τ' for intramolecular electron transfer (IET). Recent ultrafast spectroscopic measurements of ET discussed in previous sections have provided evidence for the solvent-controlled kinetics of adiabatic ET. Qualitative agreement between experimental ET rate constants and solvation constants $(1/\tau')$ have been observed. However, consideration of features which are specific to each system may be necessary to obtain better quantitative agreement between the two sets of parameters, as has been shown in the more detailed studies of BA (12) and ADMA (7).[28,65]

Some of the central issues associated with biaryl DA systems are now delineated, albeit incompletely resolved. For some systems, ET rates will be closely coupled to solvation dynamics; if so, a more complicated confluence of solvent motion (e.g., translational as well as rotational components[79]) may be important and deviations from simple exponential relaxation behavior will persist. Even for hindered biaryls, the role of excited state twisting (importance of an internal coordinate) is not to be discounted, particularly for more flexible systems and those in which more of an ET barrier must be achieved (Marcus inverted region). Models of solvation tend to fall into two categories involving: (1) specific solvation of multiple conformational states leading to an ensemble of ET rates at ultrafast times;[28,29b] or (2) according to a new paradigm,[29d,60b] solvent control of the energetic spacing of formal LE an TICT states (e.g, Figure 8), the level of state mixing and control over the position of (nonadiabatic) potential surface crossing points (ET barriers).

Further experimental investigation of the earliest times in the photochemical electron transfer (PET) process will be useful in determining whether adiabatic ET reactions are indeed completely controlled by the dynamics of the solvent molecules surrounding the DA system, and whether such processes are in turn good probes of solvation dynamics.[27b]

The kinetics of excited state ET in directly linked DA systems is considered to depend more clearly on the twisting motion between the D and A groups in some cases (e.g., DMABN). On starting with a planar chromophore, the energetics of the rearrangement of local state (LE) to charge-transfer excited state (TICT) will be governed by the twist angle (Figure 3). The driving force in the dynamics of twisting–with–charge–transfer for initially planar excited species is readily understood in terms of the proximity of ground and excited potential surfaces achieved at high twist angles that reduce DA orbital overlap.[45,47,48] DA molecules which are already twisted towards a perpendicular geometry in the ground state (e.g., 5,6,9,10), do not depend on conformational relaxation to achieve the 90° geometry to satisfy the minimum overlap rule. Dual fluorescence and solvent dependence of the lower energy emission band have been two important criteria to identify compounds which undergo excited state ET with a change in internal geometry.

The importance of perpendicular geometries has been extended to the realm of long-range charge transport in biological macromolecules (the electron transport chain in reaction centers of photosynthetic bacteria).[29b] The idea is that moderately overlapping rings provide pathways for rapid ET by providing loci for discrete (full) charge separation along a transport chain.

The fundamental understanding of photoinduced ET processes is expanded by studies of systems such as the directly linked DA molecules which were discussed here. This knowledge has already been used to develop molecules with specific properties suitable for probing microscopic environment and medium conditions (e.g., polarity, viscosity, pH, presence of metal ions). Another new area of potential application is molecular electronics. Use of carefully designed molecules which may perform tasks, such as very fast electronic switching on a molecular level, may allow one to construct electronic devices on a molecular scale. Some molecular switches based on TICT properties of certain DA systems have already been proposed.[80] In the future, development of structures with device-like characteristics, the tuning of properties—forward and reverse electron transfers that correspond to switching times—and subtle substituent influences such as those found for ADMA and analogues (**7–11**), and the acridiniums, **24–25**, will be important.

ACKNOWLEDGMENT

Support for this work by the Department of Energy, Office of Basic Energy Sciences, is gratefully acknowledged.

REFERENCES

1. Boxer, S. G. *Biochim. Biophys. Acta*, **1983**, *726*, 265.
2. (a) Marcus, R. A. *Ann. Rev. Phys. Chem.*, **1964**, *15*, 155; (b) Marcus, R. A. *Discuss. Faraday Soc.*, **1960**, *29*, 21.
3. Hush, N. S. *Trans. Faraday Soc.*, **1961**, *57*, 155.
4. Levich, V. G. *Adv. Electrochem. Eng.*, **1966**, *4*, 249.
5. (a) Dogonadze, R. R.; Kuznetsov, A. M.; Levich, V. G. *Electrochim. Acta*, **1968**, *13*, 1025; (b) Levich, V. G.; Dogonadze, R. R. *Coll. Czech. Commun.*, **1961**, *26*, 193.
6. (a) Rips, I.; Jortner, J. *J. Chem. Phys.*, **1987**, *87*, 2090; (b) Rips, I.; Klafter, J.; Jortner, J. *J. Phys. Chem.*, **1990**, *94*, 8557; (c) Brunschwig, B. S.; Ehrenson, S.; Sutin, N. *J. Phys. Chem.*, **1987**, *91*, 4714; (d) Kim, H. J.; Hynes, J. T. *J. Phys. Chem.*, **1990**, *94*, 2736.
7. (a) Gould, I. R.; Young, R. H.; Moody, R. E.; Farid, S. *J. Phys. Chem.*, **1991**, *95*, 2068; (b) Mataga, N. *Pure. Appl. Chem.*, **1984**, *56(9)*, 1255.
8. (a) Ojima, S.; Miyasaka, H.; Mataga, N. *J. Phys. Chem.*, **1990**, *94*, 4147; (b) Fleming, G. R.; Wolynes, P. G. *Physics Today*, May **1990**, 36.
9. (a) Penner, G. H. *J. Mol. Struct. (Theochem)*, **1986**, *137*, 191; (b) Suzuki, H. *Bull. Chem. Soc. Jpn.*, **1959**, *32*, 1340.
10. Colter, A. K.; Clemens, L. M. *J. Phys. Chem.*, **1964**, *68*, 651.
11. Irie, M.; Yoshida, K.; Hayashi, K. *J. Phys. Chem.*, **1977**, *81*, 969.

12. Werst, D. W.; Gentry, W. R.; Barbara, P. F. *J. Phys. Chem.*, **1985**, *89*, 729.
13. Wasielewski, M. R. *Chem. Rev.* **1992**, *92*, 435.
14. Gust, D.; Moore, T. A. *Topics in Current Chemistry*. **1991**, Vol. 159, p. 103.
15. Connolly, J. S.; Bolton, J. R. *Photoinduced Electron Transfer* (Fox, M. A.; Chanon, M., Eds.), Part D. Elsevier, Amsterdam, **1988**, p. 303.
16. Vaulthey, E.; Suppan, P.; Haselbach, E. *Helvetica Chim. Acta*, **1988**, *71*, 93.
17. Johnson, M. D.; Miller, J. R; Green, N. S.; Closs, G. L. *J. Phys. Chem.*, **1989**, *93*, 2093.
18. (a) Wasielewski, M. R.; Niemczyk, M. P.; Johnson, D. G.; Svec, W. A.; Minsek, D. W. *Tetrahedron*, **1989**, *45*, 4785; (b) Verhoeven, J. W. *Pure Appl. Chem.*, **1990**, *62*, 1585; (c) Liu, J. Y.; Schmidt, J. A.; Bolton, J. R. *J. Phys. Chem.*, **1991**, *95*, 6924.
19. Sakata, Y.; Tsue, H.; Goto, Y.; Misumi, S.; Asahi, T.; Nishikawa, S.; Okada, T.; Mataga, N. *Chem. Lett.*, **1991**, *81*, 1307.
20. Oliver, A. M.; Craig, D. C.; Padden-Row, M. N; Kroon, J.; Verhoeven, J. W. *Chem. Phys. Lett.*, **1988**, *150*, 366.
21. Zeng, Y.; Zimmt, M. *J. Am. Chem. Soc.*, **1991**, *113*, 5107.
22. (a) Rehm, D.; Weller, A. *Ber. Bunsenges. Phys. Chem.*, **1969**, *73*, 834; (b) Weller, A. *Z. Phys. Chem.*, **1982**, *133*, 93.
23. (a) Jones, G., II; Santhanam, M.; Chiang, S. H. *J. Am. Chem. Soc.* **1980**, *102*, 6088; (b) Jones, G., II; Chiang, S. H.; Becker, W. G.; Welch, J. A. *J. Phys. Chem.*, **1982**, *86*, 2805.
24. (a) Marcus, R. A.; Sutin, N. *Biochimica et Biophysica Acta*, **1985**, *811*, 265; (b) Van der Zwan, G. Hynes, J. *Chem. Phys.*, **1984**, *90*, 21; (c) Sparpaglione, M.; Mukamel, S. *J. Phys. Chem.*, **1987**, *91*, 3938; (d) Onuchic, J. N.; Beratan, D. N.; Hopfield, J. J. *J. Phys. Chem.*, **1986**, *90*, 3707.
25. (a) Kosower, E. M.; Huppert, D. *Ann. Rev. Phys. Chem.*, **1986**, *37*, 127; (b) Su, S. G.; Simon, J. D. *J. Phys. Chem.*, **1986**, *90*, 6475; (b) Su, S. G.; Simon, J. D. *J. Chem. Phys.*, **1988**, *89*, 908.
26. Newton, M. *Chem. Rev.*, **1991**, *91*, 767.
27. (a) Sumi, H.; Marcus, R. A. *J. Chem Phys.*, **1986**, *84*, 4894; (b) Bagchi, B.; Chandra, A.; Fleming, G. R. *Ultrafast Phenomena VII* (Harris, L. B.; Ippen, E. P.; Mouron, G. A.; Zewail, A. H., Eds.). Springer–Verlag, Berlin, 1990, p. 408.
28. Mataga, N., Nishikawa, S., Asahi, T.; Okada, T. *J. Phys. Chem.*, **1990**, *94*, 1443.
29. (a) Kang, T. J.; Kahlow, M. A.; Giser, D.; Swallen, S.; Nagarajan, V.; Jarzeba, W.; Barbara; P. F. *J. Phys. Chem.*, **1988**, *92*, 6800; (b) Lueck, H.; Windsor, M. W.; Rettig, W. *J. Phys. Chem.*, **1990**, *94*, 4550; (c) Barbara, P. F.; Jarzeba, W. *Adv. Photochem.*, **1990**, *15*, 1; (d) Tominaga, K.; Walker, G. C.; Jarzeba, W.; Barbara, P. F. *J. Phys. Chem.*, **1991**, *95*, 10475.
30. (a) Mulliken, R. S. *J. Am. Chem. Soc.*, **1950**, *72*, 600; (b) Mulliken, R. S. *J. Am. Chem. Soc.*, **1952**, *74*, 811; (c) Mulliken, R. S. *J. Chem. Phys.*, **1955**, *23*, 397.
31. Foster, R. *Organic Charge-Transfer Complexes*. Academic Press, London, 1969.
32. Jones, G., II. *Photoinduced Electron Transfer* (Fox, M. A.; Chanon, M., Eds.), Part A. Elsevier, Amsterdam, 1988, p. 245.
33. Ojima, S.; Miyasaka, H.; Mataga, N. I. *Phys. Chem.*, **1990**, *94*, 5834 and 7534.
34. (a) Hilinski, E. F.; Masnovi, J. M.; Amatore, C.; Kochi, J. K.; Rentzepis, P. M. *J. Am. Chem. Soc.*, **1983**, *105*, 6167; (b) Goodman, J. L.; Peters, K. S. *J. Am. Chem. Soc.*, **1985**, *107*, 1441 and 6459.
35. Beens, H.; Weller, A. *Organic Molecular Photophysics* (Birks, J. B., Ed.). Wiley, London, 1975, Vol. 2, p. 159.
36. Orbach, N.; Ottolenghi, M. *Chem. Phys. Lett.*, **1975**, *35*, 175.
37. (a) Lim, B. T.; Okajima, S.; Lim, G. C. *J. Chem. Phys.*, **1986**, *84*, 1937; (b) Jones, G., II; Haney, W. A.; Phan, X. T. *J. Am. Chem. Soc.* **1988**, *110*, 1922.
38. (a) Niimara, N.; Yoshihara, K.; Hosoya, H.; Nagakura, S. *J. Phys. Chem.*, **1969**, *73*, 2670; (b) Masuhara, H.; Mataga, N. *Z. Phys. Chem. (Munich)*, **1972**, *80*, 113.
39. Mataga, N.; Kubota, T. *Molecular Interactions and Electronic Spectra*. Marcel Press, New York, 1970, p. 411.

40. Okada, T.; Fujita, T.; Kubota, M.; Masaki, S.; Mataga, N.; Ide, R.; Sakata, Y.; Misumi, S. *Chem. Phys. Lett.*, **1972**, *14*, 563.

41. (a) Hermant, R. M.; Bakkar, N. A. C.; Sherer, T.; Krijnen, B.; Verhoeven, J. W. *J. Am. Chem. Soc.*, **1990**, *112*, 1214; (b) Overing, H.; Padden-Row, M. N.; Heppener, M.; Oliver, A. M.; Cotsaris, E.; Verhoeven, J. W.; Hush, N. S. *J. Am. Chem. Soc.*, **1987**, *109*, 3258; (c) Overing, H.; Verhoeven, J. W.; Padden-Row, M. N.; Warman, J. M. *Tetrahedron*, **1989**, *45(15)*, 4751.

42. Wasielewski, M. R.; Minsek, D. W.; Niemczyk, M. P.; Svev, W. A.; Yang, N. C. *J. Am. Chem. Soc.*, **1990**, *112*, 1512.

43. (a) Rehm, D.; Weller, A. *Israel J. Chem.* **1970**, *8*, 259; (b) Eberson, L. *Electron Transfer Reactions in Organic Chemistry.* Springer Verlag, New York, 1987, pp. 27, 157.

44. (a) Lippert, E.; Rettig, W.; Bonacic-Koutecky, V.; Heisel, F.; Miehe, J. A. *Adv. Chem. Phys.*, **1987**, *68*, 1; (b) Rettig, W. *Angew. Chem. Int. Ed.*, **1986**, *75*, 971; (c) Zander, M.; Rettig, W. *Chem. Phys. Lett.*, **1984**, *110(6)*, 602.

45. Grabowski, Z. R.; Rotkiewicz, K.; Siemiarczuk, A.; Cowley, D. J.; Baumann, W. *Nouv. J. Chim.*, **1979**, *3(7)*, 443.

46. Lippert, E.; Ludor, W.; Boos, H. *Advances in Molecular Spectroscopy.* (Mangin, A., Ed.). Pergamon Press, Oxford, 1962, p. 443.

47. (a) Rettig, W.; Bonacic–Koutecky, V. *Chem. Phys. Lett.*, **1979**, *62*, 115; (b) Grabowski, Z. R.; Rotkiewicz, K.; Siemiarczuk, A. *J. Lumin.*, **1979**, *18–19*, 420; (c) Majumdar, D.; Sen, R.; Bhattacharyya, K.; Bhattacharyya, S. P. *J. Phys. Chem.*, **1991**, *95*, 4324.

48. (a) Bonacic–Koutecky, V.; Koutecky, J.; Michl, J. *Angew. Chem. Int. Ed.*, **1987**, *26*, 170; (b) Michl, J.; Bonacic-Kontecky, V. *Electronic Aspects of Organic Photochemistry.* John Wiley and Sons, New York, 1990, p. 338.

49. (a) Jones, G., II; Jackson, W. R.; Choi, C.; Bergmark, W. R. *J. Phys. Chem.*, **1985**, *89*, 294; (b) Jones, G., II; Jackson, W. R.; Kanoktanaporn, S.; Halpern, A. M. *Optics Commun.*, **1980**, *33*, 315; (c) Jones, G., II; Jackson, W. R.; Kanoktanaportn, S.; Bergmark, W. R., *Photochem and Photobiol.*, **1985**, *42*, 477.

50. (a) Howell, R.; Petek, H.; Phillips, D.; Yoshihara, Keitaro, *Chem. Phys. Lett.*, **1991**, *183*, 249; (b) Peng, W. L.; Dantus, M.; Zewail, A. H.; Kemnitz, K.; Hicks, J. M.; Eisenthal, K. B. *J. Phys. Chem.*, **1987**, *91*, 6162; (c) Herbich, J.; Salgado, F. P.; Rettschnick, R. P. H.; Grabowski, Z. R.; Wogtowicz, C. *J. Phys. Chem.*, **1991**, *95*, 3491.

51. Weisenborn, P. C. M.; Huizor, A. H.; Varma, C. A. G. O. *Chem. Phys.*, **1989**, *133*, 437.

52. Leinhos, V.; Kuhle, W.; Zachariasse, K. A. *J. Phys. Chem.*, **1991**, *95*, 2013.

53. Mataga, N.; Yao, H.; Okada, T.; Rettig, W. *J. Phys. Chem.*, **1989**, *93*, 3383.

54. (a) Okada, T.; Fujika, T.; Mataga, N. *Z. Phys. Chem.*, **1976**, *101*, 57; (b) Masaki, S.; Okada, T.; Mataga, N.; Sakata, Y.; Misumi, S. *Bull. Chem. Soc. Jpn.*, **1976**, *49*, 1277.

55. Detzer, N.; Bauman, W.; Schwager, B.; Frohling, J. C.; Brittinger, C. *Z. Naturforsch*, **1987**, *42a*, 395.

56. Okada, T.; Mataga, N.; Bauman, W.; Siemiarczuk, P. *J. Phys. Chem.*, **1987**, *91*, 4490.

57. Siemiarczuk, A.; Grabowski, Z. R.; Krowczynski, A; Asher, M.; Ottolenghi, M. *Chem. Phys. Lett.*, **1977**, *51*, 315.

58. Okada, T.; Kawai, M.; Ikemachi, T.; Mataga, N.; Sakata, Y.; Misumi, S.; Shionoya, S. *J. Phys. Chem.*, **1984**, *88*, 1976.

59. (a) Maroncelli, M.; Fleming, G. *J. Chem. Phys.*, **1987**, *86*, 6221; (b) Jarzeba, W.; Walker, G. C.; Johnson, A. E.; Barbara, P. F. *Chem. Phys.*, **1991**, *152*, 57.

60. (a) Fonseca, T. *J. Chem. Phys.*, **1989**, *91*, 2869; (b) Tominaga, K.; Walker, G. C.; Kang, T. J.; Barbara, P. F.; Fonseca, T. *J. Phys. Chem.*, **1991**, *95*, 10492.

61. Birks, J. B. *Photophysics of Aromatic Molecules.* John Wiley and Sons, London, 1970, p. 86.

62. Van der Auweraer, M.; Grabowski, Z. R.; Rettig, W. *J. Phys. Chem.* **1991**, *95*, 2083.

63. Rettig, W.; Zander, M. *Ber. Bunsenges. Phys. Chem.*, **1983**, *87*, 1143.

64. Nakashima, N.; Murakawa, M.; Mataga, N. *Bull. Chem. Soc. Jpn.*, **1976**, *49(4)*, 854.

65. Kang, T. J.; Jarzeba, W.; Barbara, P. F.; Fonseca, T. *Chem. Phys.*, **1990**, *149*, 81.

66. Okada, T.; Nishikawa, S.; Kanaji, K.; Mataga, N. *Ultrafast Phenomena VII* (Harris, L. B.; Ippen, E. P.; Mouron, G. A.; Zewail, A. H., Eds.). Springer-Verlag, Berlin, 1990, 397.

67. Hara, K.; Takuya, A.; Jiro, O. *J. Am. Chem. Soc.*, **1984**, *106*, 1968.

68. (a) Al–Hassan, K. A.; Azumi, T. *Chem. Phys. Lett.*, **1988**, *150*, 344; (b) Al–Hassan, K. A. *Chem. Phys. Lett.*, **1991**, *179*, 195.

69. Tseng, J. C. C.; Singer, L. *J. Phys. Chem.*, **1989**, *93*, 7092.

70. (a) Rettig, W.; Marschner, F. *Nouv. J. Chim.*, **1983**, *7*, 425; (b) Rettig, W.; Marschner, F. *New J. Chem.*, **1990**, *14*, 819.

71. Cowley, D. J.; O'Kane, E.; Todd, R. S. *J. J. Chem. Soc. Perkin Trans. 2*, **1991**, 1495.

72. (a) Jonker, S. A.; Van Dijk, S. I.; Goubitz, K.; Reiss, C. A.; Schuddeboom, W.; Verhoeven, J. W. *Mol. Cryst. Liq. Cryst.*, **1990**, *183*, 273; (b) Jonker, S. A.; Verhoeven, J. W.; Reiss, C. A.; Goubitz, K.; Heijdedrijk, D. *Recl. Trav. Chim. Pays–Bas*, **1990**, *109*, 154; (c) Jonker, S. A.; Ariese, F.; Verhoeven, J. W. *Recl. Trav. Chim. Pays–Bas*, **1989**, *108*, 109.

73. Bailey, M. L. *Acridines*, 2nd ed. (Acheson, R. M., Ed.). John Wiley and Sons, New York, 1970, p. 631.

74. Jones, G., II; Farahat, M. S. (To be published).

75. Kakitani, T.; Mataga, N. *J. Phys. Chem.*, **1986**, *90*, 993.

76. Akesson, E.; Walker, G. C.; Barbara, P. F. *J. Chem. Phys.*, **1991**, *95*, 4188.

77. (a) Dimroth, K.: Reichardt, C.; Siepmann, T.; Bohlmann, F. *Liebig Ann. Chem.*, **1960**, *38*, 1590; (b) Reichardt, C. *Solvents and Solvent Effects in Organic Chemistry*, 2nd ed. VCH Publishers, Weinheim, 1988, p. 363.

78. Jortner, J.; Bixon, M. *Cem. Phys.*, **1988**, *88*, 167.

79. Bagchi, B.; Chandra, A.; Fleming, G. R. *J. Phys. Chem.*, **1990**, *94*, 5197.

80. (a) Rettig, W. *Appl. Phys. B*, **1988**, *45*, 145; (b) Launay, J. P.; Sowinska, M.; Leydier, L.; Gourdon, A.; Amouyal, E.; Boillot, M. L.; Heisel, F.; Miche, J. A. *Chem. Phys. Lett.*, **1989**, *160*, 89.

LIGHT-INDUCED ELECTRON TRANSFER IN INORGANIC SYSTEMS IN HOMOGENEOUS AND HETEROGENEOUS PHASES

Nick Serpone, Rita Terzian,

Darren Lawless, and Jean-Marie Herrmann

Advances in Electron Transfer Chemistry
Volume 3, pages 33–166.
Copyright © 1993 by JAI Press Inc.
All rights of reproduction in any form reserved.
ISBN: 1-55938-320-8

1. INTRODUCTION

In the field of organic chemistry, several light-induced electron transfer reactions have been described.[1] Since the publication of the series of volumes on *Photoinduced Electron Transfer* by Fox and Chanon[2] containing articles germane to the present work on light-induced electron transfer in hexacoordinate transition metal complexes,[3] on ion pairs and supramolecular systems,[4] on intramolecular electron transfer between weakly coupled redox centers,[5] and on photocatalyzed oxidations and reductions on semiconductors,[6] these activities in the inorganic sphere have continued unabated in both homogeneous and heterogeneous phase. In the former phase, the earlier work (ca. pre-1987) was mostly concerned with bimolecular processes involving, to a large extent, the six-coordinate complex, $Ru(NN)_3^{2+}$ (where NN is 2,2′-bipyridine or 1,10-phenanthroline and their related derivatives). This complex and its homologs[7] have continued to provide the core for more complex systems (polynuclear, both homo and hetero, supramolecular devices) designed to mimic the functions of the photosynthetic center to achieve light energy migration and, of particular interest here, to achieve charge separation towards producing an artificial photosynthetic device[8–16] for the conversion and storage of solar energy via formation of high-energy chemicals. Past progress in this regard has been impeded in homogeneous phase in general, and for monomeric metal complexes in particular, by the rapid exoergonic and energy-wasting back electron transfer following the photoinduced electron transfer event (Eq. 1):

$$ML_6^{n+} + h\nu \rightarrow {}^*ML_6^{n+} \tag{1a}$$

$${}^*ML_6^{n+} + MY_6^{m+} \rightarrow ML_6^{(n-1)+} + MY_6^{(m+1)+} \text{ (reductive quenching)} \tag{1b}$$

$${}^*ML_6^{n+} + MX_6^{p+} \rightarrow ML_6^{(n+1)+} + MX_6^{(p-1)+} \text{ (oxidative quenching)} \tag{1c}$$

$$ML_6^{(n-1)+} + MY_6^{(m+1)+} \rightarrow ML_6^{n+} + MY_6^{m+} \text{ (back electron transfer)} \quad (1d)$$

$$ML_6^{(n+1)+} + MX_6^{(p-1)+} \rightarrow ML_6^{n+} + MX_6^{p+} \text{ (back electron transfer)} \quad (1e)$$

Clearly, efficient light energy conversion and storage necessitates the curtailment, if not the suppression of Eqs. 1d and 1e. To achieve this, approaches taken have utilized sacrificial reagents (electron relays) to scavenge the oxidized and reduced forms of ML_6^{n+} to couple one-electron excited state chemistry,[17] or to photochemically generate reactive intermediates which will undergo subsequent oxidation and reduction.[18-20] Additionally, the above reaction sequence consists of one-electron events, rather inefficient since most processes require multi-electron transfer to achieve conversion and storage. A good case in point is the water-splitting process (Eq. 2) where reduction of water is a two-electron transfer event, while oxidation of water is a four-electron transfer event. A recent novel approach[21] designed to achieve multi-electron events has explored and exploited the complementary redox function of two metals of electron-rich metal-metal bonds in quadruply bonded (M—4—M) dimers.

$$2 H_2O + 2 e^- \rightarrow H_2 + 2 OH^- \quad (2a)$$

$$2 OH^- - 4 e^- \rightarrow O_2 + 2 H^+ \quad (2b)$$

Recognizing that many biological reactions, including photosynthesis, occur in and are a consequence of the heterogeneity of the system,[22] success in artificial photosynthesis will also necessitate mimicking the heterogeneity if not the complexity of the photosynthetic apparatus. A key point in this direction is efficient charge separation following the light absorption act, suppression of charge recombination, and creation of a pool of these separated redox equivalents to achieve efficient oxidation and reduction reactions. A variety of approaches have been reported that employ semiconductors, membranes, vesicles, and molecular systems.[23] These heterogeneous systems have been shown to not only improve charge separation relative to systems in homogeneous media, but have also achieved accumulation of electron density, sufficient to drive multi-electron transfer processes.[24-36] Another recent approach has placed the transition metal photosensitizer, $Ru(bpy)_3^{2+}$, onto a solid support (porous Vycor glass[37]) to impose some kinetic control on the photoinduced events.

Much of the recent light-induced electron transfer events in supramolecular systems have been examined by the groups of Balzani,[8,9,12-15] Meyer,[10] and Scandola.[9,11,16,38] Additional studies have also appeared that update our knowledge in light-induced electron transfer in ion pairs[39,40] and in the utilization of metal complexes as light-absorption and light-emission sensitizers.[41] Two recent books embody much of the background material relevant to the present article; the reader might explore these for greater details on processes in homogeneous-phase electron transfer,[8] and in heterogeneous electron transfer in molecular assemblies (e.g., semiconductors).[42]

2. SPECTRAL AND PHOTOPHYSICAL PROPERTIES

2.1. Transition Metal Complexes

Transition metal complexes, ML_6^{n+}, consisting of the metal M and the aromatic organic ligands L, possess the requisite properties to be employed in light-induced electron transfer processes.[41] They normally exhibit relatively low-energy metal-to-ligand charge transfer absorption bands in the visible spectral region that make them especially useful as light-energy harvesters (Figure 1).

The presence of the metal atom in ML_6^{n+} causes spin-orbit coupling to appear in the complex. This breaks down the selection rules for communication between formally spin-forbidden excited states (singlet to triplet, for example). As a result, light energy cascade down to the lowest energy excited state (formally a triplet in $Ru(NN)_3^{2+}$) is fairly rapid, the quantum yield of formation of this state is unity, and its lifetime is relatively short (hundreds of nanoseconds to several microseconds) such that low concentrations of impurities and/or other adventitious agents are of little consequence to its longevity. Of greater import, these complexes possess distinct redox centers and are often luminescent. The redox potentials and the photophysical properties of TM complexes can be fine-tuned by judicious choice of ligands[7] and metals.[3]

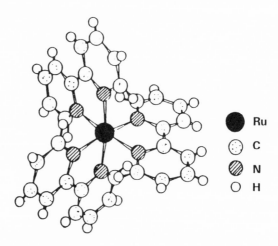

Ru
C
N
H

Figure 1a.   Structure of a ruthenium(II) polypyridine complex: $Ru(bpy)_3^{2+}$, where bpy is 2,2′-bipyridine. From ref. 43.

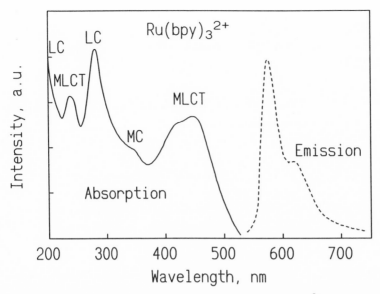

Figure 1b. Electronic absorption and emission spectra of Ru(bpy)$_3$$^{2+}$ in aqueous and in ethanol/methanol, respectively at ambient temperature. Adapted from ref. 41.

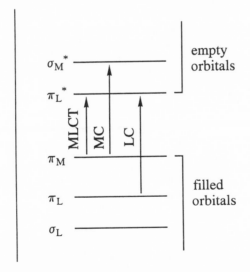

Figure 1c. Schematic representation of a molecular orbital energy level diagram for a ruthenium(II) polypyridyl complex, Ru(NN)$_3$$^{2+}$, in octahedral symmetry, illustrating the various possible electronic transitions: MLCT denotes an electron transition from an orbital predominantly metal in character to one that is predominantly ligand (metal-to-ligand charge-transfer); MC denotes metal centered electron transition in which the electron is simply redistributed amongst metal like orbitals; LC referes to ligand centered electron transitions in which electrons migrate between orbitals largely ligand in character. From ref. 7.

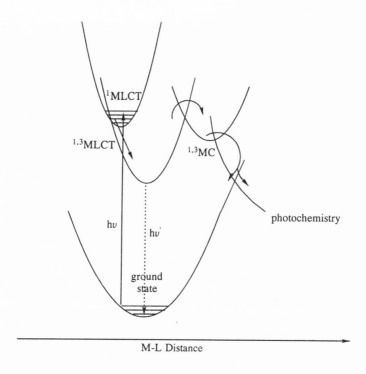

Figure 1d. Scheme showing the possible energetic positions of the MLCT and MC states in a complex like Ru(bpy)$_3$$^{2+}$, together with the possible deactivation channels of the optical electron transition energy. Adapted from ref. 7.

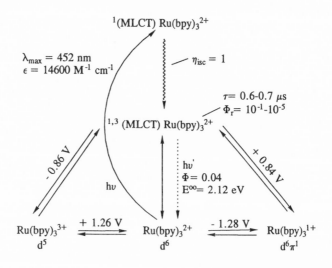

Figure 1e. Scheme summarizing some of the photophysical, photochemical, and electrochemical properties of the Ru(bpy)$_3$$^{2+}$ complex. Adapted from refs. 7 and 41.

38

Light is an essential reagent in photoinduced electron transfer (Eqs. 1a–c). Absorption of light by a TM complex populates rapidly and very efficiently, in most instances, the lowest energy excited state of a given multiplicity. As noted in Figure 1, this state is the ^{1,3}(MLCT) state for Ru(bpy)$_3^{2+}$ which is luminescent, is relatively long-lived (ca. 600 ns[7]), and possesses a very rich redox chemistry.[7] These excited states are simultaneously better oxidants and reductants than is the ground state complex. The origin of this enhanced redox behavior is best understood considering the one-electron energy level diagram of Figure 2. The redox ability of the excited state depends on the ionization potential and the electron affinity of the new electronic isomer of the ground state species. Thus light excitation of an electron from the HOMO to the LUMO orbital decreases the ionization potential; the subsequent generation of the *hole* in the HOMO orbital increases the electron affinity. In the thermally relaxed state, electron transfer events can be treated by thermodynamic and kinetic considerations analogous to electron transfer processes of ground state species.[41]

To a first approximation, where the entropy differences between ground and excited state species are negligible, the reduction and oxidation potentials of an excited state are given by (Eq. 3):

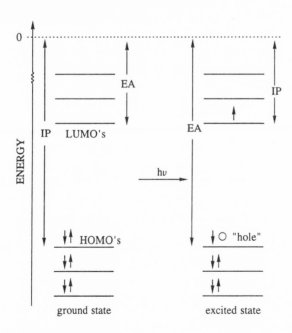

Figure 2. One-electron orbital energy level diagram illustrating the changes in ionization potentials and electron affinities of a molecule under light irradiation. Adapted from ref. 3.

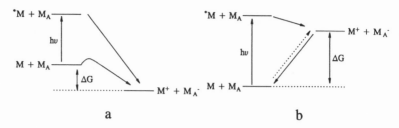

Figure 3. Scheme showing two possible energetic situations for electron transfer processes implicating excite state species. Adapted from ref. 41.

$$E^{o}(^{*}M/M^{-}) = E^{o}(M/M^{-}) + E^{oo}(^{*}M/M) \qquad (3a)$$

$$E^{o}(^{*}M/M^{+}) = E^{o}(M/M^{+}) + E^{oo}(^{*}M/M) \qquad (3b)$$

where $E^{o}(M/M^{-})$ and $E^{o}(M/M^{+})$ are the reduction and oxidation potentials of the ground state complex ML_6^{n+} (herein denoted as M), and $E^{oo}(^{*}M/M)$ is the 0–0 spectroscopic energy of the excited state. The enhanced redox properties of the excited state species is not a sufficient condition for efficient electron transfer, since it must compete with additional deactivation channels as exemplified by the excited state lifetime τ $\{= 1/k_{obs} = 1/(k_{el.tr.} + \Sigma k_i)\}$.

Absorption of light by the light harvester ML_6^{n+} (photosensitizer) in the presence of another TM complex or organic species (oxidative quencher, M_A) will give rise to two possible situations[41] (Figure 3) which are entirely analogous to those occurring in semiconductors (see below). The first (Figure 3a) corresponds to a process whereby light acts as a catalyst, accelerating a kinetically slow (high activation energy) exergonic dark reaction ($\Delta G < 0$). Note that the exergonicity of the light-induced electron transfer event is increased ($\Delta G < {}^{*}\Delta G$). In the second situation, light transforms M into a better reductant such that a thermodynamically unfavorable ($\Delta G > 0$) electron transfer event in the ground state M becomes allowed. Here, light is used to drive $M + M_A$ to $M^+ + M_A^-$ via $^{*}M + M_A$, and a fraction of the light absorbed is converted into *chemical energy* of the products that will be released in the back electron transfer step (Eqs. 1d,e). It must be reemphasized here that, in some instances, only a fraction of the light absorbed produces the reactive excited state (dependent, as it were, on the internal conversion processes and on the intersystem crossing efficiency), and then only a fraction of such reactive excited state species will undergo electron transfer as it must compete with other competitive processes (Eq. 4). Where M and M_A are covalently (weakly) linked to give a dinuclear species, $M-M_A$, and M remains as the light-absorbing chromophore, the intramolecular electron transfer process in this case would also have to compete with excited state decay of this excited chromophore (Eq. 5).

$$^*M + M_A \rightarrow M^+ + M_A^- \qquad k_q \qquad (4a)$$

$$^*M \rightarrow M \qquad 1/\tau \qquad (4b)$$

$$*M\text{-}M_A \rightarrow M^+\text{-}M_A- \qquad k_{intra} \qquad (5a)$$

$$^*M\text{-}M_A \rightarrow M\text{-}M_A \qquad 1/\tau \qquad (5b)$$

2.2. Semiconductors

Before undertaking a description of heterogeneous electron transfer between a solid phase and, of relevance here, a solution phase, it will be instructive to consider some of the diversities of semiconductor particulates that are unlike the molecular systems with which many of us are more familiar. In so doing, we shall consider the distinctiveness of semiconductor particulates in terms of *structural and electronic properties*. The reader is referred to three recent books[42,44] which deal with these aspects at some great length. Herein we treat only those properties of direct relevance to electron transfer.

Structural Properties of Semiconducting Particulates

Particulate systems are distinguished by their size. Thus, particles whose size is less than 1000 Å are referred to as *colloids*, and above 1000 Å are known as macroparticles. The interest here is in colloids in which we also distinguish those semiconductor particulates which display quantum size effects (for TiO_2, radius < 3 Å; for CdS, radius < 25 Å[42]). These effects are displayed when the Bohr radius of the first exciton is similar to or larger than that of the particle. Semiconductor colloids have become the focus of much interest in heterogeneous photocatalysis and electron transfer reactions owing to their particular properties: (1) excellent light harvesters; (2) charge carriers (electrons and holes) generated subsequent to the light absorption event can be separated in space and time; (3) particle surface can be modified to increase the light absorption characteristics in lower energy spectral regions, particularly important for large bandgap metal oxide semiconductors, as well as increase the catalytic activity; (4) particle surface can provide a template for more efficient redox reactions; (5) relatively easy to fabricate; and (6) relatively inexpensive. For those colloids whose size is less than the deBroglie wavelength, the usual spectroscopic techniques can be employed to characterize their properties, not least of which are those processes (energy and electron transfer) that are particle-size dependent.

During the preparation of semiconductor colloid particles, lattice defects are unavoidable even though the material may possess a high degree of crystallinity. We consider metal oxide semiconductors (n-type) which are anion deficient. An

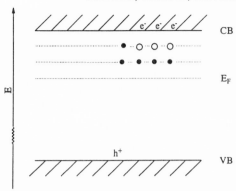

Figure 4. Energy scheme illustrating the oxygen vacancies in a metal oxide semi-conductor in which they act as electron donors. After ref. 42.

example of such a material is TiO_2. Figure 4 illustrates[42] the concept of anion vacancies and their ionization to populate the conduction band. The ionization can also be described using the Kroger notation. Thus:

$$
\begin{array}{ccc}
O & M & O \\
M & O & M \\
O & M & O
\end{array}
\rightleftharpoons
\begin{array}{ccc}
O & M & O \\
M & \textcircled{2-} & M \\
O & M & O
\end{array}
+ 1/2\, O_2 \tag{6}
$$

or more simply:

$$O_o \rightleftharpoons V_O + 1/2\, O_2 \tag{7}$$

where loss of oxygen leaves the lattice neutral. The anion vacancy can now be ionized because of the proximity (a few meV) of the defect level to the conduction band. The first and second ionizations are described by Eqs. 8 and 9. Clearly, anion vacancies in a metal oxide lattice, which are regarded as electron donors, generate an n-type semiconductor. It is worth noting that cation vacancies (electron acceptors from valence band) lead to p-type conduction. The overall process is given by Eq. 10 (adding Eqs. 7, 8, and 9):[42]

$$V_O \rightleftharpoons V_O^{\bullet} + e_{CB}^{-} \tag{8}$$

$$V_O^{\bullet} \rightleftharpoons V_O^{\bullet\bullet} + e_{CB}^{-} \tag{9}$$

$$O_o \rightleftharpoons V_O^{\bullet\bullet} + 1/2\, O_2 + 2e_{CB}^{-} \tag{10}$$

and the corresponding mass action expression (Eq. 11) is:

$$K = [V_O^{\bullet\bullet}]\, [\, e_{CB}^{-}\,]^2 P_{O_2}^{1/2} \tag{11}$$

The first ionization (Eq. 8) is achieved at reasonably high temperatures, where $2[V_O^{\bullet\bullet}] = [e_{CB}^-]$; the electron concentration then becomes $[e_{CB}^-] = (2K)^{1/3} P_{O_2}^{-1/6}$. This latter expression has important consequences in the electrical conductance (a measure of the free electron concentration) of the material, and suggests that the conductance should increase or decrease as minus one-sixth of the pressure of oxygen. This notion has been verified experimentally in one of our laboratories. Some of the relevant work is treated in Section 6 of this chapter.

Optical and Electronic Properties

Semiconductors absorb light at some threshold wavelength, λ_{th} (nm), sometimes referred to as the fundamental band edge; it is related to the bandgap energy E_{bg} by the relation $\lambda_{th} = 1240/E_{bg}$ (eV).[42] By analogy to molecular systems, which follow the Beer-Lambert law for light absorption, a similar expression is had for the absorption of light by a semiconductor. Thus, the absorption coefficient α (also described as the reciprocal absorption length in cm^{-1} units) is related to the transmitted light intensity I_{trans} and the incident light intensity I_0 by the exponential law (Eq. 12):

$$I_{trans} = I_0 \exp\{-\alpha\, l\} \tag{12}$$

where l is the penetration depth of light in cm. For TiO_2, $\alpha = 2.6 \times 10^4\ cm^{-1}$ at 320 nm.[42] The magnitude of α increases with the energy of the photons near threshold according to Eq. 13, where n is 1/2 for a direct transition and n = 2 for an indirect transition. This distinction between *direct* and *indirect* semiconductors, based on

$$\alpha = \frac{\{const[h\nu - E_{bg}]^n\}}{h\nu} \tag{13}$$

the type of transition, is illustrated in Figure 5.[42,45–47] Selection rules for photon absorption dictate that only transitions where there is zero net momentum $\mathbf{k}$ change are allowed; the absorption process is dependent on the band structure of the semiconductor solid. Hence, where electron transition occurs from the highest energy level of the valence band (analogous to the HOMO's in molecular systems) to the lowest energy level of the conduction band (equivalent to the LUMO's) and involves no momentum change, the transition is said to be a *direct transition* and the solid is referred to as a *direct bandgap semiconductor* (Figure 5a). Where the transition occurs from the valence band maximum which is located at a position different from the conduction band minimum on the $\mathbf{k}$ coordinate, and involves a momentum change in order to satisfy momentum conservation, the transition is said to be an *indirect transition* and the solid is referred to as an *indirect bandgap material.*[47]

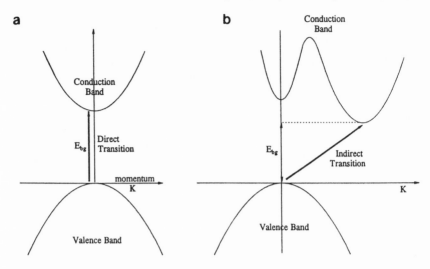

Figure 5. (a) Plot showing the energy of the valence and conduction bands as a function of the momentum coordinate **k** for a direct bandgap semiconductor; (b) plot showing the energy of the two bands for an indirect bandgap material. After ref. 45.

Figure 6 illustrates bandgaps and band edge positions (at pH 1) for a selected number of semiconductor materials which have been useful in heterogeneous photocatalysis. These are TiO_2, ZnO, CdS, and WO_3. Also shown are redox potentials for a series of redox couples to illustrate the thermodynamic limitations

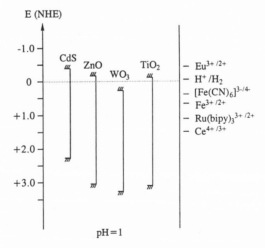

Figure 6. Band-edge positions for several semiconductors in contact with an aqueous electrolyte at pH 1. After ref. 44.

of the type of photoreactions that can be carried out with the photogenerated charge carriers. Thus, reduction of species in the electrolyte medium requires that their redox level be below the conduction band position of the semiconductor used. For example, Eu^{3+} cannot be reduced by TiO_2 photogenerated electrons. An aside to this discussion, but a relevant one in practical terms when these semiconductors are used in aqueous media, is that if the redox levels for the reduction and the oxidation of water lie within the bandgap of the semiconductor, the material will photocorrode (anodically and/or cathodically). It should be noted that the band edge positions reflect flatband potentials determined from capacity measurements.[42] However, when a semiconductor bulk or colloid material comes into contact with an aqueous electrolyte, changes occur (space charge layers and band bending) and these are now described below for an n-type semiconductor.

The formation of a space charge layer necessitates an interfacial transfer of electrons between the semiconductor and an electroactive species in the electrolyte medium. Within this space charge layer, the valence and conduction bands are bent. Four possible situations arise; they are depicted in Figure 7.[46] The *flatband condition* is that at which there is no space charge layer. If electrons accumulate on the semiconductor side, an *accumulation layer* obtains and the bands are bent downward. If, however, electrons accumulate in the electrolyte side, a *depletion layer* is formed and the bands are bent upward. Depletion of the majority carriers (here the electrons) can continue to a point such that their concentration at the surface decreases below the intrinsic level.[42] With the electronic equilibrium maintained, the concentration of the minority carriers (here the holes) in this region of the space charge layer will exceed that of the electrons, and an *inversion layer* is created. The consequence of this is that the Fermi level is closer to the valence

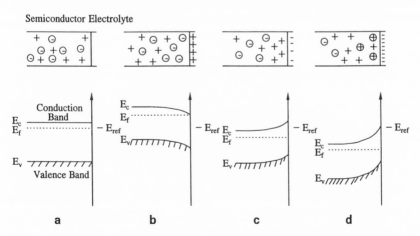

Figure 7. Formation of a space charge layer on an n-type semiconductor at the interface with a solution: (**a**) flatband condition; (**b**) accumulation layer; (**c**) depletion layer; (**d**) inversion layer. From ref.46.

band than the conduction band and the semiconductor then acquires p-type semi-conductivity at the surface but remains n-type in the bulk. The depletion layer case is the one that is most useful in relation to photocatalysis and electron transfer processes. For TiO_2 in particular, and metal oxide semiconductors in general, the colloid particle surface is amphoteric with sites that can be charged positively and negatively (Eqs. 14, 15) depending upon the point of zero charge (pH at which the surface charge is 0). The potential difference between the semiconductor and solution changes by 0.059 V per pH unit.

$$\equiv Ti\text{-}O^- + H^+ \rightleftharpoons \equiv Ti\text{-}OH \tag{14}$$

$$\equiv Ti\text{-}OH + H^+ \rightleftharpoons \equiv Ti\text{-}OH_2^+ \tag{15}$$

Flatband potential measurements have shown that for TiO_2 colloidal particles, the redox potential of the conduction band changes in a cathodic direction as (Eq. 16):

$$E_{CB} = -0.1 - 0.059 pH \tag{16}$$

The importance of the above Nernstian expression is that photoredox reactions occurring on the surface of semiconductor materials can be fine-tuned by varying the pH of the solution. A case in point is the reduction of H^+ to H_2 for which according to Figure 6 the reaction is thermodynamically allowed over the whole pH range.

Because the photogenerated electrons and holes are formed only in the region where light has been absorbed by the semiconducting material, it is instructive to consider now the width of the space charge layer. For a planar electrode, the width W of the space charge layer is given by:

$$W = L_D(\Delta\phi \, e/kT)^{1/2} \tag{17}$$

where L_D is the Debye length given by the relation:

$$L_D = (\varepsilon_0 \, \varepsilon \, kT/2e^2 N_0)^{1/2} \tag{18}$$

Here, N_0 is the number of ionized dopant molecules per cm^3 and $\Delta\phi$ is the potential drop within the layer.[46] The potential distribution, $\Delta\phi$, for a spherical n-type semiconductor particle in equilibrium with an electrolyte containing a redox couple whose Fermi level is E_F has been derived by Albery and Bartlett;[48] it is illustrated in Figure 8 for large and small particles.[46] The difference in the potential at $r = 0$ and at a distance r is given by:

$$\Delta\phi = \{kT/6e\}\{[r - (r_0 - W)]/L_D\}^2\{1 + 2(r_0 - W)/r\} \tag{19}$$

Two limiting cases arise from this expression that bear great significance for light-induced electron transfer in semiconductor suspensions. The first case occurs

n-TYPE SEMICONDUCTOR

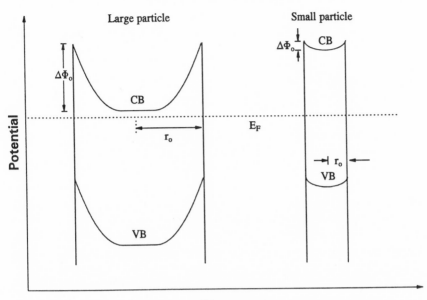

Distance

Figure 8. Plots illustrating the formation of a space charge layer in a large and small semiconductor particle in equilibrium with a solution redox system, whose Fermi level is denoted E_F. Note that the small particle is nearly depleted of charge the consequence of which is that its E_F is positioned at the middle of the bandgap and the band bending is negligibly small. From ref. 46.

when $r_0 \gg W$; that is, when the particle size is much greater than the depletion layer width. Where $r = r_0$, one obtains:

$$\Delta\phi_0 = \{kT/2e\}\{W/L_D\}^2 \qquad (20)$$

Equation 20 considers the total potential drop $\Delta\phi_0$ within the semiconductor particle. The congruence of Eq. 20 (after rearrangement) with Eq. 17 suggests that the width of the depletion layer for a large spherical particle is identical to that for a planar electrode. In the case of very small particles, the condition $r_0 = W$ holds and the corresponding relation for the potential drop becomes:

$$\Delta\phi_0 = \{kT/6e\}\{r_0/L_D\}^2 \qquad (21)$$

Equation 21 infers that the electric field in colloidal semiconductors is small and that high dopant levels are needed to produce a significant potential difference between the surface and the center of the particle. As an example,[42] to get a 50 mV potential drop in a colloidal TiO_2 particle with $r = 6$ nm requires the concentration of ionized donor impurities to be 5×10^{19} cm^{-3}. In undoped TiO_2 particles, this concentration is much smaller so that band bending is negligible. Where an electron transfer process occurs from very small semiconductor particles to a solution species which depletes the majority carriers, the electrical potential difference drops in the Helmholtz layer, diffuse layer contributions being neglected.[46] The consequence of this is that the band edge position will shift. Similarly, if photogenerated holes on small particles rapidly migrate from n-type semiconductors to a solution hole acceptor and the majority carriers remain in the particle, the conduction band edge at the surface will shift cathodically. This same process can be induced by applying a negative bias on the semiconductor electrode.

The depletion layer at the semiconductor/solution interface, also referred to as a Schottky barrier, plays an important role in light-induced charge separation (Figure 9).[42] The electrostatic field generated in the space charge layer causes the spatial separation of photoformed electrons and holes subsequent to the light absorption event. The direction of the field in a large particle (as in powders) of an n-type semiconductor is such that electrons migrate into the particle bulk to the *dark* side of the particle, while the holes migrate to the illuminated side of the particle surface. Both charge carriers present (and trapped) at the surface are poised to undergo redox reactions. In the case of a semiconductor electrode, the electrons having migrated to the back contact of the electrode move through the external circuit to the counter-electrode of a photoelectrochemical cell.

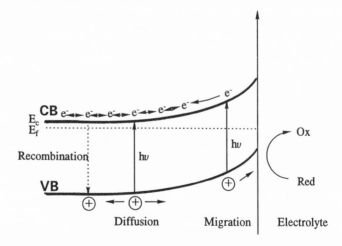

Figure 9. Photoinduced charge carrier separation assisted by the local electrostatic fields present in the depletion layer of a semiconductor electrode (or large particle) in contact with a redox system. Adapted from ref. 42.

Where the semiconductor particles are small (of colloidal dimensions), the separation of electrons and holes occurs primarily via diffusion because of the small electric fields generated in such particles (small, if any band bending). The average transit time for the charge carriers from the particle interior to the surface is given by Eq. 22.[42] For colloidal particles, this transit time is of the order of a few picoseconds (ps). For TiO_2 colloid particles, average radius of 6 nm and a diffusion coefficient for electrons $D_{e-} = 2 \times 10^{-2}$ cm^{2}s^{-1}, the transit time is $\sim$ 3 ps.[42] Once formed and trapped on surface sites, charge carriers can recombine by radiative (luminescence) and nonradiative pathways, and can interact with adsorbed solutes or catalysts on the particle surface. It is also worth noting that where diffusion of carriers is faster than recombination, as might occur in particles of very small dimensions, the quantum yield of photoredox processes can in principle reach unity. For this to be the case, however, also necessitates that at least one of the carriers be removed fairly rapidly in order to suppress surface recombinations. This fact calls attention to the important role played by interfacial charge transfer kinetics.[42]

$$\tau_d = r_o^2/\pi^2 D \qquad (22)$$

We now examine the neighboring environment of a semiconductor particle that might have an effect on interfacial charge transfer dynamics. For this we consider qualitatively the potential distribution in an electrolyte surrounding a *spherically*

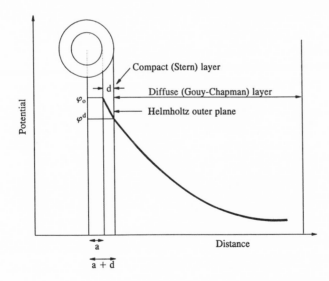

Figure 10. Compact layer and diffuse layer surrounding a charged spherical particle together with the variation of the potential gradient as a function of distance. From ref. 46.

charged colloidal semiconductor particle.[42] The solution environment that surrounds a particle consists of two distinct layers: (1) a compact layer, often referred
to as the Stern-Graham layer, having a width of about 1 to 5 Å depending on whether
or not the adsorbed ions are solvated; and (2) a diffuse layer sometimes referred to
as the Gouy-Chapman layer, whose width can extend from the outer Helmholtz
plane (see Figure 10) to about several hundred(s) nanometers into the electrolyte
space. The surface potential is denoted φ_o while the potential within the Stern layer
decreases almost linearly to a value φ^d at the Helmholtz plane. Within the diffuse
layer, the potential decreases exponentially and is distance-dependent as described
by Eq. 23 for low potentials:

$$\varphi = \varphi^d \exp[-\kappa(x - d)] \tag{23}$$

where d is the distance of closest approach of the ions to the colloid surface and
κ is the reciprocal of the Debye length.[42]

3. KINETIC ASPECTS OF ELECTRON TRANSFER

3.1. Electron Transfer in a Homogeneous Phase

Transfer of an electron from one species to another (*intermolecular*) or within
one part of a molecular system to another part of the same system (*intramolecular*)
might appear at first view to be a simple process. Yet the theoretical description of
such a process has posed great challenges. There are two principal processes that
are usually encountered: (1) in the first, the electron is exchanged between two
species (or between two regions of the same system) *without net chemical change*,
while in the second (2) electron transfer *leads to net chemical change*. The latter
are the commonly known oxidation/reduction reactions and are not dealt with in
this chapter.

In the classical theoretical model for electron transfer, the potential surfaces of
the reactants and products are connected by a ground state/excited state relationship
(Figure 11).[38,42] Initially, a *precursor* complex, M...M$_A$, forms between the electron
donor, M, and the electron acceptor, M$_A$. Subsequently, electron transfer takes place
along the reaction coordinate via the transition state to form the *successor* complex,
M$^+$...M$_A^-$, which can be considered as an electronically excited state of M...M$_A$.[38]
Additionally, this electronic excited state can be populated by light absorption
(E_{op}). This process is known as *optical electron transfer*, not to be confused with
photoinduced electron transfer;[38] such processes have also become known as
intervalence charge transfer (IVCT) transitions. These variations are illustrated in
Figure 12 which shows a schematic representation of different types of electron
transfer processes in a binary system. It is worth noting that optical electron transfer
and photoinduced electron transfer lead to the same product state for an endoergic
reaction.

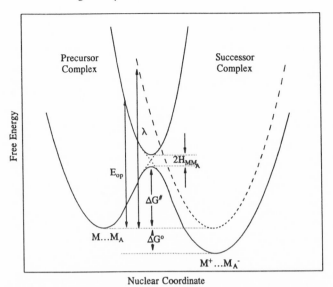

Figure 11. Free energy diagram for an adiabatic electron transfer process from the metal donor, M, to the electron acceptor, M_A; the $M...M_A$ denotes the nuclear configuration of the precursor complex, while $M^+...M_A^-$ refers to the nuclear configuration of the successor complex; ΔG^0 is the standard thermodynamic free energy for the process, and $\Delta G^{\#}$ is the free energy of activation. Adapted from refs. 38 and 42.

Important kinetic information can be accessed on thermal electron transfers from studies of the corresponding optical IVCT transitions.[38] Where concentrations of the reactants M and M_A for a binary system are low, such transitions are usually not observed, unless M and M_A form ion pairs (here IVCT transitions are also known as *outer-sphere*). Further, where the reactants are covalently linked (weakly) as in dinuclear or polynuclear complexes, observation of IVCT transitions is facilitated (intramolecular electron transfer). These are witnessed in the near-infrared region for isoergonic reactions, while IVCT are seen in the visible or UV regions for endoergonic reactions.[38]

$$M...M_A \xrightarrow{k} M^+...M_A^- \qquad (24)$$

In its most simplistic form (Eq. 24), description of the electron transfer step between the precursor complex, $M...M_A$, and the successor complex, $M^+...M_A^-$ can be understood in terms of conventional transition state theory.[49] As illustrated in Figure 11 and in the cartoon of Figure 13, in addition to the activation energy $\Delta G^{\#}$, the nuclear factor ν_N, that describes the changes in bond lengths and bond angles in M and M_A (inner nuclear coordinates), and solvent reorganization around the

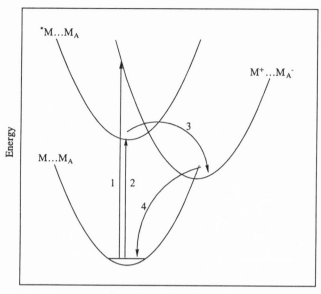

nuclear configuration

Figure 12. Scheme showing the potential surfaces of a binary system, M---M$_A$, together with the possible types of electron transfer processes: (**1**) optical electron transfer; (**2 + 3**) photoinduced electron transfer; and (**4**) thermal electron transfer. Adapted from ref. 38.

two reactants (outer nuclear coordinates), together with the electronic factor k_{el}, which describes the coupling of the precursor complex surface with that of the successor complex, have a direct consequence on k of Eq. 25. Thus:

$$k = \nu_N \, k_{el} \exp\{-\Delta G^{\#}/RT\} \tag{25}$$

As well, the Franck–Condon principle requires that prior to electron transfer, both reactants must undergo distortion (changes in inner and outer nuclear coordinates) to achieve a geometry such that the electron transfer occurs at an isoenergetic crossing point of the two surfaces. The term k_{el} in Eq. 25, sometimes also denoted as the transmission coefficient (κ) of the reaction, expresses the probability that the precursor complex converts into the successor complex at the crossing point. By contrast, the nuclear term ν_N in Eq. 25 expresses the maximum possible value for the rate constant k. For most compounds, the value of this nuclear frequency factor is taken as 10^{13} s^{-1}.[38]

The activation free energy $\Delta G^{\#}$ reflects both the energy needed to effect the distortion of M and M$_A$ in the precursor complex and the driving force for the

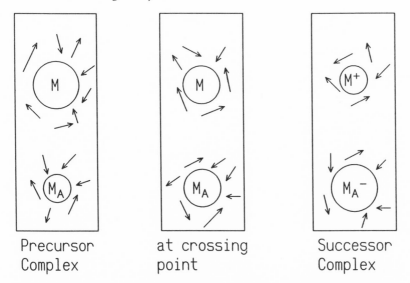

Precursor   at crossing   Successor
Complex   point   Complex

Figure 13. Diagram illustrating the nuclear/structural changes necessitated in an electron transfer process from the precursor complex to the successor complex. Note the identity in the structure of the two reactants, M and M_A, at the crossing point of their respective potential energy surfaces.

reaction. In this sense, the free energy of activation $\Delta G^{\#}$ can be expressed, according to the Marcus theory, by the parabolic free energy expression (Eq. 26):

$$\Delta G^{\#} = (\lambda/4)\{1 + \Delta G^{o}/\lambda\}^2 \tag{26}$$

$$\Delta G^{\#} = \{(\lambda/4)^2 + (\Delta G^{o}/2)^2\}^{1/2} \tag{27}$$

A different but empirical relationship (Eq. 27) based on experimental data has been given by Rehm and Weller.[50,51] In Eqs. 26 and 27, ΔG^{o} is the standard free energy change for the electron transfer process, and λ is the reorganizational energy ($\lambda = \lambda_i + \lambda_o$), and represents the intrinsic barriers to electron transfer corresponding to changes in bond lengths and bond angles (λ_i) and in solvent reorganization (λ_o). In the diagram of Figure 11, λ corresponds to the difference between the precursor (minimum in the solid curve) and successor (dashed curve) complex surfaces for an hypothetical isoergonic reaction and for a given fixed nuclear coordinate (identical distortions). This reorganizational parameters can be estimated using Marcus theory[49,52–55] from a knowledge of structural parameters, vibrational frequencies of M and M_A reactants and solvent dielectric constants. However, it is more common to determine λ from the self-exchange reactions of the couples

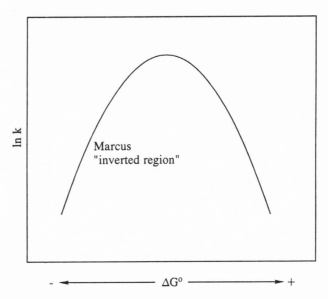

Figure 14. Plot showing the dependence of the electron transfer rate constant on the standard free energy for the reaction (the driving force). The region to the left of the maximum is the so-called "Marcus inverted region".

M/M$^+$ and M$_A$/M$_A^-$.[52] The merits of the relationships of Eqs. 26 and 27 have been critically reviewed.[56]

The Marcus expression for the free energy of activation (Eq. 26) is quite informative regarding the connectivity between $\Delta G^\#$ and the standard free energy of reaction; that is, the driving force for the electron transfer process, as to whether ΔG° is zero, is equal to λ, or is equal to $-\lambda$. Where $\Delta G^\circ = 0$, the free energy of activation $\Delta G^\#$ equals $\lambda/4$, but is equal to 0 when $\Delta G^\circ = -\lambda$; as the driving force for the reaction increases (ΔG° more negative), $\Delta G^\#$ increases again, thus describing a right-side-up parabolic curve. In view of Eq. 25, the rate constant for electron transfer k increases for small ΔG° values, reaches a plateau, and then expectedly decreases for thermodynamically very favorable reactions (high negative ΔG°). The latter is the well known "Marcus inverted region" (Figure 14), which can be understood considering the three possible ways that the precursor and successor complex surfaces of Figure 11 can cross. This is depicted schematically in Figure 15. In the *normal region*, for small positive free energies of reactions, the rate constant is relatively small; it increases significantly as the successor complex surface is shifted vertically lower (keeping the precursor surface fixed) at some point of which the crossing is essentially *activationless* ($\Delta G^\#$ is 0), following which further vertical lowering leads to an increase in the driving force and brings the crossing to a position to the left of the precursor surface minimum such that the

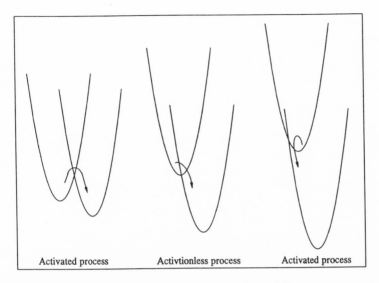

Figure 15. Potential energy curves for electron transfer in different energy regimes: $\Delta G > -\lambda$ (*normal region*); $\Delta G = -\lambda$ (*activation-free process*); and $\Delta G < -\lambda$ (*inverted region*). From ref. 38.

electron transfer process becomes an activated process once again. It should be noted that the maximum value of k will depend largely on the value of ν_N and to some extent on that of k_{el} in Eq. 25.

The various theoretical descriptions of electron transfer have approached electron exchange as either *adiabatic* or *nonadiabatic* electron transfer. In the adiabatic electron transfer case, the system(s) is thought to pass smoothly through an activated state to the final state; the electron is not viewed as a discrete particle, but rather the system(s) may pass through a series of intermediate stages. In a more theoretical sense, the adiabaticity of electron transfer is characterized by a relatively strong electronic interaction between the reactants M and M_A ($H_{MMa} \geq 0.025$ eV or ~ 200 cm^{-1} at ambient temperature).[42] Also, where the electronic interaction is strong, the transmission coefficient k_{el} (or κ) is unity and k for electron transfer is then a function only of the nuclear frequency factor ν_N and the free energy of activation $\Delta G^{\#}$. That is, the rate-determining step is the nuclear motion that leads to the transition state geometry, and the process is insensitive to factors that affect electronic interactions (center to center distances, solvent medium, orientational geometry, among others).[38]

In nonadiabatic electron transfer, the electron is viewed as a discrete particle, and considerations are given to potential energy changes when the electron is exchanged between the two interacting systems (or regions). Moreover, electronic coupling between reactant and product states is relatively weak; that is, H_{MMa} is very small and the corresponding transmission coefficient k_{el} is $\ll 1$. Here, the electron transfer step at the transition state is viewed as rate-determining, and such factors (see above) that affect electronic interactions have a direct consequence on the magnitude of k of Eq. 25.

Bimolecular (Intermolecular) Processes

In the case of a bimolecular electron transfer process between reactant M and reactant M_A, the overall reaction may be viewed as:

$$M + M_A \xrightarrow{k_{obs}} M^+ + M_A^-$$ (28)

Considering the detailed steps that might occur to bring the two reactants together via a diffusive step (k_{diff}) to form the encounter (precursor) complex, $M...M_A$, and separation step (k_{-diff}) that breaks up this precursor complex, the actual electron transfer step (k_{el}) that yields the successor complex, $M^+...M_A^-$, is shown in Eq. 29:

$$M + M_A \underset{k_{-diff}}{\overset{k_{diff}}{\rightleftharpoons}} M...MA \xrightarrow{k_{el}} M^+...M_A^-$$ (29)

and the corresponding expression for the experimentally determined rate constant k_{obs} for electron transfer is given by:

$$k_{obs} = k_{diff}k_{el}/(k_{el} + k_{-diff})$$ (30)

where the two diffusional rate constants can be calculated using expressions available in the literature.[57] Where this becomes possible, the electron transfer process for a bimolecular reaction can then be related to the analogous unimolecular process occurring within weakly covalently bonded systems.

Excited state electron transfer events add another dimension to the simple reaction above for a process involving only ground states. The overall scheme becomes somewhat more complicated if the detailed steps are accounted for. These are schematically illustrated in Figure 16. In this scheme, k_{diff}, k_{-diff}, k'_{diff}, and k'_{-diff} represent the rate constants of formation and dissociation of the outer-sphere encounter (precursor to forward and back electron transfer, respectively) complex.If we consider the electron donor M to be excited by a suitable light source and the electron acceptor, M_A, to act as the quencher, then from simple photochemical considerations the lifetime of the excited donor, τ_M^o, in the absence of quencher is:

$$\tau_M^o = 1/(k'_{prod} + k_{rad} + k_{nrad})$$ (31)

where k_{rad} and k_{nrad} are the radiative and nonradiative rate constants for decay of the excited state species, M^*, and k'_{prod} is the unimolecular rate constant for product

products P'

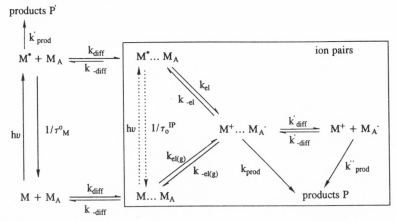

Figure 16. Schematic representation of the various steps in a photoinduced intermolecular electron transfer process between the electron donor/chromophore M (*solid lines*); the dashed line shows light absorption for the case where the electron donor/chromophore and the electron acceptor form a tight ion-pair. After refs. 4, 7, and 9.

formation by M^*. Where the quencher species, M_A, is present, the lifetime of M^* will be given by:

$$\tau_M = 1/(k'_{prod} + k_{rad} + k_{nrad} + k_q[M_A]) \qquad (32)$$

where k_q is the quenching rate constant, here assumed to be related to oxidative quenching of the excited state species M^*. From the Stern–Volmer quenching model one obtains:

$$\tau_M^{\,o}/\tau_M = 1 + \tau_M^{\,o}\, k_q\, [M_A] \qquad (33)$$

The bimolecular quenching rate constant, k_q, depends on several factors which are embodied in the scheme of Figure 16. Within this context, k_q is equated to k_{obs} above. Thus,

$$k_q = k_{obs} = \frac{k_{diff}}{1 + (k_{-diff}/k_{el}) + k_{-diff}k_{-el}/k_{el(g)}k_{el}} \qquad (34)$$

For substantially exoergic processes, where $k_{el} \gg k_{-el}$, Eq. 34 becomes identical to Eq. 30.[58]

Unimolecular (Outer Sphere) Processes: Ion Pairs

Where the reactants M and M_A heretofore discussed are oppositely charged transition metal complexes (cations and anions), they may form ion pairs. In the scheme of Figure 16, it is tacitly assumed that the formation constant for the encounter complex, M...M_A, is negligibly low ($K_{IP} \ll 1$). However, where K_{IP} is 10^1 to 10^3 M^{-1}, normally found for species that form ion pairs, any photoinduced electron transfer process in such an ion pair system will necessitate consideration of not only excitation of the "free" ions, but also of the ion pairs now available in non-insignificant quantities in the solution medium. This is illustrated also in Figure 16; the dashed arrows between M...M_A and *M...M_A depict light excitation and relaxation of the ion pair.

Before we consider the electron transfer process in these systems, it is worthwhile that we examine briefly the electronic changes that occur from the (weak) electrostatic interactions that give rise to ion pairs. If the electrostatic interactions between M and M_A are negligibly small, such that $K_{IP} \ll 1$, the electronic spectrum of the (M + M_A) combination will be the simple sum of the electronic spectra of the individual components M and M_A. However, if the coulombic forces between M and M_A are non-negligible (weak), the resulting spectrum of the (M + M_A) systems is the superposition of the individual spectra; in addition, there appears a band that is unique to the (M + M_A) ion pair. This band originates from a charge-transfer transition (that we herein will denote simply as IPCT) between the donor (in our case, M) and the acceptor (here M_A). The transition has been described in the literature as outer-sphere charge-transfer (OSCT), or as ion-pair charge-transfer (IPCT) which can be of the metal-to-ligand (MLCT) or ligand-to-metal (LMCT) charge transfer type, as intervalence charge-transfer (IVCT, i.e., metal-to-metal charge-transfer MMCT), and as second-sphere charge-transfer (SSCT, i.e., ligand-to-ligand charge-transfer LLCT).[4] This optical IPCT transition occurs if M and M_A are reducible and oxidizable ions (or molecules), and if, when in close contact, there is a small orbital overlap between the two species.[40]

Light absorption into the IPCT band is in essence a photoinduced electron transfer event if the transfer is irreversible, and to an optical electron transfer if the event is reversible (Figure 17). In the latter case, formation of $M^+M_A^-$ is followed by a rapid back electron transfer necessitating a rather small activation energy; such a process is often referred to as a nonradiative deactivation.[40] No net change occurs. In the former case, the ion pair, $M^+M_A^-$, undergoes further secondary rearrangement, the kinetics of which must be such as to compete with back electron transfer. One such example is the $\{[Co(NH_3)_6]^{3+}(I^-)_3\}$ ion pair which upon light excitation into the $I^- \rightarrow Co^{3+}$ charge transfer band produces irreversibly the subsequent ion pair $\{[Co(NH_3)_6]^{2+} I\bullet(I^-)_2\}$ which, after cage escape the reduced cobalt(II) complex undergoes rapid decomposition to give Co_{aq}^{2+} and NH_4^+ ions in acid media. It is also evident from Figure 17 that the optical energy for the charge transfer transition (E_{op}) depends on the potential difference (ΔG) between the redox couples M/M^+ and M_A/M_A^- and on the reorganizational energy λ; thus, $E_{op} = \Delta G + \lambda$. It should be noted that electron transfer from M...M_A to M^+...M_A^- can also occur thermally requiring ΔG_{th} to reach the crossing point of the two potential curves (Figure 17). As well, the vibrationally relaxed redox electronic isomer M^+...M_A^- may also undergo a reverse thermal electron transfer whose free energy of activa-

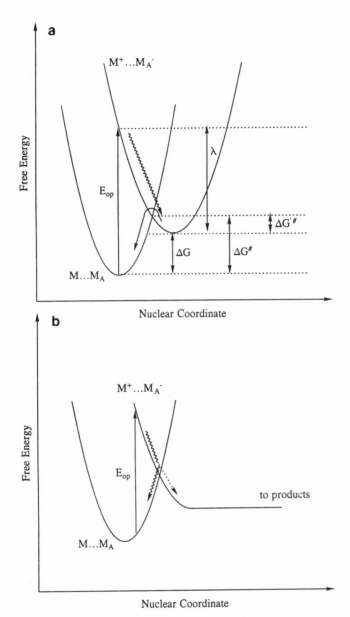

Figure 17. (a) Potential energy diagram for electron transfer from a donor (M) to an acceptor (M_A) in an outer-sphere complex: ΔG° is the free energy of the reaction; $\Delta G^{\#}$ is the free energy of activation for the thermal electron transfer from the precursor complex to the successor complex; $\Delta G'^{\#}$ is the free activation energy for the reverse electron transfer from the successor complex to the precursor complex; λ is the reorganizational energy; and E_{op} is the energy for the optical electron transfer from the precursor complex to a Franck-Condon state of the successor complex. (b) Plot illustrating the reverse electron transfer as a non-radiative electron transition. After ref. 40.

tion will be given by: $\Delta G^{\#\prime} = \Delta G^{\#} - \Delta G$.[40] An example of a system where photoinduced electron transfer is followed by reverse electron transfer is the $\{[Co(sep)]^{3+}\}\{[Ru(CN)_6]^{4-}\}$ system; the reduced product of electron transfer, $[Co(sep)]^{2+}$, being a caged compound, cannot undergo ligand dissociation, and so back electron transfer occurs with unitary efficiency.[4]

The expression for the quenching rate constant (Eq. 35), k_q, is expectedly more complicated than Eq. 34 and involves both the intermolecular component (dynamic quenching) and the intramolecular component (static quenching).[4] Consideration of the latter can lead to an estimate of the association constant of the ion pair. Equation 35 is simplified under certain conditions:[4] (1) where k_{el} in Figure 16 is much less than the rate of relaxation k_o $(= 1/\tau_o)$ of the excited species, no quenching takes place; (2) where $k_{-diff} >> k_{el}$, the presence of ion pairs is of no consequence to quenching since the precursor ion pair $^*M...M_A$ obtained by direct excitation of $M...M_A$ dissociates before electron transfer occurs (here, quenching follows the typical dynamic quenching whose efficiency depends on the relative values of k_{el} and $1/\tau_o$); (3) where $k_{el} >> (1/\tau_o^{IP} + k_{-diff})$, the $^*M$ species in the excited ion pair $^*M...M_A$ undergoes rapid static quenching, rate constant $= k_{el}$, while the "free" excited M^* species are dynamically quenched via encounters with the acceptor species M_A; and (4) where dynamic quenching is negligible, $\tau_o k_{diff}[M_A] << 1$, and static quenching predominates (item 3), the ratio of lifetimes τ_o/τ (or emission intensities, I_o/I) yields an estimate of the association constant for ion-pair formation, $\{= 1 + K[M_A]\}$.[4]

$$k_q = \frac{k_{diff}k_{el}}{(k_{el} + k_{-diff})} \bullet \frac{1 + 1/(\tau_o^{IP}k_{el})}{1 + 1/\{\tau_o^{IP}(k_{el} + k_{-diff})\}} \tag{35}$$

Intramolecular (Inner Sphere) Processes: Supermolecules

The interactions between M and M_A may be such that they may be considered to be covalently (weakly) linked to give rise to supermolecules of the type $M\text{-}(BL)\text{-}M_A$, where (BL) is a bridging ligand and in which the parent species retain their chemical identities and properties. The photoinduced intramolecular electron transfer events can then be described by the scheme shown in Figure 18. A hypothetical complex that might serve as an example is $\{(CN)_5Ru^{II}\text{-}(BL)\text{-}Co^{III}(NH_3)_5\}$ implicated in reactions 36 and 37:

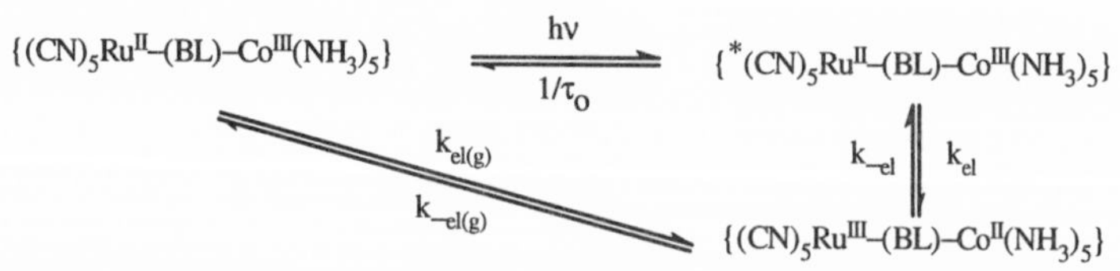

$$\tag{36}$$

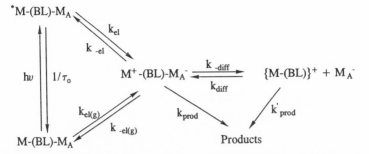

Figure 18. Mechanistic steps in an electron transfer process in a binuclear system connected by a bridging ligand (BL). After ref. 4.

$$\{(CN)_5Ru^{III}\text{-}(BL)\text{-}Co^{II}(NH_3)_5\} \underset{k_{diff}}{\overset{k_{-diff}}{\rightleftarrows}} [(CN)_5Ru^{III}(BL)]^{2-} + [Co^{II}(NH_3)_5]^{2+}$$

$$k_{prod} \searrow \qquad \swarrow k'_{prod}$$

$$Co_{aq}^{2+} + 5NH_3 \tag{37}$$

Several excellent examples of photoinduced electron transfer in ion pairs and in supramolecular systems have recently been reviewed by Balzani and Scandola,[4,9] by Vogler,[5,40] and by Hennig.[39]

3.2. Interfacial Electron Transfer in a Heterogeneous Phase

An event that is primordial in effecting conversion of (sun)light energy to useful chemical energy (fuels) is photoinduced charge separation (e.g., Eq. 1b,c), followed by suppression of the charge recombination event (Eq. 1d,e). In natural photosynthesis, membranes of bilayer dimension are exploited to house the conversion centers and to keep the oxidized and reduced products separated. To mimic the natural process, various assemblies (micelles, microemulsions, vesicles,...) have been examined in artificial photosynthesis to effect the function(s) of the membranes. The charged lipid/water interface has been exploited to put a kinetic control on the electron transfer events. Although the light conversion efficiency may be high in homogeneous phase systems, where suitable donor/acceptors are available, it is equally true that where the light-induced charge separation event is endoergonic, the ability of the charged species to back-react by diffusion controlled rates becomes deleterious to overall conversion efficiencies. By contrast, a heterogeneous system based on semiconductor particulates can inhibit this undesirable back electron transfer between reduced and oxidized primary products. The physical (vectorial) separation of charged species is indeed a prerequisite for building an artificial device that will efficiently convert the (sun)light energy into useful chemical energy. Semiconductor particulates are especially attractive in this regard, since they possess some of the requisite attributes to achieve efficient charge

separation and energy conversion: they absorb light efficiently and produce minimal entropy; by their very nature, they produce charged entities (electrons and
holes) which in turn can lead to charged separated products via the unidirectional
photoredox reactions that occur at the semiconductor/solution interface; their redox
potentials can be fine-tuned to the redox reactions that store the light energy; and
finally, of some importance, they are chemically and photochemically stable for
periods longer than their homogeneous counterparts.

In artificial devices that utilize small semiconductor particles, the rapid movement of charge carriers (conduction band electrons, e_{CB}^-, and valence band holes,
h_{VB}^+) in their respective bands and the presence of local electrostatic fields at the
particles/water interface make possible the separation of oxidizing and reducing
equivalents. Our initial work (1983) on effecting and demonstrating vectorial
displacement of charges when two different semiconductors are coupled, and
recently summarized,[59] has since been confirmed by others by spectroscopic,[60]
photoelectrochemical,[61] photocatalytic,[62] and pulsed laser spectroscopic methods.[63] This new strategy, which we have coined *Inter-Particle Electron Transfer
(IPET)*, has been exploited in the photocleavage of H_2S in aqueous alkaline
media,[64] and in the photodehydrogenation of alcohols[65] where relatively high
photochemical efficiencies (~ 30%) have been attained. It has also been examined
by photoelectrochemical[66] and by photoconductance studies.[67]

On an Electrode (Metal or Semiconductor)

We begin by noting an important difference between homogeneous and heterogeneous electron transfer processes. While in the former case the rate constants are
usually denoted in units of $M^{-1}s^{-1}$ for bimolecular events and in units of s^{-1} for
unimolecular processes, in heterogeneous phase the rate constants are normally
reported in units of cm s^{-1}. To grasp this difference, we examine the details of the
process(es) occurring at the interface between a solid electrode (could be a metal
or a semiconductor) and a redox species in an electrolyte medium taken in this
discussion to be a charged or uncharged electron acceptor, A.[42] We also presume
that A is solvated; if the solvent is water, we have A_{aq}. Note that the electrode surface
may also be highly solvated by chemisorbed or physisorbed water and the approach
of A_{aq} to the surface is restricted to within the region between the Helmholtz outer
plane and the surface, R_{het} (Figure 19).

Before interfacial electron transfer can occur between A_{aq} and the surface redox
equivalents within the reaction layer thickness, diffusion of A_{aq} from the bulk to
the surface must first take place; typically heterogeneous diffusion rate constants
are ~ 5×10^2 cm s^{-1}.[42] These events are summarized by Eqs. 38 and 39:

$$A_{aq}(bulk) \underset{k_{-diff}}{\overset{k_{diff}}{\rightleftharpoons}} A_{aq}(surface) \qquad (38)$$

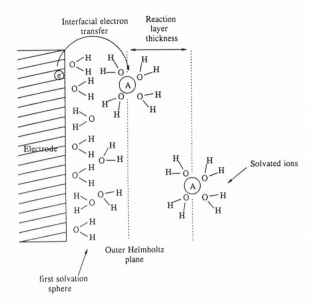

Figure 19. Schematic representation of the various steps implicated in the interfacial electron transfer from an electrode (metal or semiconductor) to a solvated electron acceptor. From ref. 42.

$$A_{aq}(\text{surface}) + e^-_{\text{electrode}} \xrightarrow{\text{k}_{ct}} A_{aq}^-(\text{surface}) \qquad (39)$$

where k_{-diff} is diffusion of the unreacted acceptor species away from the surface. The equilibrium constant, K, is given by the ratio of the concentration of A_{aq} (surface) in mols cm^{-2}, and the concentration of A_{aq} in the solution bulk $[A_{aq}(\text{bulk})]$, or simply by k_{diff}/k_{-diff}. The kinetic analysis of reactions 38 and 39 yield for the observed interfacial electron transfer rate constant, k_{obs}, the expression:

$$(k_{obs})^{-1} = (k_{ads})^{-1} + (Kk_{ct})^{-1} \qquad (40)$$

Where Kk_{ct} is $<< k_{diff}$, that is, where the adsorption equilibrium is established during the electron transfer event, k_{obs} is given by:

$$k_{obs} = K\,k_{ct} \qquad (41)$$

Another distinction between homogeneous and heterogeneous electron transfers is the meaning of *reorganizational energy* in a heterogeneous medium, λ_{het}. In homogeneous phase, electron transfer involves two reactants and their corresponding solvation shells which rearrange in forming the activated complex. In heterogeneous phase, electron transfer involves only one reactant and its solvation shell. Thus the inner sphere component of the reorganizational energy, λ_i, for the heterogeneous phase system is intuitively expected to be: $\lambda_i^{ket} = 0.5\ \lambda_i^{hom}$.[42] This expectation appears to have been confirmed. Insofar as the outer sphere components of λ are concerned, however, the relationship is complicated by the fact that these are dependent on the distance R between the reacting species in the activated complex of the homogeneous process, and on the distance R_{het} between species A and the electrode in the heterogeneous process. This relationship follows the expression:[42]

$$\lambda_o^{het} = 0.5\ \lambda_o^{homo} + (e^2/2)\ \{1/R - 1/2R_{het}\}\ \{1/n^2 - 1/D_S\} \qquad (42)$$

If the A_{aq} species penetrates the solvent layer at the electrode surface, $2R_{het} = R$; if it cannot penetrate the solvent shell, then $2R_{het} > R$. Under these conditions, $\lambda_o^{het} \geq 0.5\ \lambda_o^{hom}$, and in general Eq. 43 is valid. This equation can be used to compare the rate constant for a homogeneous exchange reaction $(A' + A^- \rightarrow A + A'^-)$ with that for an interfacial electron transfer involving the same redox couple.[42]

$$\lambda^{het} \geq 0.5\ \lambda^{homo} \qquad (43)$$

On a Semiconductor Colloid Particle

The above treatment has considered interfacial electron transfer event between an acceptor species in the solution bulk and a fixed electrode, be it a metal or a semiconductor crystal. We now examine similar electron transfer events where the semiconductor *microelectrode* is a small particle of colloidal dimensions. In this case, both A_{aq} and the particle are free to diffuse in solution, albeit the rates of diffusion are no doubt different. Additionally, any efficient redox process—thus efficient redox dynamics at the interface—that involves semiconductor particles will necessitate the rapid removal (scavenging) of the photogenerated charge carriers (conduction band electrons and/or valence band holes) at the particle surface, since the process will have to compete with charge carrier recombination. For example, recombination is fairly rapid (less than 30 ns) in naked TiO_2 colloids,[68] but methods are available (substitutional doping) to retard this process.[69]

To the extent that both the particle and the redox reactant are able to migrate within the solution bulk, some of the requirements encountered in bimolecular redox reactions in homogeneous phase also apply in heterogeneous phase. Thus, the overall heterogeneous redox process consists of the following two steps:[42]

1. Formation of an *encounter "complex"* between the particle and the electron (or hole) acceptor; the process dynamics are diffusion limited and will therefore depend on the medium viscosity and on the distance between the particle center and the center of the acceptor species.
2. The interfacial charge transfer step occurring at the particle/solution interface will be characterized by a rate parameter k_{ct} (in units of cm s^{-1}).

The various stages are simplistically summarized in reactions 44–46:

1. Excitation and charge carrier diffusion to the particle surface.

$$(particle) + h\nu \rightarrow (e^-....h^+) \qquad (44)$$

2. Encounter "complex" formation with acceptor via diffusion.

$$(e^-....h^+) + A_{aq} \rightarrow (h^+....e^-) A_{aq} \tag{45}$$

3. Interfacial electron (or hole) transfer.

$$(h^+....e^-) A_{aq} \xrightarrow{k_{ct}} (h^+.....) + A_{aq}^- \tag{46}$$

The experimentally observed rate constant, k_{obs}, for the bimolecular events is given by:

$$(k_{obs})^{-1} = (1/4\pi R^2)\{1/k_{ct} + R/D\} \tag{47}$$

where R is the sum of the radii of the particle and of the acceptor species, and D is the sum of the diffusion coefficients of the particle and the acceptor in the given solvent. Noteworthy here is the evident relationship between the usual and more familiar second-order rate constant, k_{obs}, with the electrochemical rate constant for electron transfer in heterogeneous media, k_{ct}.

Equation 47 establishes certain relationships between the magnitude of the various parameters. If $1/k_{ct}$ is $>>$ R/D, that is if the interfacial charge transfer rate is much slower than diffusion of the reactants and is rate-determining, then:

$$k_{obs} = 4\pi R^2 k_{ct} \tag{48}$$

However, if $1/k_{ct}$ is $<<$ R/D; that is, if the interfacial charge transfer is much faster than diffusion, then Eq. 47 becomes (the Smoluchowski expression):

$$k_{obs} = 4\pi RD \tag{49}$$

For the cases where the electron (or hole) acceptor is ligated to the particle surface, either via chemisorption or physisorption, the rate of interfacial charge transfer, k_{ct}, is then given by:[42]

$$k_{ct} = \delta\, k_{ct}^{ave} \tag{50}$$

where δ is the reaction layer thickness, and k_{ct}^{ave} is the average rate for the charge carrier to tunnel through the interface.

The techniques used to determine the charge transfer rates, k_{ct}, in heterogeneous media have been examined extensively by the group of Gratzel,[42] and by Albery and his group,[70] among others, and elegantly summarized in a recent book.[42] The reader is referred to this book and to the original literature for further details.

4. EXCITED STATE ELECTRON TRANSFER IN TRANSITION METAL COMPLEXES: SELECTED EXAMPLES

4.1. Intermolecular

We saw earlier (Eqs. 28 to 30) that for a simple bimolecular electron transfer between an excited electron donor M and an electron acceptor M_A necessitates the two reactants to diffuse towards each other, form a (encounter) precursor complex, followed by electron transfer to form the successor complex $M^+...M_A^-$, after which cage escape can occur, either completely to yield the two redox products, M^+ and M_A^-, or partially (or not at all) to return both reactants to their original ground state. However, as we shall see below, back electron transfer may also return one of the species to its excited state from which electron transfer originated. We consider now some selected examples from the large body of literature available on ruthenium(II) polypyridine complexes.

Oxidative quenching of the metal-to-ligand charge-transfer excited state of $Ry(bpy)_3^{2+}$ (bpy is 2,2'-bipyridine) by such electron transfer acceptors as methylviologen (MV^{2+}) and $Ru(NH_3)_5py^{3+}$ has been examined in aqueous media as a function of concentration and type of added electrolyte ($NaCl$, $NaClO_4$, and $CaCl_2$) to: (1) assess the yield of cage escape products; (2) examine the influence of added NaCl on the rate of back electron transfer from these products; and most important (3) assess the ionic strength dependence of the diffusional parameters, k_{diff} and k_{-diff}, of Eq. 30.[70] There exist three diffusion related regimes:

1. *Diffusional pre-equilibrium* when $k_{el} \ll k_{-diff}$, which simplifies Eq. 30 to $k_{obs} = (k_{diff}/k_{-diff})k_{el}$ such that the observed rate is smaller than diffusion and is mostly determined by k_{el}. Most bimolecular electron transfer reactions belong to this class.
2. The *diffusion-controlled* limit occurs when $k_{el} \gg k_{-diff}$, reducing Eq. 30 to $k_{obs} = k_{diff}$; here the rate of reaction is insensitive to k_{el}.
3. Finally, there may be cases for which $k_{el} \approx k_{-diff}$, termed *nearly diffusion controlled* and which requires that Eq. 30 be used; the reaction rate is moderately sensitive to k_{el}.

Scheme 51 can be used to rationalize the events:[71]

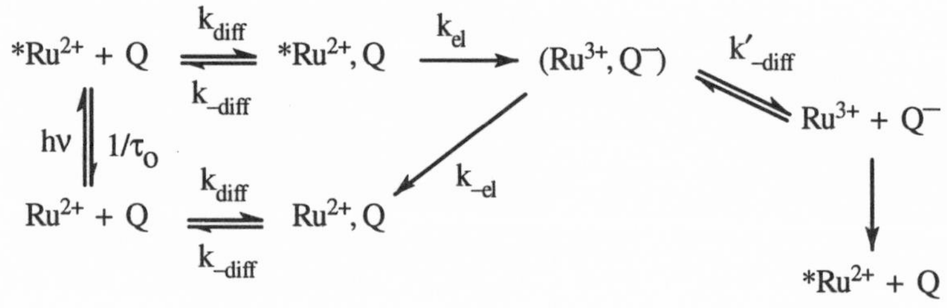

$$*Ru^{2+} + Q \underset{k_{-diff}}{\overset{k_{diff}}{\rightleftharpoons}} *Ru^{2+}, Q \xrightarrow{k_{el}} (Ru^{3+}, Q^-) \underset{}{\overset{k'_{-diff}}{\rightleftharpoons}} Ru^{3+} + Q^-$$

$$hv \updownarrow 1/\tau_0$$

$$Ru^{2+} + Q \underset{k_{-diff}}{\overset{k_{diff}}{\rightleftharpoons}} Ru^{2+}, Q \qquad k_{-el}$$

$$*Ru^{2+} + Q \tag{51}$$

$$*Ru(bpy)_3^{2+} + MV^{2+} \xrightarrow{k_{obs}} Ru(bpy)_3^{3+} + MV^+ \tag{52}$$

$$*Ru(bpy)_3^{2+} + Ru(NH_3)_5py^{3+} \xrightarrow{k_{obs}} Ru(bpy)_3^{3+} + Ru(NH_3)_5py^{2+} \tag{53}$$

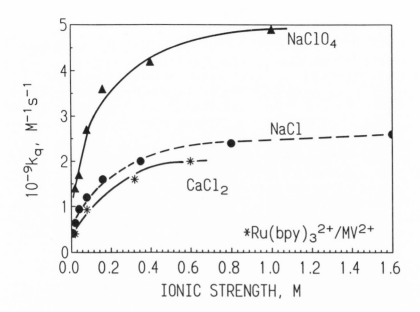

Figure 20. Plots showing the dependence of the oxidative electron transfer quenching of $^*Ru(bpy)_3^{2+}$ by MV^{2+} on the ionic strength for different added electrolytes in aqueous media. Adapted from ref. 71.

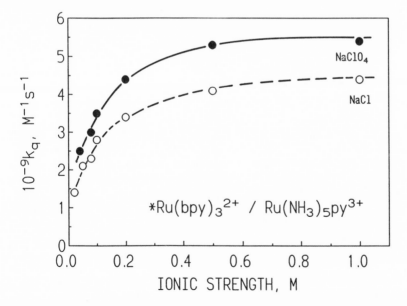

Figure 21. Plots illustrating the dependence of the oxidative electron transfer quenching of $^*Ru(bpy)_3^{2+}$ by $Ru(NH_3)_5py^{3+}$ on the ionic strength for NaCl and $NaClO_4$ as added electrolytes in aqueous media. Adapted from ref. 71.

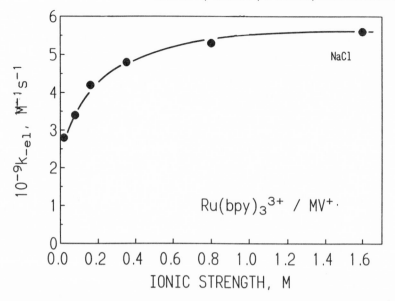

Figure 22. Plot showing the dependence of the back electron transfer from MV^+ to $Ru(bpy)_3^{3+}$ on the ionic strength (adjusted with NaCl) following forward oxidative quenching of the excited state of the ruthenium(II) complex by methylviologen in aqueous media. Adapted from ref. 71.

The dependence of the quenching constants, k_{obs}, on the ionic strength, μ, is illustrated in Figures 20 and 21 for MV^{2+} and for $Ru(NH_3)_5py^{3+}$, respectively, (reactions 52 and 53). The influence of the electrolyte NaCl on the rate of the back electron transfer (reverse of reaction 52, Eq. 54) is portrayed in Figure 22.

$$Ru(bpy)_3^{3+} + MV^+ \xrightarrow{k^b_{obs}} Ru(bpy)_3^{2+} + MV^{2+} \tag{54}$$

Added ClO_4^- electrolyte shows a pronounced acceleration, relative to Cl^- of the quenching rate constant for the $^*Ru(bpy)_3^{2+}/MV^{2+}$ system, an effect also observed in the quenching of $^*Ru(bpy)_3^{2+}$ by cobalt(III) sepulchrate.[72]

Variations in quenching constants, k_{obs}, with increase in ionic strength are rather small (Figures 20 and 21) in both Cl^- and ClO_4^- media, suggesting that these variations between the two counterions used cannot arise from differences in k_{diff} and/or k_{-diff} as a result of different extents of ion pairing, since such effect is expected to be ionic strength-dependent. Rather, the data are understandable if the same diffusional rate constants are used: k_{el} is 1.9×10^9 s^{-1} for the Cl^- media and 7.6×10^9 s^{-1} for the ClO_4^- media.[71] A specific counterion effect arises from the influence of the counterion on the unimolecular reactive step (k_{el} in scheme 51), thereby suggesting that the counterion may be intimately implicated in the precursor complex. The differences in k_{el} between chloride and perchlorate media are due to the following differences in the hydrophilicity of the two counterions:[71] (1) the hydration shell around the anion provides a more or less rigid solvent environment to the precursor complex, requiring a more or less larger outer-sphere reorganizational energy; or (2) the hydration shell allows a more or less closer approach of the two reactants, $^*Ru(bpy)_3^{2+}$ and MV^{2+}, both of which are cationic and the process may follow a more or less adiabatic behavior.

Reaction 53 also shows a specific anion effect, albeit the difference between the two sets of data in Figure 21 is much smaller than when MV^{2+} is the quencher. The cage escape yields determined for the $^*Ru(bpy)_3^{2+}/MV^{2+}$ system decrease markedly as the ionic strength is increased: $\eta_{cage\ escape}$ is 0.38 when the ionic strength (μ) is 0.01 M; it is 0.22 when μ is 1.60 M.[71]

Comparison of the quenching rate data between NaCl and $CaCl_2$ as the added electrolytes is of interest. On the basis of ionic strength effects, the quenching rate constants in aqueous $CaCl_2$ media are consistently smaller than in media-containing NaCl. However, if the quenching constants are compared as a function of concentration of chloride (the so-called Olson–Simonson effect[73]), both sets of data fall on a single curve (see Figure 8 of ref. 71).

For electron transfer reactions in the nearly diffusion controlled regime, specific counterion effects and the Olson–Simonson effect could lead to substantial erroneous conclusions if rate constants k_{obs} are compared at constant ionic strength for different so-called inert supporting electrolytes.[71] Salt effects can be parametrically accounted for with reasonable accuracy in terms of diffusional rate constants estimated from the Debye–Huckel model under certain conditions (cf. ref. 71).

The quantum yields of redox products (or of radical species) in the bulk, formed subsequent to the electron transfer step (see scheme 51) depends on the rate by which the solvent-caged species undergo back electron transfer and on the rate of cage escape.[74] When the cage escape yields are low, the rate of back electron transfer which competes with cage escape will be comparable to that of the diffusion process.

Early on we remarked that (in condensed phase) the Marcus theory of electron transfer predicts that the rates of electron transfer increase with increasing exergonicity ($-\Delta G^o$) towards a maximum; further increase in the thermodynamic driving force for electron transfer subsequently leads to a *decrease* in the rates. This decrease in rates in the *inverted region* is caused by a reduction in the thermally averaged vibrational overlap integral between initial and final states (the so-called Franck–Condon factor). Systematic examinations designed to observe these dramatic changes in electron transfer rates with variations in exergonicity, from low to highly negative free energies (ΔG^o), have been hampered by the fact that such rates are limited by the diffusion process for the forward electron transfer (see Scheme 51). However, this diffusion-limiting factor need not affect the rates of back electron transfer between species in the successor complex which is solvent-caged, $\{Ru^{3+},Q^-\}$ in Scheme 51 (or between radical species), since such rates will be unimolecular. Such reductions in rates of highly exergonic electron transfer for back electron transfer processes, occurring within geminate radical ion pairs produced in luminescence quenching experiments, has been evidenced for the complexes $Ru(bpy)_3^{2+}$ and $Ru(dp\text{-}phen)_3^{2+}$ (dp-phen is 4,7-diphenyl-1,10-phenanthroline).[75]

Addition of an aromatic amine such as N,N,N',N'-tetramethyl-1,4-phenylenediamine (TMPD) to a solution of $Ru(bpy)_3^{2+}$ (mixed solvent mixture: CH_3CN/ water) quenches the luminescence from this ruthenium(II) complex via reductive electron transfer (reaction 55). The quenching rate constants for this amine and others {PD, 1,4-phenylenediamine; TMB, N,N,N',N'-tetramethylbenzidine; DPPD, diphenyl-1,4-phenylenediamine; 3,3'-DMB, 3,3'-dimethylbenzidine; phenothiazine; 1,4-anisidine; DPA, diphenylamine; and 1,4-toluidine} of the luminescence from $^*Ru(NN)_3^{2+}$, where NN is bpy, phen, or dp-phen, are summarized in Table 1.[74] Except for 1,4-anisidine, DPA, and 1,4-toluidine, bimolecular electron transfer quenching of the luminescent state of ruthenium(II) complexes occurs efficiently in a solvent-separated collision pair or a contact collision pair. Electrostatic repulsion between the geminate cation radicals in a polar solvent enhances the rate of separation of the radical, defined in Scheme 51 as k'_{-diff} (or as k_{dis}[74]). Prior to its separation, the rate of reverse electron transfer within the solvent cage depends on the inter-radical interaction(s) and the energy gap between the cation radical pair and the original species in their ground states.[74]

$$^*Ru(bpy)_3^{2+} + TMPD \rightarrow Ru(bpy)_3^+ + TMPD^+ \tag{55}$$

The concentration of the reduced transition metal complex, $Ru(bpy)_3^+$, is given by the expression:

Table 1. Redox Potentials of Donors, Quenching Rate Constants, and Fractions of Radical Formation of the Quenching of $^*Ru(NN)_3^{2+}$ in a Mixed Solvent of CH_3CN and H_2O[74]

Donors	E^0, V vs. SCE	$Ru(bpy)_3^{2+}$ $10^{-8}k_q$, $M^{-1}s^{-1}$	F_1	$Ru(phen)_3^{2+}$ $10^{-8}k_q$, $M^{-1}s^{-1}$	Fl	$Ru(dp\text{-}phen)_3^{2+}$ $10^{-8}k_q$, $M^{-1}s^{-1}$	Fl
TMPD	0.16	50	0.84	74	0.51	79	0.36
PD	0.17	36	0.81	32	0.70	34	0.46
TMB	0.32	57	0.59	—	—	92	0.16
DPPD	0.35	55	0.60	76	0.46	84	0.20
3,3'-DMB	0.45	34	0.56	65	0.33	84	0.20
Phenothiazine	0.53	53	0.66	80	0.47	89	0.32
1,4-Anisidine	0.71	39	0.81	13	0.65	16	0.40
DPA	0.78	—	—	0.17	0.65	—	—
1,4-Toluidine	0.78	0.18	0.91	—	—	0.54	0.46

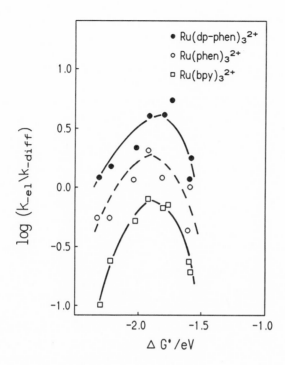

Figure 23. ΔG° dependence of k_{-el}/k_{-diff}, $1/(F_1 - 1)$, for the electron transfer quenching of $^*Ru(NN)_3^{2+}$ by the aromatic amines: NN = bpy [□], NN = phen [○], and NN = dp-phen [●]. From ref. 74.

$$[Ru(NN)_3^+] = \{(k_{decay} - k^{\circ}_{decay})/k_{decay}\}F_1[^*Ru(NN)_3^{2+}] \tag{56}$$

where the { } term represents the quenching efficiency, F_1 denotes the efficiency of formation of the reduced product $Ru(NN)_3^+$ in the bimolecular quenching process. Representing the time-average back electron transfer as k_b (or as k_{-el} in Scheme 51), the fraction of cation radical pairs that do not undergo reverse electron transfer is given by F_1 in Eq. 57:

$$F_1 = k_{-diff}/(k_{-diff} + k_{-el}) \tag{57}$$

and thus

$$k_{-el}/k_{-diff} = 1/(F_1 - 1) \tag{58}$$

To the extent that k_{-diff} depends mostly on the reaction radii of the species in the solvent cage, for analogous species[76] any variation in the ratio k_{-el}/k_{-diff} reflects mostly variations in the rate of reverse electron transfer, k_{-el}.[74] Figure 23 illustrates these variations as a function of the exergonicity of the reverse electron transfer process. In all three cases, the maximum reverse electron transfer rate occurs at an exergonicity of $-\Delta G^0 = 1.7$ eV, and the predicted reduction in these rates at more negative ΔG^0 is evident.

It is instructive at this time to consider some of the steps implicated in the electron transfer between two reactant species in some appropriate medium. What are the changes that take place? We have already alluded to some of these earlier: changes in bond lengths and bond angles of the reactants M and M_A, and diffusion of the two species to some appropriate distance for electron transfer to occur. Clearly, diffusion will be highly dependent on the properties of the solvent medium, the charges on the two species, and on the nature and the concentration of the counterions present (ionic strength). As well, it is expected that the closer the two reacting partners approach each other, the faster the electron transfer will be. Thus the reaction distance will be a significant factor. Finally, as in any activated process, the temperature will also have a significant effect.

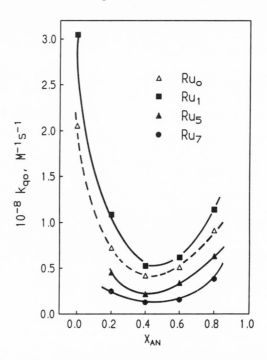

Figure 24. Corrected quenching constants k_{qo} for the $Ru\{(C_nH_{2n+1})_2bpy\}_3^{2+}/MV^{2+}$ system for $n = 0, 1, 5,$ and 7 shown as a function of solvent composition. From ref. 77.

In a recent report, Greiner and co-workers[77] examined some of these factors. They examined the solvent dependence (acetonitrile/water) of the quenching (forward electron transfer) of hydrocarbon-substituted Ru$\{(C_nH_{2n+1})_2bpy\}_3^{2+}$ complexes (at the 4,4′ position) by MV^{2+} together with the recombination (back electron transfer) of the free ions that escaped from the solvent cage. The distance dependence was probed by varying n ($= 0, 1, 5$, and 7); the quenching rate is smaller for the complexes containing the longer hydrocarbon residues, while the chain length seems to have little effect on the back electron transfer. There appear to be no variations between charged quencher, such as MV^{2+}, and an uncharged viologen such as the neutral propylviologensulfonate (PVS).[77] The dependence of electron transfer on ionic strength was also examined using ClO_4^- salts.

Quenching rate constants for the Ru$\{(C_nH_{2n+1})_2bpy\}_3^{2+}/MV^{2+}$ system, corrected for diffusion using Eq. 30, are illustrated in Figure 24 for $n = 0, 1, 5$, and 7 as a function of solvent composition, χ_{AN}; the dependence of these quenching rate constants for Ru(bpy)$_3^{2+}/MV^{2+}$ on the ionic strength is portrayed in Figure 25.[77] Figure 26 illustrates the effect of charge on the electron transfer quenching at constant ionic strength (0.4 M) for the Ru(bpy)$_3$(ClO$_4$)$_2$ /MV(ClO$_4$)$_2$ system, and with no added salt when PVS is the electron acceptor quencher, at various solvent compositions.

The forward electron transfer process is clearly influenced by the chain length of the hydrocarbon attached to the bpy ligand; this process follows an exponential

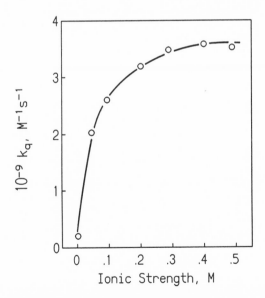

Figure 25. Dependence of the quenching constant k_q for the Ru(bpy)$_3^{2+}$/ MV^{2+} system on the ionic strength adjusted with NaClO$_4$. From ref. 77.

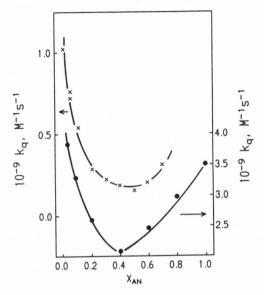

Figure 26. Solvent dependence of the quenching constants k_q for the $Ru(bpy)_3^{2+}/MV^{2+}$ system at an ionic strength of 0.4M ($NaClO_4$) [●], and for the $Ru(bpy)_3^{2+}/PVS$ system without added salt [x]. From ref. 77.

decrease in the quenching constants with distance, as expected (ionic strength was 0.2 M).[78] By contrast, reverse electron transfer is virtually independent of chain length (Figure 27).[77] Electron transfer is diffusion controlled at relatively high ionic strengths (0.4 M); at zero ionic strength, the quenching constants are less than the diffusion control limit.

An interesting feature in Figures 24 to 27 is the minimum in the curves which always occurs at a solvent composition of $\chi_{AN} = 0.4$. Calculations of the electron transfer rate constant, k_{el}, for nonadiabatic and adiabatic transfer cannot reproduce this minimum in the curve(s) of rate constant *versus* solvent composition, the $\Delta G^{\#}$ values so estimated from the reorganizational parameter λ and $\Delta G°$ being too small.[77] As seen earlier, the classical approach to adiabatic reactions within the framework of the absolute rate theory gives for the rate constant for electron transfer:

$$k = k°_{el} \exp(-\Delta G^{\#}/RT) = \kappa \, \nu_N \exp(-\Delta G^{\#}/RT) \qquad (59)$$

where the transmission coefficient is unity for adiabatic reactions and $\Delta G^{\#} = (\lambda + \Delta G°)^2/(4 \, \lambda \, RT)$ is the activation energy necessary to reach the crossing point of the potential energy surfaces for reactants and products (see Figure 11), and ν_N is the effective nuclear frequency factor related to the vibrational frequencies of the reactants and solvent. As the electron transfer event occurs, the sudden constant

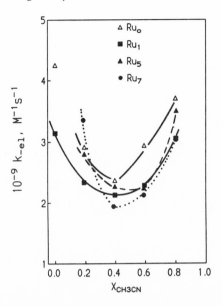

Figure 27. Solvent dependence of the back electron transfer rate constants k_b for the system $Ru\{(C_2H_{2n+1})_2bpy\}_3^{3+}/MV^{+\bullet}$ for $n = 0$, 1, 5, and 7 at an ionic strength of 0.2 M (NaCl). From ref. 77.

changes in charge redistribution between M and M_A must be accompanied by solvent relaxation; the longitudinal relaxation time τ_L becomes the more suitable property to consider: $v_N = 1/\tau_L$.[77] A simple Arrhenius equation using the experimentally determined values of the activation free energy, $\Delta G^{\#}$, and the experimental values of τ_L as the pre-exponential factor gives a satisfactory agreement between the calculated and experimental values of the quenching constants.[77]

At high exergonicity, bimolecular electron transfer quenching, in which transfer is assumed to occur by collision between freely diffusing donor and acceptor, becomes and remains diffusion controlled. These observations were first reported by Rehm and Weller.[79] In the systems $Ru\{(C_nH_{2n+1})_2bpy\}_3^{2+}/MV^{2+}$, the diffusion of donor and acceptor is retained, but does allow adjustment of the electron transfer distance.[77] The forward transfer (quenching) occurs with $\Delta G^\circ \sim -0.4$ eV, while the reverse electron transfer between the totally separated and freely diffusing electron donor and acceptor takes place with $\Delta G^\circ = -1.7$ eV. The transfer distance appears to vary with the degree of exergonicity; that is, with the degree with which the freely diffusing partners attempt to find the optimum distance for the electron transfer event to take place.[78] The optimum distance at small exergonicity is smaller than the contact distance of the two partners; thus, one observes a distance dependence in the forward electron transfer (quenching). By contrast, at high exergonicities, the optimum transfer distance is greater than the contact distance

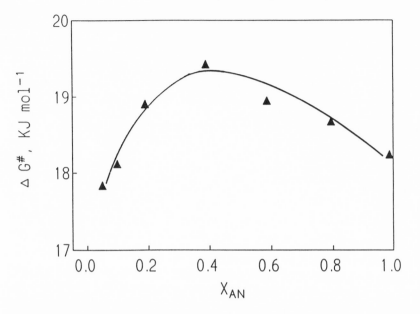

Figure 28. Solvent dependence of the free energy of activation for the oxidative quenching reaction of $^*Ru(bpy)_3^{2+}$ by MV^{2+} in perchlorate media (ionic strength, 0.4 M). From ref. 77.

and can be reached by diffusion; no distance dependence is therefore expected for the $Ru\{(C_nH_{2n+1})_2bpy\}_3^{2+}/ MV^{2+}$ systems and none is found (see Figure 27). The reverse electron transfer is diffusion controlled and shows the Rehm-Weller behavior.[77]

The free energy of activation for quenching $^*Ru(bpy)_3^{2+}$ by MV^{2+} in acetonitrile is virtually independent of ionic strength from 0.05 to 0.5 M; $\Delta G^\#$ is about 18.5 ± 0.6 KJ mol^{-1}. However, at a constant ionic strength of 0.4 M, the free energy of activation does show dependence on the solvent composition (Figure 28); a maximum is evident at $\chi_{AN} = 0.4$.[77]

Reductive electron transfer quenching of $Ru(NN)_3^{2+}$ systems [NN = bpy (C_0); 4,4'-dimethyl-2,2'-bipyridine (C_1); and 4,4'-diheptadecyl-2,2'-bipyridine (C_{17})] by various organic electron donors [TMPD; p-methoxy-N,N-dimethylaniline (DMA); and N,N-dimethyl-p-toluidine (DMT)] in acetonitrile, tetrachloroethylene, and chloroform show a decrease in the quenching constant as the hydrocarbon chain length is increased; the data are summarized in Table 2.[80] The sudden drop in k_q from the C_0 to the C_1 system appears to arise from steric effects, while the decrease in k_q on going from the C_1 to the C_{17} system arises from a decrease in the transmission coefficient.[80]

Table 2. Luminescence Quenching Rate Constants for the $Ru(NN)_3^{2+}$ Systems by Electron Donors[80]

		$10^{-7}k_q, M^{-1}s^{-1}$		
NN (E½, V)	*Quencher (E½, V)*	CH_3CN	$C_2H_2C_4$	$CHCl_3$
C_0 (0.75)	TMPD (0.12)	1200	240	450
C_1(0.64)		780	130	210
C_{17} (~0.64)		730	110	120
C_0(0.75)	DMA (0.55)	440	94	150
C_1(0.64)		190	21	15
C_{17} (~ 0.64)		150	8.7	4.6
C_0(0.75)	DMT (0.71)	130	23	26
C_1 (0.64)		4.9	0.65	0.52
C_{17}(~0.64)		36	0.85	0.55

Enhancement of the efficiencies of photoinduced electron transfer and suppression of the subsequent dark (reverse) electron transfer to the ground state species has been central in efforts to achieve efficient charge separation to photoredox products.[81] Unequivocal evidence that back electron transfer to the excited state also has to be considered has been presented,[82] and the importance of this in the oxidative electron transfer quenching of $Ru(bpy)_3^{2+}$ by neutral electron acceptors has been demonstrated.[83] Tazuke and co-workers[84] have also recently established the significance of reverse electron transfer to the excited state in the reductive quenching of $Ru(phen)_2(CN)_2$ by neutral organic quenchers (Table 3). Except for the TMPD quencher, reductive quenching of this system exhibits a negative temperature dependence in plots of k_q versus $1/T$ (so-called Eyring plots). This abnormal negative temperature dependence is accompanied by large and negative

Table 3. Reductive and Oxidative Quenching of Excited $Ru(phen)_2(CN)_2$ by Organic Quenchers in Acetonitrile at 298 K and Ionic Strength of 0[84]

Quencher (E½ V vs. SCE)	$10^{-9} k_q, M^{-1}s^{-1}$
methyl m-nitrobenzoate (−1.04)	4.5
m-nitroanisole (−1.14)	1.5
p-nitrotoluene (−1.21)	0.34
N,N,N′,N′-tetramethylphenylenediamine (0.12)	6.9
p-aminodiphenylamine (0.27)	1.1
N,N,N′,N′-tetramethylbenzidine (0.43)	0.14

activation entropies, and phenomenologically these control the reaction path.[84] In the case of $Ru(bpy)_3^{2+}$, the negative temperature dependence is witnessed only for the oxidative quenching pathway.[82]

To interpret the negative temperature dependence in both reductive and oxidative electron transfer quenching in the dicyano system, the pathways summarized in Scheme 60 are worth considering:

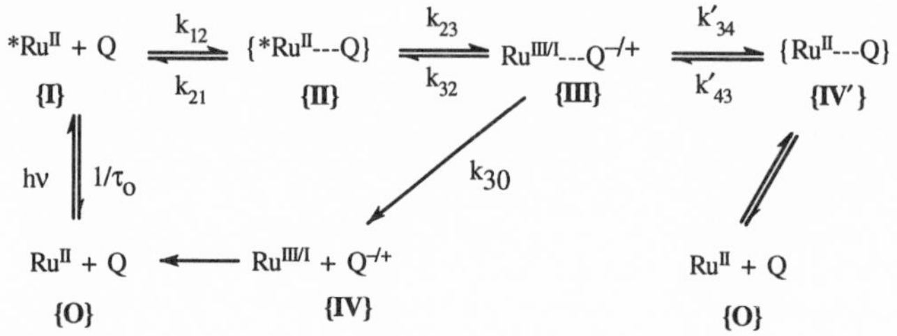

$$
\begin{array}{c}
{}^*Ru^{II} + Q \underset{k_{21}}{\overset{k_{12}}{\rightleftharpoons}} \{{}^*Ru^{II}\text{---}Q\} \underset{k_{32}}{\overset{k_{23}}{\rightleftharpoons}} Ru^{III/I}\text{---}Q^{-/+} \underset{k'_{43}}{\overset{k'_{34}}{\rightleftharpoons}} \{Ru^{II}\text{---}Q\} \\
\{I\} \qquad\qquad \{II\} \qquad\qquad \{III\} \qquad\qquad \{IV'\} \\[6pt]
hv \Big\updownarrow 1/\tau_o \qquad\qquad k_{30} \\[6pt]
Ru^{II} + Q \longleftarrow Ru^{III/I} + Q^{-/+} \qquad\qquad Ru^{II} + Q \\
\{O\} \qquad\qquad \{IV\} \qquad\qquad \{O\}
\end{array}
\tag{60}
$$

from which the rate constant k_q (or k_{obs}) is given by:

$$k_q = K_{12}\, k_{23}\, k_{30} / (k_{30} + k_{32}) \tag{61}$$

Two cases can be envisaged from this expression depending on the relative magnitudes of k_{32} and k_{30}; thus,

Case I: if $k_{30} \gg k_{32}$ then $k_q = K_{12}\, k_{23}$
Case II: if $k_{30} \ll k_{32}$ then $k_q = K_{12}\, k_{23}\, k_{30} / k_{32}$

To the extent that K_{12} is independent (diffusion) of temperature, the observed negative temperature dependence is only expected for Case II.[84]

A comparison of the results from electron transfer quenching of ${}^*Ru(bpy)_3^{2+}$ and of ${}^*Ru(phen)_2(CN)_2$ proves instructive. Electron transfer reactions that produce photoredox products of opposite charges (Eqs. 62 to 64) exhibit a negative temperature dependence of the quenching constant, while the activation enthalpies are always positive when the photoredox products are positively charged (Eq. 65).[84]

$$
{}^*Ru(phen)_2(CN)_2 + \text{Electron Acceptor} \rightarrow [-] + [+] \tag{62}
$$

$$\omega_p = -0.85 \text{ kcal mol}^{-1}$$

$$
{}^*Ru(phen)_2(CN)_2 + \text{Electron Donor} \rightarrow [+] + [-] \tag{63}
$$

$$\omega_p = -0.85 \text{ kcal mol}^{-1}$$

$$
{}^*Ru(bpy)_3^{2+} + \text{Electron Acceptor} \rightarrow [3+] + [-] \tag{64}
$$

$$\omega_p = -2.34 \text{ kcal mol}^{-1}$$

$$^*\text{Ru(bpy)}_3{}^{2+} + \text{Electron Donor} \rightarrow [+] + [+] \tag{65}$$

$$\omega_p = +0.78 \text{ kcal mol}^{-1}$$

where the electrostatic interaction within the product ion pairs (repulsive or attractive) can be expressed by the sign of the electrostatic work, ω_p, needed to bring the two product ions to the electron transfer distance, as it would be in species {III} of Scheme 60. The sign of ω_p dictates the quenching pathway (Case I or Case II above) and consequently the overall temperature dependence of k_q.[84] Clearly, the reductive quenching of $^*\text{Ru(bpy)}_3{}^{2+}$ (Eq. 65) which produces electrostatically repulsive ions will favor step k_{30}; that is, dissociation (or separation) of {$\text{Ru}^{\text{III/I}}$--- $Q^{-/+}$} in Scheme 60, thus pointing to the route described in Case I and for which $\Delta H^\#$ are expected to be positive. By contrast, the product ions formed in Eqs. 62 to 64 will form ion radical pairs, and consequently the fate of the product ions will be back electron transfer to the excited state (Case II and $\Delta H^\#$ are negative) and/or back to the ground state. If the latter is the major deactivating pathway for the system {III} in Scheme 60, then $\Delta H^\#$ will be positive and Case I applies. The changes in the electrostatic entropy accompanied by back electron transfer to the excited state, $\Delta S_{32}{}^{\text{es}}$, are all positive for Eqs. 62 to 64 (+4.8, +4.8, +8.9 eu, respectively). Thus the quenching reaction embodied in Eqs. 62 to 64 favor back electron transfer to the excited state and show a negative temperature dependence of k_q.

In the presence of peroxydisulfate, $S_2O_8{}^{2-}$, the *triplet* metal-to-ligand charge-transfer state of $\text{Ru(bpy)}_3{}^{2+}$ is oxidatively quenched. Early studies by Bolletta and co-workers[85] showed that two ruthenium(III) species were formed per mole of peroxydisulfate (Eqs. 66–68). However, a closer examination of this process

$$\text{Ru(bpy)}_3{}^{2+} + h\nu \rightarrow \rightarrow {}^3(\text{MLCT})\text{Ru(bpy)}_3{}^{2+} \tag{66}$$

$$^3(\text{MLCT})\text{Ru(bpy)}_3{}^{2+} + S_2O_8{}^{2-} \xrightarrow{k_q} \text{Ru(bpy)}_3{}^{3+} + SO_4{}^{-\bullet} + SO_4{}^{2-} \tag{67}$$

$$\text{Ru(bpy)}_3{}^{2+} + SO_4{}^{-\bullet} \xrightarrow{k_1} \text{Ru(bpy)}_3{}^{3+} + SO_4{}^{2-} \tag{68}$$

indicates that the quantum yield of formation of $\text{Ru(bpy)}_3{}^{3+}$ species is less than 2, because of side reactions of the sulfate radical (reactions 69 and 70),[86]

$$SO_4{}^{-\bullet} + \text{Ru(bpy)}_3{}^{n+} \rightarrow \text{Ru(bpy)}_2(\text{bpy-OH})^{n+} H^+ + SO_4{}^{2-} \tag{69}$$

and if organic substrates (or impurities) are also present.

$$SO_4^{-\bullet} + RH \rightarrow SO_4^{2-} + R\bullet + H^+ \tag{70}$$

The maximum electron transfer quenching efficiency is obtained at pH 3, while the maximum overall quantum yield of the ruthenium(III) complex ($\phi_{Ru(III)} \sim 1.2$) is obtained at pH 5.[86,87] In 1N H_2SO_4 media, the quenching constant k_q is 4.87×10^8 $M^{-1}s^{-1}$;[84] in argon-purged aqueous media with no added solutes, $k_q = 3.2 \times 10^9 M^{-1}s^{-1}$,[86] while with solutes present, k_q is somewhat smaller owing to increased ionic strength ($k_q = 1.85 \times 10^9$ $M^{-1}s^{-1}$ at $\mu = 0.02$ M and 1.30×10^9 $M^{-1}s^{-1}$ at $\mu = 0.06$ M).[87] The cage escape efficiency, η_{esc}, for the sulfate radical is unity for the system $\{Ru^{III}\text{---}S_2O_8^{3-}\}$ and back electron transfer, although thermodynamically permissible, is not plausible.[85] External applied magnetic fields have no effect on oxidative quenching,[87] contrary to the results of oxidative quenching of $^*Ru(bpy)_3^{2+}$ by MV^{2+} which showed a drop of about 10%.[88]

$$H_3AsO_3 + SO_4^{-\bullet} \rightarrow As(IV) + SO_4^{2-} \tag{71}$$

$$H_3AsO_3 + Ru(bpy)_3^{3+} \rightarrow As(IV) + Ru(bpy)_3^{2+} \tag{72}$$

$$As(IV) + Ru(bpy)_3^{3+} \rightarrow As(V) + Ru(bpy)_3^{2+} \tag{73}$$

The sulfate radical produced in Eq. 67 is useful since it can be used to oxidize a variety of other added solutes as noted in Eq. 70. Thus, As(III) is oxidized to As(V) (reactions 71 to 73).[89] The mechanism of oxidation of added solutes to a system containing the ruthenium(II) bipyridyl complex and peroxydisulfate, upon irradiation, is summarized in Scheme 74:

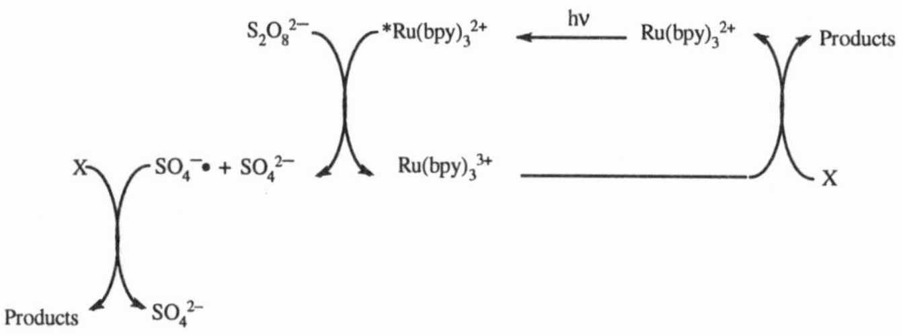

$$\tag{74}$$

where X may be such reducing substrates as H_3AsO_3, oxalate, formate, and $Co(edta)^{2-}$.[89] The kinetic expression unifying the events in Scheme 74 (Eq. 75) shows that:

$$k_q[^*Ru(bpy)_3^{2+}][S_2O_8^{2-}] + k_1[Ru(bpy)_3^{2+}][SO_4^{-\bullet}] = k_x[X][Ru(bpy)_3^{3+}] \tag{75}$$

If this equation holds for these substrates and the concentration of $Ru(bpy)_3^{2+}$ remains constant, then the ruthenium(II) complex acts as a photocatalyst.

Photoinduced oxidative electron transfer quenching of $^*Ru(bpy)_3^{2+}$ by suitable electron acceptors also finds application in inorganic synthesis in ligand substitution reactions.[90] Thus $Co^{II}EDTA$ can be formed by irradiating a system containing $Ru(bpy)_3^{2+}$, $Co(NH_3)_5Br^{2+}$, and EDTA (reactions 76 to 81).

$$k_q = 1.1 - 1.7 \times 10^9 \ M^{-1}s^{-1}$$

$$^*Ru(bpy)_3^{2+} + Co^{III}(NH_3)_5Br^{2+} \rightarrow Co^{II}(NH_3)_5Br^+ + Ru(bpy)_3^{3+} \qquad (76)$$

$$Co^{II}(NH_3)_5Br^+ \xrightarrow{H^+} Co^{2+}_{aq} + 5NH_4^+ + Br^- \qquad (77)$$

$$Co^{2+}_{aq} \ \{or \ Co(NH_3)_5Br^+\} + (EDTA)^{4-} \rightleftharpoons Co^{II}(EDTA)^{2-} \qquad (78)$$

$$Co^{II}(EDTA)^{2-} + Ru(bpy)_3^{3+} \rightarrow Co^{II}(EDTA)^- + Ru(bpy)_3^{2+} \qquad (79)$$

$$k'_q = 2 \times 10^9 \ M^{-1}s^{-1}$$

$$^*Ru(bpy)_3^{2+} + Co^{II}(EDTA)^- \rightarrow Ru(bpy)_3^{3+} + Co^{II}(EDTA)^{2-} \qquad (80)$$

The photosensized reduction of a Co^{III}-Schiff base complex, $Co^{III}(dop)(OH_2)_2^{2+}$, by irradiated $Ru(bpy)_3^{2+}$ occurs efficiently in a network of a gelatin hydrogel and in aqueous gelatin solution under anaerobic conditions.[91] Initially, a reduced intermediate of the Co^{III} complex forms; on termination of the irradiation, this intermediate turned slowly to the diaqua Co^{II}-Schiff base complex in the hydrogel system, but was gradually oxidized in the aqueous gelatin to the Co^{III}-Schiff base complex. It appears that the gelatin acts as a ligand for the Co^{III} complex, and as both an electron donor and an electron acceptor.[91] The process is summarized in Scheme 81.

$Co^{III}(dop)(OH_2)_2^{2+}$

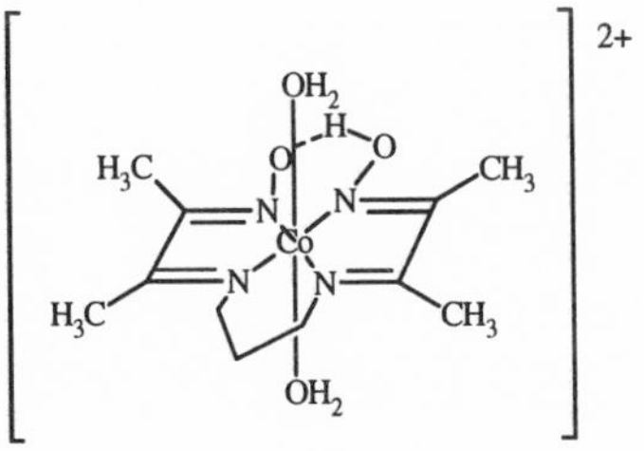

Reduction of CO_2 to formate is a difficult task by a monoelectronic process as it requires a potential of at least -2 V.[92] However, the reduction can be achieved by

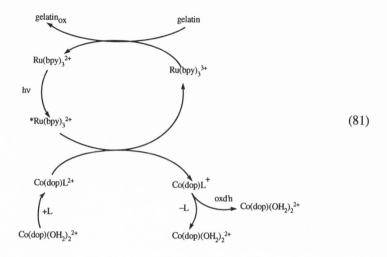

$$(81)$$

a photoassisted process which implicates the excited state of $Ru(bpy)_3^{2+}$ and other species—either $Co(bpy)_3^{2+}$ or cis-$Ru(bpy)_2(CO)X^{n+}$ where $X = Cl$ or H and $n = 1$, or where $X = CO$ and $n = 2$ or $Ru(bpy)(CO)_2Cl_2$—as homogeneous catalysts.[93] The efficiency of formate production depends on the presence of water and excess ligand but is independent of CO_2 pressure. The process involves reductive quenching of the $^*Ru(NN)_3^{2+}$ (NN is bpy or a derivative; or phen) excited state by a tertiary amine (e.g., triethanolamine) to the corresponding Ru^I complex which reduces the carbon dioxide catalyst to Ru^I and then further to Ru^0.[93] For the system composed of $Ru(bpy)_3^{2+}$ and cis-$Ru(bpy)_2(CO)H^+$, the maximum quantum yield of formate produced is 15%. The various steps in the process are summarized in reactions 82 to 88:

$$^*Ru(bpy)_3^{2+} + TEOA \rightarrow Ru(bpy)_3^+ + TEOA^+ \qquad (82)$$

$$2Ru(bpy)_3^+ + Ru(bpy)_2(CO)Cl^+ \rightarrow Ru(bpy)_2(CO)^0 + 2Ru(bpy)_3^{2+} + Cl^- \quad (83)$$

or

$$2Ru(bpy)_3^+ + H^+ + Ru(bpy)_2(CO)H^+ \rightarrow Ru(bpy)_2(CO)^0 + 2Ru(bpy)_3^{2+} + H_2 \ (84)$$

Thermally, formate is formed via reaction 85:

$$Ru(bpy)_2(CO)^0 + Cl^- + CO_2 + H^+ \rightarrow Ru(bpy)_2(CO)Cl^+ + HCOO^- \qquad (85)$$

while under catalytic conditions, formate is formed via Eqs. 86–88:

$$Ru(bpy)_3(CO)^0 + CO_2 \rightarrow Ru(bpy)_2(CO)(CO_2)^0 \xrightarrow{H^+} Ru(bpy)_2(CO)(COOH)^+ \quad (86)$$

$$\xrightarrow{H^+} Ru(bpy)_2(CO)(HCOO)^+ \quad (87)$$

$$Ru(bpy)_2(CO)(HCOO)^+ + Cl^- \rightarrow Ru(bpy)_2(CO)Cl^+ + HCOO^- \quad (88)$$

The overall process is presented in Figure 29.[93]

The photoinduced reduction of phenacyl halides ($YC_6H_4COCH_2X$ where X is Br, or Cl, and Y is p-CN, p-Br, H, p-Me, or p-MeO) by 9,10-dihydro-10-methylacridine ($AcrH_2$) in both the absence and presence of perchloric acid in acetonitrile is an efficient process by visible light irradiation (450 nm; ambient temperature) of $Ru(bpy)_3^{2+}$ as a photocatalyst to give 10-methylacridinium ion ($AcrH^+$) and the corresponding acetophenone derivatives.[94] Without $HClO_4$, quenching of $^*Ru(bpy)_3^{2+}$ by $AcrH_2$ occurs by reductive electron transfer to give $Ru(bpy)_3^+$ which reduces the phenacyl halide derivatives; in the presence of perchloric acid at low pH (strongly acid media) the quenching event takes place via oxidative electron transfer from the phenacyl halides to $^*Ru(bpy)_3^{2+}$ to give the corresponding ruthenium(III) complex, which subsequently oxidizes $AcrH_2$. The oxidative and reductive paths are summarized in Schemes 89 and 90.[94] The related quenching rate constants are presented in Table 4.

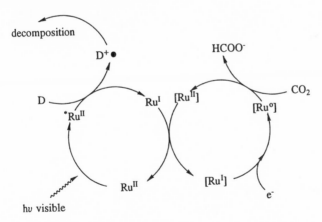

Figure 29. Schematic representation of the processes occurring in the generation of formate by photoinduced reduction of CO_2 using the mixed catalytic system and following a pathway of reductive quenching of the excited state of the ruthenium(II) photosensitizer; it involves a photosensitizer cycle {ruthenium(II) tris chelate, *left*} and a CO_2 catalytic reduction cycle {ruthenium bis- or mono-chelate, *right*}; D denotes TEOA, the electron donor; the ligands of the metal ions have not been included for simplicity. From ref. 93.

Table 4. Electron Transfer Rate Constants, k_q, for the
Acid-Catalyzed Reductive Quenching of **Ru(bpy)$_3$$^{2+}$ by
Phenacyl Halides in the Presence of HClO$_4$ in
Acetonitrile at Ambient Temperature[94]

Phenacyl Halide	$10^{-7}\ k_q,\ M^{-1}s^{-1}$ *(in the presence of HClO$_4$)*	
	0.30 M HClO$_4$	*2.0 M HClO$_4$*
p-CNC$_6$H$_4$COCH$_2$Br	2. 6	15
p-BrC$_6$H$_4$COCH$_2$Br	0.87	7.3
PhCOCH$_2$Br	0.68	7.8
p-MeC$_6$H$_4$COCH$_2$Br	2.1	12
p-MeOC$_6$H$_4$COCH$_2$Br	7.9	33
PhCOCH$_2$Cl	0.29	1.6

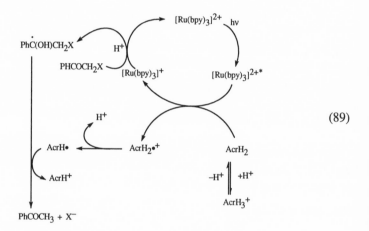

(89)

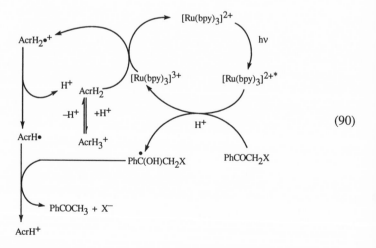

(90)

A novel ruthenium(II) photosensitizer, Ru{S(–)-PhEt*bpy}$_3^{2+}$, possessing a predominant Λ-configuration reduces Co(acac)$_3$ {acac = anion of acetylacetone} catalytically (turnover number, 40) with high enantioselectivity (k_Λ / k_Δ = 1.54) under irradiation by visible light (> 400 nm) in 9:1 ethanol/water mixture.[95] Scheme 91 illustrates the oxidative quenching of the excited state of the ruthenium(II) photosensitizer by the Co(acac)$_3$ complex. Table 5 summarizes the kinetics of the photoreduction of the Co^{III} complex.

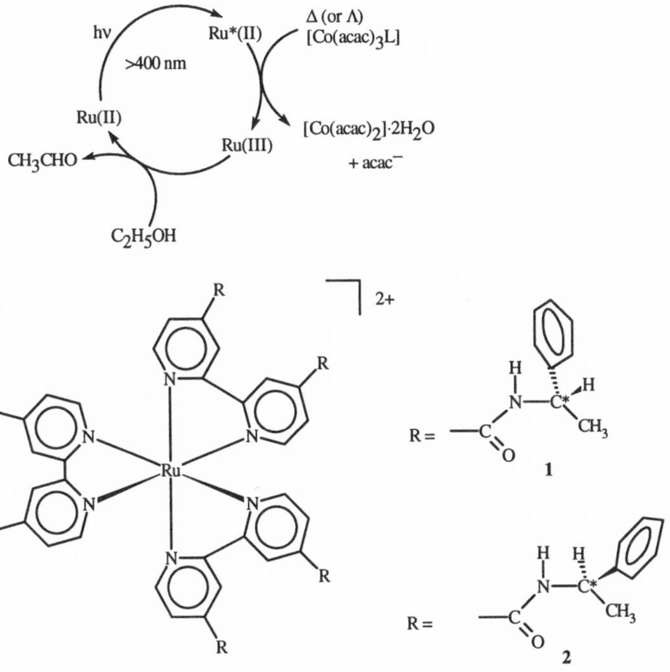

(91)

Table 5. Photoreduction of Co(acac)$_3$ Catalyzed by the Chiral Ruthenium(II) Photosensitizer[95]

Catalyst[a]	Ethanol Content, %	$10^2 k_{obs}$, min^{-1}	Selectivity	
1	90	0.23 ± 0.01	Λ/Δ	1.54
	70	1.79 ± 0.01		1.05
	50	2.25 ± 0.01		1.03
2	97	0.04 ± 0.01	Δ/Λ	1.54

Notes: [a]See Scheme 91.

Table 6. Solvent Effects on the Reductive Quenching of
*Ru(bpy)$_3{}^{2+}$ by Polychlorophenolate Anions at 25 °C, $\mu = 0.01$
M, and 0.01 NaOH[97]

[Methanol] in water (vol.%)	$10^{-9}k_q$, $M^{-1}s^{-1}$		
	2,4-DCP	2,6-DCP	2,4,6-TCP
20	0.53	0.28	0.30
40	1.3	0.62	0.50
60	1.8	1.4	1.1
80	3.4	3.4	2.2
100	5.1	7.2	4.9

Effective photoassisted reduction of *vic*-dibromo and keto compounds proceeds in acetonitrile media by irradiated Ru(bpy)$_3{}^{2+}$ in the presence of the electron donor triethylamine, TEA. The resulting Ru(bpy)$_3{}^+$ species acts as an electron transfer mediator (and as a catalyst) in the debromination of the substrates, reactions 92–94.[96]

$$^*Ru(bpy)_3{}^{2+} + TEA \rightarrow Ru(bpy)_3{}^+ + TEA^{+\bullet} \tag{92}$$

$$Ru(bpy)_3{}^+ + R_1{-}CHBrCHBr{-}R_2 \rightarrow Ru(bpy)_3{}^{2+} + R_1{-}\overset{\bullet}{C}HCHBr{-}R_2 + Br^- \tag{93}$$

$$Ru(bpy)_3{}^+ + R_1{-}\overset{\bullet}{C}HCHBr{-}R_2 \rightarrow Ru(bpy)_3{}^{2+} + R_1{-}CH = CH{-}R_2 + Br^- \tag{94}$$

Such phenolics as the anionic forms of 2,4-dichlorophenol, 2,6-dichlorophenol, and 2,4,6-trichlorophenol are excellent electron donors and can quench the excited state of Ru(bpy)$_3{}^{2+}$ by reductive electron transfer.[97] The quenching rate constants in methanol/water solvent mixtures are nearly diffusion controlled (Table 6). Following the formation of the corresponding phenoxyl radicals, further reaction with a phenolate anion leads to partial dechlorination and to dimerization.

4.2. Intramolecular

In luminescence quenching (static and dynamic) and excited state lifetime quenching (dynamic) experiments, one measures an experimental quenching rate constant, k_q, which, as we saw earlier (Eq. 35), is made up of several other microrate constants. One of the goals in examining photoinduced electron transfer reactions in ion pairs is the possibility of measuring directly the microrate constant for the electron transfer event, k_{el} and possibly for the step k_{-el}. Unfortunately, although this may sound simple, only a few cases have been documented.[4,40,98,99]

Ion Pairs

We can use the scheme illustrated in Figure 16 as a basis for the discussion on electron transfer events taking place in ion pairs. If they are taken to be formed by coupling M and M_A species in the ground state, electron transfer can then be viewed as an intramolecular process. Following the light absorption act to produce the excited ion pair species, M^*---M_A, electron transfer can take place by a reductive or oxidative process to give M^----M_A^+ or M^+---M_A^- intermediate products, respectively, which may or may not be photostable. Two principal fates await these species (we consider the case for oxidative electron transfer): (1) back electron transfer; and (2) if one of the partners of the ion pair is unstable (kinetically labile), pathological product formation occurs. The relative competition between these two events is the decisive factor. In case (2) a net photochemical change takes place and the reaction becomes irreversible. Some examples where (1) predominates and where (2) is predominant will now be considered.

A good example that illustrates case (1) and the measurement of k_{el} is provided by the ion pairs formed between a ruthenium(II) polypyridine complex, $Ru(NN)_3^{2+}$ or $Ru(NN)_2(NN')^{2+}$ (NN = bpy and NN' = 4,4'-Cl$_2$-bpy or 2,2'-biquinoline), and the polytungstate species $Mn(OH)PW_{11}O_{39}^{6-}$ or $Co(H_2O)$-$SiW_{11}O_{39}^{6-}$ (denoted here MW_{11}^{6-}).[99,100] Two factors have rendered the determination of k_{el} feasible: (1) the relatively long lifetimes of the excited ruthenium(II) chromophore (about 125 to 400 ns),[4] and (2) the high charge on the polytungstate anions which ensures a high association constant ($\sim 10^4$ M^{-1}) in the ground state with rates of separation of the M---M_A ion-pair species, k_{-diff}, around 10^6 s^{-1}, and k_{diff} around 10^{10} M^{-1}s^{-1}.[4] The relevant systems and the corresponding data are presented in Table 7. Oxidative quenching in $Ru(NN)_2(NN')^{2+}$ –MW_{11}^{6-} occurs only for those systems for which the process is exergonic; in the last case, the driving force is endergonic and no quenching of the excited state of or the luminescence from the ruthenium(II) chromophore is observed.

The formation of stable photoproducts, M^+---M_A^-, or the dissociated photoredox products M^+ and M_A^- in the scheme of Figure 16, depend on the rate of secondary processes which compete with back electron transfer. As noted in the example illustrated in reactions 36 and 37 (where BL is CN), cobalt(III) ammine complexes are suitable electron acceptors in an irreversible photoreaction, since the photoredox product cobalt(II) ammine species is kinetically labile and undergoes a very rapid decay to form Co^{II}_{aq} and free NH$_3$ ligands.[101] The example depicted by Eqs. 36 and 37 represents the more common type of photoinduced electron transfer reactions which implicate charge transfer excitation, namely metal-to-metal charge

Table 7. Oxidative Electron Transfer Quenching of Excited $^*Ru(NN)_2(NN')^{2+}$ Complexes by Polytungstate Anions[4,40]

$^*Ru(NN)_2(NN')^{2+}$	MW_{11}^{6-}	ΔG, eV	k_q, M^{-1}s^{-1}	k_{el}, s^{-1}
NN = NN' = bpy	M = Mn	−0.34	2.1×10^{10}	$\geq 5 \times 10^8$
NN = NN' = bpy	M = Co	−0.20	2.3×10^{10}	8.5×10^7
NN = bpy, NN' = 4,4'-Cl$_2$bpy	M = Co	−0.09	—	4×10^6
NN = bpy NN' = 2,2'-quin	M = Co	+0.35	$< 3 \times 10^8$	$< 5 \times 10^{-5}$

transfer (MMCT). If ethylenediamine (en) replaces ammonia as the ligand in the cobalt(III) electron acceptor complex, a different photoinduced electron transfer chemistry obtains via MMCT excitation (Eq. 95).[102,103] Thus,

$$Ru^{II}(CN)_6^{-4}\text{----}Co^{III}(en)_3^{3+} \xrightarrow{h\nu} Ru^{III}(CN)_6^{3-}\text{----}Co^{II}(en)_3^{2+}$$

$$+ H_2O \downarrow -en$$

$$\{(CN)_5Ru^{II}\text{--}CN\text{--}Co^{III}(en)_2(H_2O)\}^- \leftarrow \{(CN)_5Ru^{III}\text{--}CN\text{--}Co^{II}(en)_2(H_2O)\}^-$$

(95)

Although the tris-en cobalt(III) complex is also substitutionally labile, its decay is slower than that of the ammine analog.[104,105] The stable cyanide-bridged Ru^{II}-CN-Co^{III} complex is formed via an inner sphere pathway.

Where the cobalt(II)-ammine complex is replaced by the analogous cobalt(III) sepulchrate complex, $Co(sep)^{3+}$ {six N donor atoms}, no permanent photochemical change obtains by MMCT excitation, inasmuch as the primary reduced photoproduct $Co^{II}(sep)^{2+}$ is kinetically inert owing to the nature of the cage.[102,103] Back electron transfer is now the only secondary process possible.[40]

Electron transfer reactions photoinduced by outer sphere ligand-to-metal or ligand-to-ligand charge transfer excitation, MLCT or LLCT, are lacking; one example of photoinduced electron transfer by outer-sphere metal-to-ligand charge transfer is known.[40] This is depicted in reactions 96:

$$Rh^{III}(bpy)_3^{3+}\text{----}Fe^{II}(CN)_6^{4-} \xrightarrow{h\nu} Rh^{III}(bpy)_2(bpy^-)^{2+}\text{---}Fe^{III}(CN)_6^{3-} \quad (96a)$$

$$Rh^{III}(bpy)_2(bpy^-)^{2+}\text{----}Fe^{III}(CN)_6^{3-} + 2\ H_2O\ \rightarrow Rh^{II}(bpy)_3^{2+} + Fe^{III}(CN)_6^{3-} \quad (96b)$$

$$Rh^{II}(bpy)_3^{2+} \rightarrow Rh^{II}(bpy)_2^{2+} + bpy \quad (96c)$$

$$Rh^{II}(bpy)_2^{2+} + Fe^{III}(CN)_6^{3-} + 2H_2O \rightarrow Rh^{III}(bpy)_2(H_2O)_2^{3+} + Fe^{II}(CN)_6^{4-} \quad (96d)$$

The low quantum yield of the overall photoinduced (irreversible) reaction is due to a rapid thermal reversal of the optical charge transfer transition (cf. Figure 12). Here, the back electron transfer step competes effectively with the separation step (Eq. 96b; k_{-diff}).

Additional examples are treated in the recent review articles by Scandola and co-workers[38] and by Vogler and Kunkley.[40]

Molecular Systems (Chromophore/Quencher Complexes)

One of the principal goals of photoinduced electron transfer research is the desire to achieve efficient separation of the redox equivalents following the light energy conversion step, and subsequently to prevent recombination of the charge carriers (suppress back electron transfer) so as to achieve energy storage as stored chemical potential. This in effect would mimic one of the functions of the photosynthesis apparatus.

Examples of complexes where both electron donors and electron acceptors are integral part are known, for which MLCT excitation (in complexes) or $\pi \rightarrow \pi^*$ excitation (in a porphyrin chromophore) leads to oxidative or reductive electron transfer quenching of the excited state formed.[106,107] One such example from a porphyrin-based system has revealed a relatively long-lived (3 μs) photoinduced charge separation onto the peripheral donor and acceptor redox sites; the stored chemical potential was $\Delta G^o > 1.0$ eV.[107]

An interesting system in metal complexes which begins to show similar prospects in terms of charge separation is the ruthenium(II) complex [(Mebpy-3DQ^{2+})Ru^{II}(Mebpy-PTZ)$_2$]$^{4+}$, where Mebpy-3DQ^{2+} (abbreviated as DQ^{2+}b in the scheme below) and Mebpy-PTZ (abbreviated as bPTZ) refer to the ligands:[106]

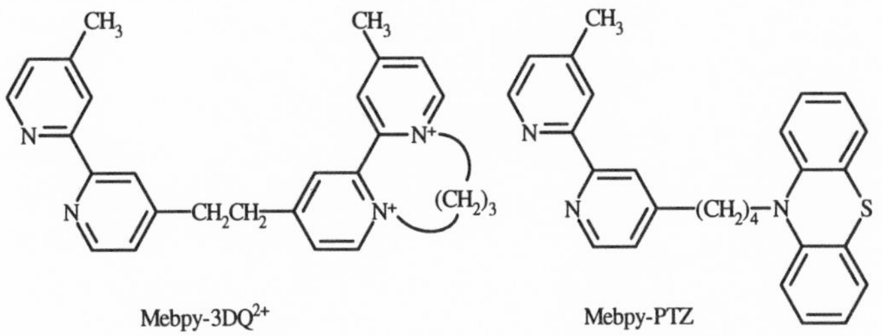
CH3
CH3
CH3
N
N
N+
N+
(CH2)3
CH2CH2
N
N
(CH2)4
N
S
Mebpy-3DQ2+
Mebpy-PTZ

MLCT excitation of this complex is followed by a sequence of intramolecular events (illustrated in Scheme 97) which lead to the highly efficient formation of a charge separated state based on the PTZ and DQ^{2+} based redox sites (an interligand charge transfer state, ILCT) whose stored chemical potential is about 1.3 eV.

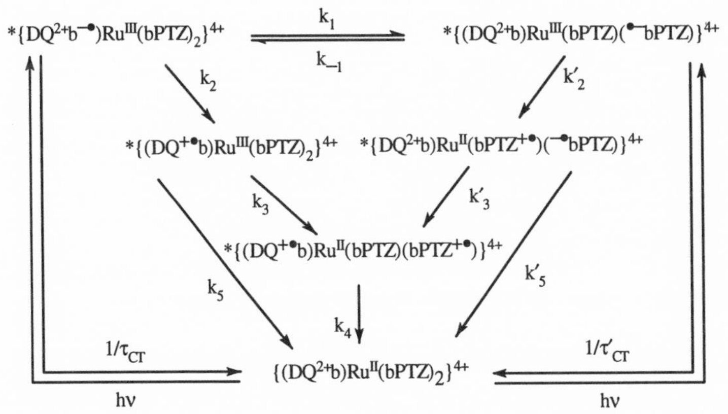
*{DQ2+b−•)RuIII(bPTZ)2}4+
k1
k−1
*{(DQ2+b)RuIII(bPTZ)(•−bPTZ)}4+
k2
k'2
*{(DQ+•b)RuIII(bPTZ)2}4+
*{DQ2+b)RuII(bPTZ+•)(−•bPTZ)}4+
k3
k'3
*{(DQ+•b)RuII(bPTZ)(bPTZ+•)}4+
k5
k'5
k4
1/τCT
1/τ'CT
{(DQ2+b)RuII(bPTZ)2}4+
hν
hν

(97)

Sensitization of the ILCT occurs either or both via step k_2 (oxidative quenching) and k'_2 (reductive quenching). The quantum yield of formation of the ILCT state is about 0.26 suggesting the importance of the intramolecular recombination events k_5 and/or k'_5.[106] The longevity of the charge separated state (ILCT) is 165 ns, sufficiently long that redox chemistry from this ILCT state can take place using Cu^{2+} (reactions 98 and 99).

$$\text{fast}$$
$$PTZ^+-Ru^{II}-DQ^+ + Cu^{2+} \rightarrow PTZ^+-Ru^{II}-DQ^{2+} + Cu^+ \qquad (98)$$

$$k_q = 4.3 \times 10^8 \text{ M}^{-1}\text{s}^{-1}$$

$$PTZ^+-Ru^{II}-DQ^{2+} + Cu^+ \rightarrow PTZ-Ru^{II}-DQ^{2+} + Cu^{2+} \qquad (99)$$

Recent attempts to understand the mechanics and the factors involved in forming long-lived charge separated states have been undertaken by Meyer and his group[108] in their systematic investigations into: (1) examining well-defined MLCT excited states to explore the fundamental details of electron and energy transfer pathways; (2) preparing molecular assemblies in which long-range energy and electron transfer can occur and thereby mimic the reaction center in photosynthesis; (3) preparing assemblies that can act as interfaces between simple excited state electron transfer and the multi-electron catalysts needed to carry out such redox reactions as the oxidation of water to dioxygen, or the reduction of carbon dioxide to methanol (or to methane); and (4) translating the solution based photochemistry to polymeric films such that devices can be fabricated for possible applications.

To illustrate the search for long-lived charge separation, it is worth stepping back to consider the intermolecular system made up of the often used excited chromophore $^*Ru(bpy)_3^{2+}$ and the electron acceptor methylviologen (MV^{2+}) used in oxidative quenching (reactions 100):

$$^*Ru(bpy)_3^{2+} + MV^{2+} \rightleftharpoons \{^*Ru(bpy)_3^{2+}, MV^{2+}\} \rightarrow \{Ru(bpy)_3^{3+}, MV^{+\bullet}\} \quad (100a)$$

$$\overset{k_1}{\{Ru(bpy)_3^{3+}, MV^{+\bullet}\} \rightarrow Ru(bpy)_3^{3+} + MV^{+\bullet}} \qquad (100b)$$

$$\overset{k_2}{\{Ru(bpy)_3^{3+}, MV^{+\bullet}\} \rightarrow \{Ru(bpy)_3^{2+}, MV^{2+}\}} \qquad (100c)$$

where the efficiency for cage escape is given by $\eta_{esc} = k_1/(k_1 + k_2)$ and which depends on the charges on the initially formed redox pair, the quantity of energy released in the back electron transfer step, k_2, and on spin effects.[109] The fundamental problem in energy storage is how to use the separated photoredox products in reaction 100b before the energy wasting reaction 100c takes place; that is, it is imperative that reaction 100c be suppressed and that the back electron transfer from the separated pair also be circumvented. One approach has been to control intramolecularly the excitation event and the quenching event(s) by, for example, attaching the quencher on the chromophore to produce a chromophore/quencher complex; such a system is illustrated in Scheme 101:[109]

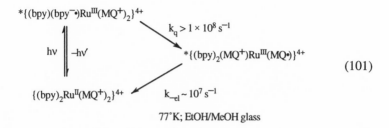

$$(101)$$

where MQ^+ is N-methyl-4,4'bipyridine, a better electron acceptor than bpy by about 0.5 V.[109] The excited state $^*\{(bpy)_2(MQ^+)Ru^{III}(MQ\bullet)\}^{4+}$ can subsequently be intercepted by electron donors and electron acceptors. In the above system, back electron transfer is about an order of magnitude slower than the intramolecular forward electron transfer event.

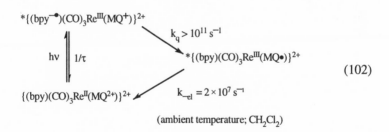

$$(102)$$

In a related rhenium(II) system, the back electron transfer process, k_{-el}, is slower by a factor of $\sim 10^4$ (Scheme 102).[109] The corresponding rhenium(I) complex has also been examined in some detail; the relevant rate constants are depicted in Scheme 103. Here, k_{-el} is at least one order of magnitude slower than the intramolecular forward electron transfer.[108]

The role of free energy change, ΔG, on intramolecular electron transfer has been examined for a series of chromophore/quencher complexes, $\{(4,4'-(X)_2-bpy)Re^I(CO)_3(MQ^+)\}^{2+}$ (where X = C(O)OEt, H, NH_2), in EtOH/MeOH glasses at 77K;

$$*\{(bpy^{-\bullet})(CO)_3Re^{II}(MQ^+)\}^+$$

$$k_q > 2 \times 10^8 \, s^{-1}$$

$$hv \quad 1/\tau \qquad\qquad *\{(bpy)(CO)_3Re^{II}(MQ\bullet)\}^+ \qquad\qquad (103)$$

$$\{(bpy)(CO)_3Re^I(MQ^{2+})\}^+ \qquad k_{-el} = 1.9 \times 10^7 \, s^{-1}$$

(295K; dichloromethane)

$$*\{(4,4'-(X)_2bpy^{-\bullet})Re^{II}(CO)_3(MQ^+)\}^{2+} \quad \rightarrow$$

$$*\{(4,4'-(X)_2bpy)Re^{II}(CO)_3(MQ\bullet\}^{2+} \qquad (104)$$

intramolecular electron transfer quenching via reaction 104 occurs only if the process is sufficiently driven thermodynamically.[110] The driving force for intramolecular quenching is $\Delta G = -0.1$ eV for X = C(O)OEt, -0.49 eV for X = H, and -1.0 eV for X = NH_2. While the MLCT emission is totally quenched at ambient temperature, emission quenching still occurs in the low-temperature glasses, not the case for the other two analogs. This calls attention to the fact that the ability of frozen solvent dipoles to inhibit intramolecular electron transfer depends on the magnitude of ΔG.[110] The events are illustrated in Figure 30 which shows on an energy versus solvent librational coordinate Q of the two excited states involved, $*\{(4,4'-(X)_2bpy^{-\bullet}Re^{II}(CO)_3(MQ^+)\}^{2+}$ and $*\{(4,4'-(X)_2bpy^{-\bullet})Re^{II}(CO)_3(MQ\bullet)\}^{2+}$ and the ground state $\{(bpy)Re^I(CO)_3(MQ^+)\}^{2+}$ for all of which the solvent dipole environments at equilibrium are all different. Following MLCT excitation and vibrational relaxation, either of the two excited states are formed with an energy content $(\Delta G_{es,1} + \lambda_{o,1})$ for the $bpy^{-\bullet}$-Re^{II} state and $(\Delta G_{es,2} + \lambda_{o,2})$ for the Re^{II}-$MQ\bullet$ state, and where $\lambda_{o,x}$ denotes the solvent dipole reorganizational energies between each of the excited states and the ground state. As displayed in case A of Figure 30, the energy condition that is required for electron transfer to occur from one excited state to the other is $(\Delta G_{es,1} + \lambda_{o,1}) > (\Delta G_{es,2} + \lambda_{o,2})$ which is the case only for the X = NH_2 complex where low energy $Re^I \rightarrow MQ^+$ excitation gives the intramolecular electron transfer state by direct absorption (process b in case A of Figure 30). This state can also be reached via $Re^I \rightarrow bpy$ excitation (process a) followed by intramolecular electron transfer. By contrast, where $(\Delta G_{es,1} + \lambda_{o,1}) < (\Delta G_{es,2} + \lambda_{o,2})$ as in the cases for X = H, and C(O)Et (Figure 30B) and nuclear tunnelling effects in the librational modes are unimportant, intramolecular electron transfer in a frozen environment does not occur.[110] In case B of Figure 30, the MLCT excited state Re^{II}-$MQ\bullet$, formed by $Re^I \rightarrow MQ^+$ excitation, decays either by emission or by reverse electron transfer to give the $bpy^{-\bullet}$-Re^{II} excited state. This reversal in the direction of the intramolecular electron transfer in the low-temperature environment is induced energetically by the greater solvent reorganizational energy, $\lambda_{o,2} > \lambda_{o,1} - (\Delta G_{o,2} - \Delta G_{o,1})$, for the lower energy

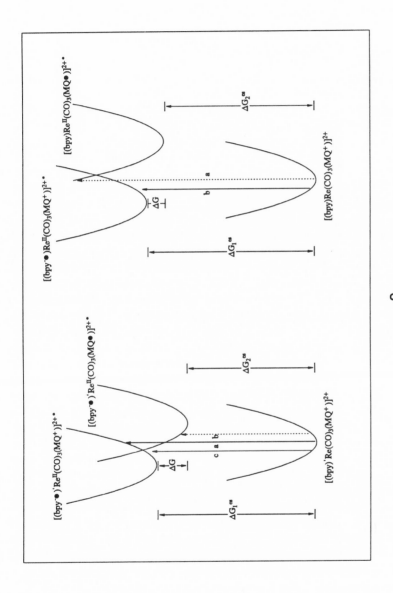

Figure 30. Energy versus solvent librational coordinate diagrams for light-induced intramolecular electron transfer. Case A applies to bpy′ being 4,4′-(NH₂)₂-bpy. From ref. 110.

a

excited state.[110] An additional factor that may complicate the intramolecular electron transfer event in Re-MQ$^+$ complexes is the *flattening* of the relative orientations of the two benzenic rings of the MQ$^+$ ligand on acceptance of an electron.

$$^*(4,4'-(X)_2bpy^{-\bullet})Re^{II}(CO)_3(MQ^+)^{2+} \rightarrow {}^*(4,4'-(X)_2bpy)Re^{II}(CO)_3MQ^\bullet)^{2+} \quad (104)$$

Conformational changes in the electron acceptor of a series of chromophore/ quencher complexes and their effects have been examined in $\{(bpy)Re^I(CO)_3(MQ^+)\}^{2+}$ systems in which the MQ$^+$ ligand is modified by addition of methyl groups at the 3,3'-positions.[111] Spectral and electrochemical results from examining a series of these complexes has demonstrated that: (1) the dihedral angle, θ, has a significant effect on the energies and intensities of the MQ$^+$-based MLCT transitions; (2) the relative ordering of the bpy- and MQ$^+$-based MLCT excited states in $\{(bpy)Re^I(CO)_3(MQ^+)\}^{2+}$ depends on the dihedral angle; and (3) changes in θ and in the solvent dipole reorientation play significant roles in the light-induced $\lambda^*(bpy) \rightarrow \lambda^*(MQ^+)$ intramolecular electron transfer (reaction 104 above). Figure 31 illustrates, on an energy/coordinate diagram, the potential energy surfaces for an averaged librational mode for the ground state complex $\{(bpy)Re^I(CO)_3(MQ^+)\}^{2+}$, and the two excited state species $^*\{(bpy^{-\bullet})Re^{II}(CO)_3(MQ^+)\}^{2+}$ and $^*\{(bpy)Re^{II}(CO)_3(MQ\bullet)\}^{2+}$. For the latter, Figure 31 displays two surfaces for two values of the dihedral angle θ. Two classical factors respond to the difference in the electronic configuration between the two excited states: solvent dipole reorientation, and rotation about the angle θ from 47° to 0°.[111] The latter factor plays an insignificant role in fluid media, as rotation about the dihedral angle occurs in sub-picosecond times. However, in glasses this rotation could be inhibited and thereby affect intramolecular electron transfer; this does not appear to be the case for the complexes where X is NH$_2$ (see above). The requirement for solvent dipole reorientation can inhibit electron transfer in glasses.[111] In low-temperature glasses, the average solvent dipole orientations are those appropriate to the electronic configuration of the ground state complex. MLCT excitation ($h\nu'$ in Figure 31) yields the excited state Re^{II}-MQ$\bullet$ in a Franck–Condon state (vertical transition) with the solvent dipole orientations of the ground state complex. In fluid media, rapid reorientation of the solvent dipoles in a time shorter than excited state decay occurs to reflect orientations suitable to the excited state electronic configuration. However, in glasses such reorientation is inhibited (solvent dipoles are frozen) and librational relaxation cannot occur on the time scale of the excited state lifetime, and the reverse intramolecular electron transfer embodied in reaction 104 above ensues.

Reductive electron transfer quenching has also been examined in related Re(I) systems, where the electron acceptor MQ$^+$ is replaced by an electron donor such as those from the phenothiazine variety (py-PTZ), following rhenium(I) $\rightarrow \pi^*(bpy)$ excitation which rapidly (< 10 ns) leads to the formation of the charge-separated

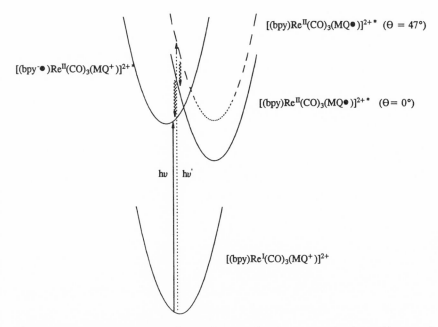

Figure 31. Energy versus coordinate diagram for an averaged solvent librational mode for the excited state and ground state of $[(bpy)Re^I(CO)_3(MQ^+)]^{2+}$. The dashed line refers to the vertical $d\pi \rightarrow \pi^*(MQ^+)$ transition with the dihedral angle between the rings at the ground state value of $\theta = 47°$. From ref. 111.

state $^*\{(bpy^-\bullet)Re^{II}(CO)_3(py\text{-}PTZ^+\bullet)\}^+$. The events are depicted in Scheme 105. Picosecond transient absorption experiments have shown that the appearance of the py-PTZ$^+\bullet$ redox site appears within ~ 200 ps.[112] The transient behavior shows that initial excitation of the MLCT chromophore is followed by rapid intramolecular electron transfer quenching with $k_q = 4.8 \times 10^9$ s^{-1} to give the charge separated state, which in turn decays back to the ground state with $k_{-el} = 4.0 \times 10^7$ s^{-1} at 295K and in acetonitrile media.[108,109,112] Two features are worth noting in Scheme 105: (1) intramolecular quenching (k_q) produces an organic donor-acceptor "exciplex"

$$^*\{(bpy^-\bullet)Re^{II}(CO)_3(py\text{-}PTZ)\}^+ \underset{k_{-q}}{\overset{k_q}{\rightleftharpoons}} {}^*\{(bpy^-\bullet)Re^I(CO)_3(py\text{-}PTZ'\bullet)\}^+$$

$$\{(bpy)Re^I(CO)_3(py\text{-}PTZ)\}^+ \qquad (105)$$

chemically linked by a Re^I bridge; and (2) the excitation-quenching sequence leads to intramolecular sensitization of a low-lying interligand charge transfer (ILCT) excited state.[112] A further example of the formation of the charge separated state is afforded by modification of these rhenium(I) complexes. Specifically, transient absorption experiments on the complex {(PTZ-bpy)(CO)$_3$Re^I(4,4'-bpy)Re^I(CO)$_3$(bpz)}$^{2+}$, where bpz is 2,2'-bipyrazine, 4,4'-bpy is 4,4'-bipyridinium cation, and PTZ-bpy is the 4-phenothiazinyl-4'methyl-2,2'-bpipyridine ligand, have demonstrated that following MLCT excitation leads to photochemical redox splitting over a distance of some 11 A^o across the 4,4'-bpy ligand bridge (see reaction below).[113] The redox separated state lies about 1.4 eV above the ground state and ~ 0.4 eV below the next lowest excited state; the charge separated state returns to the ground state with k ~ 1.8×10^7 s^{-1} which is to be compared with the analogous state *{(PTZ$^+$•-bpy$^-$•)Re^I(CO)$_3$(4-Etpy)}$^+$ which decays with k ~ 1.1×10^7 s^{-1} in acetonitrile and with k ~ 4.7×10^6 s^{-1} in dichloroethane.[113] This calls attention to the fact that in the latter charge separated state, back electron transfer to give the ground state may take place via a through-space process in view of the folding ability of the PTZ-bpy ligand and the facial geometry at the metal complex sites; the two redox sites are separated by ~ 4 A^o in the monomeric complex and by about 6 A^o in the ligand-bridged dimeric complex.

$$\{(PTZ-bpy)(CO)_3Re^I(4,4'-bpy)Re^I(CO)_3(bpz)\}^{2+} \xrightarrow[DCE, 293K]{h\nu}$$

$$\{(PTZ^+•-bpy)(CO)_3Re^I(4,4'-bpy)Re^I(CO)_3(bpz^-•)\}^{2+}$$

The influence of distance on intramolecular electron transfer from photoexcited ruthenium(II) diimine complexes to N,N'-diquaternarized bipyridines has recently been examined by Schmehl and co-workers.[114] The photochemical events illustrating intramolecular electron transfer and back electron transfer are summarized in reactions 106 through 108; the corresponding rate constants are summarized in Table 8.

Table 8. Intramolecular Electron Transfer Rate Constants and Back Electron Transfer Rate Constants for {(bpy)$_2$Ru^{II}(L^{2+})}$^{4+}$ Complexes in Acetonitrile at 298 K[114]

L	$10^{-6}k_{el}$, s^{-1}	$10^{-10}k_{-el}$, s^{-1}	No. of carbons
4.2.3-DQ	590	—	2
4.3.3-DQ	22	—	3
4.4.3-DQ	54	—	4
4.5.3-DQ	6.4	1.0	5
4.6.3-DQ	8.7	2.0	6
4.12.3-DQ	1.2	0.3	12

$$\{(bpy)_2Ru^{II}(4.x.3-DQ^{2+})\}^{4+} \xrightarrow{h\nu,k_{el}} {}^*\{(bpy)_2Ru^{III}(4.x.3-DQ^+)\}^{4+} \qquad (106)$$

$$^*\{(bpy)_2Ru^{III}(4.x.3-DQ^+)\}^{4+} \xrightarrow{k_{-el}} \{(bpy)_2Ru^{II}(4.x.3-DQ^{2+})\}^{4+} \qquad (107)$$

$$^*\{(bpy)_2Ru^{III}(4.x.3-DQ^+)\}^{4+} + (C_2H_5)_3N \xrightarrow{k_{trapping}}$$

$$\{(bpy)_2Ru^{II}(4.x.3-DQ^+)\}^{3+} + (C_2H_5)_3^{+\bullet} \qquad (108)$$

etc...[see ref.114]

where 4.x.3.-DQ^{2+} is a ligand in which a 4,4′-dimethyl-2,2′-bipyridine links to a diquaternary 2,2′-bipyridine through a methylene chain (n = 3); the ligand is pictured below:

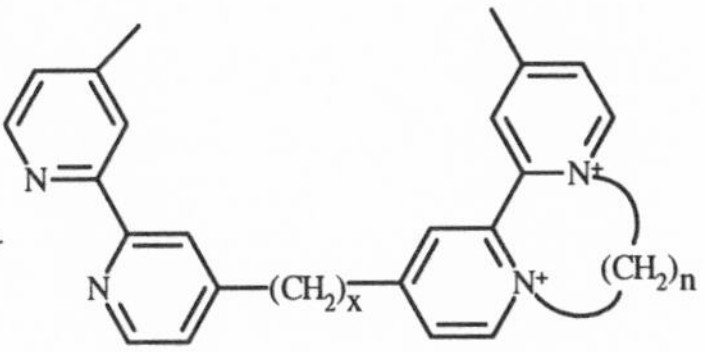

Rate constants for intramolecular electron transfer from the excited ruthenium(II) complex to the diquaternary 2,2′-bpyridine ligand decrease as the length of the bridging chain increases from x = 2 to 12. The observed electron transfer rates exhibit an even-odd chain length alternation for x = 2 to 6. The k_{-el} rates also show a dependence on distance. Addition of $Et_4N^+ClO_4^-$ to acetonitrile solutions of the 6- and 12-carbon bridged complexes increases k_{el} (reaction 106) due to strong electrostatic factors on the distribution of conformers in solution prior to electron transfer. Viscosity effects also alter the rate of intramolecular electron transfer; the nearly complete disappearance of intramolecular electron transfer in polymethyl-metacrylate for the 4.6.3-DQ complex results from a slow dielectric relaxation of the medium.[114]

Supramolecular Systems

The area of supramolecular photochemistry has received considerable attention lately. Several excellent examples of photoinduced intramolecular electron transfer in these systems are treated in a recent book (1991) by Balzani and Scandola.[9]

5. ELECTRON TRANSFER PROCESSES IN AN ORGANIZED ASSEMBLY—SEMICONDUCTOR/SOLUTION SYSTEMS

Electron transfer processes at a semiconductor/solution interface play an important role in light energy conversion devices where such interfaces are used as light-harvesting units. Irradiation of a semiconductor (SC; for example TiO_2 or CdS) with energy greater than or equal to the bandgap energy suitable to the semiconductor (E_{BG} for TiO_2, 3.0–3.2 eV, and for CdS, 2.4–2.6 eV) generates electron/hole pairs, which subsequently separate to give conduction band electrons (e^-_{CB}) and valence band holes (h^+_{VB}), reaction 109, caused by the electric field in the space charge layer for large crystals or by diffusion in the case of ultrasmall colloid particles (see Section 3.2 above). We have already alluded to the fact that these separated charge carriers rapidly migrate to the surface of the particle, where they get trapped at some defect site and become poised to recombine (radiatively and/or nonradiatively) and/or to initiate redox chemistry with surface adsorbates (or with solution substrates). This photoredox chemistry has been exploited extensively in the fields of interfacial electron transfer[6,42] and in heterogeneous photocatalysis.[44] In this section, we restrict our discussion to some early and more recent interesting aspects of interfacial electron ejection (semi-conductor to electron acceptor), and electron injection (electron donor to semiconductor). Photosensitization of semiconductor particles has important consequences in color photography, electrography, photocatalysis, solar energy conversion systems, and artificial photosynthesis. Some aspects in heterogeneous photocatalysis are treated in greater detail later in Section 6.

$$SC + h\nu \rightarrow e^-_{CB}/h^+_{VB} \rightarrow e^-_{CB} + h^+_{VB} \qquad (109)$$

$$e^-_{CB} + Ox_{ads} \rightarrow Red_{ads} \qquad (110)$$

$$h^+_{VB} + Red_{ads} \rightarrow Ox_{ads} \qquad (111)$$

5.1. Semiconductor to Acceptor Electron Transfer (Charge Ejection)

In examining the dynamics of electron ejection from colloidal semiconductor particles across the interface to an adsorbed substrate or to a solution bulk species, methylviologen has proven a useful electron acceptor for two principal reasons: (1) it undergoes a reversible one-electron reduction with a well-defined and pH-independent redox potential $\{E^\circ(MV^{2+/+}) = -0.440$ V vs. NHE$\}$; and (2) its reduced form, $MV^{+\bullet}$, is readily identifiable by its characteristic blue color ($\lambda = 602$ nm and $\varepsilon = 11,000$ $M^{-1}cm^{-1}$).[115,116] The interfacial event between excited TiO_2 and MV^{2+} is illustrated in Figure 32A. The observed rate constant for reduction of MV^{2+} to $MV^{+\bullet}$ increases linearly with the concentration of MV^{2+} (at $[MV^{2+}] \geq 2.5 \times 10^{-5}$

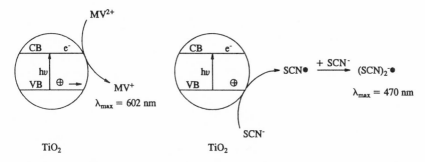

Figure 32. Graph illustrating reduction (charge ejection) of MV^{2+} and oxidation of SCN^- over irradiated TiO_2 semiconductor particles.

M); $k_{obs}\{*TiO_2 + MV^{2+}\} = 1.2 \times 10^7$ $M^{-1}s^{-1}$.[115] At low concentrations, MV^{2+} is quantitatively reduced since $[e^-/h^+] >> [MV^{2+}]$. At higher concentrations of the electron acceptor, $[MV^{+}\bullet]$ approaches the limit of 3.5×10^{-5} M. The photoinduced reduction process does not involve MV^{2+} species adsorbed or pre-adsorbed on the semiconductor particle surface for the following reasons: (1) if MV^{2+} were adsorbed on the surface, prompt appearance of $MV^{+}\bullet$ would be observed concomitantly and within the laser pulse of ~ 15–20 ns, rather than the gradual formation of the reduced species in the micro- to millisecond time frame; (2) as well, the rate of electron transfer would be expected to reach a plateau at the higher concentrations of MV^{2+}. Neither of these two cases are verified experimentally.[115]

The observed rate of formation of $MV^{+\bullet}$ is much lower than that predicted for a diffusion-controlled reaction, k_{diff} ~ 7×10^{10} $M^{-1}s^{-1}$ estimated for a reaction radius of 100 A° and for a diffusion coefficient of MV^{2+} of 9.2×10^{-6} cm^2s^{-1} using the Smoluchowsky equation 49 ($k_{diff} = 4\pi RD$). Evidently, interfacial electron transfer at pH 5 does not occur at every encounter between excited TiO_2 and MV^{2+} species.[115] The quantum efficiency of formation of $MV^{+}\bullet$ is unity under the conditions used. As reactions are dependent on the thermodynamic driving force, the position of the conduction band of colloidal TiO_2 particles is expected to greatly influence the rate of interfacial electron transfer. Because the TiO_2 surface contains acidic and basic sites (see Eqs. 14 and 15), the conduction band potential changes with pH. Figure 33 illustrates the redox potentials of the conduction and valence bands as a function of pH; also shown are the positions of the redox potentials of MV^{2+}, SCN^-, and of the reduction and oxidation of water. The k_{obs} increases by three orders of magnitude over the pH region 3 to 8; at higher pH (> 10), the rate of reduction of MV^{2+} approaches the diffusion controlled limit: $k_{obs} = 7 \times 10^9$ $M^{-1}s^{-1}$, and electron transfer across the interface is no longer rate determining.[115] From Eq. 47 and the experimental values of k_{obs} at various pH's, the rate of electron transfer k_{ct} can be estimated: for R = 100 A°, D = 9.2×10^{-6} cm^2s^{-1} for MV^{2+}, and $k_{obs} = 1.2 \times 10^7$ $M^{-1}s^{-1}$ at pH 5, the electrochemical rate constant k_{ct} is 1.6×10^{-3}

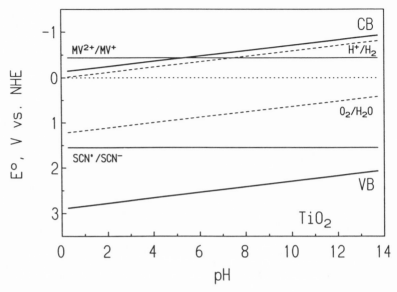

Figure 33. Plots showing the effect of pH on the redox potentials of the conduction band (CB) and valence band (VB) of TiO$_2$ together with other relevant redox couples in aqueous media.

cm s^{-1}. Recalling that the rate constant for a cathodic electron transfer process varies with the overvoltage η as per the Tafel equation, at 298 K we obtain:[117]

$$\log (k_{ct} / k^{\circ}_{ct}) = - (1 - \alpha)\eta / 0.059 \qquad (112)$$

The pH at which $\eta = 0$ is 5.34; at this pH, k°_{ct} is 4×10^{-3} cm s^{-1}. This corresponds to the rate constant of electron transfer at the standard potential of the MV$^{2+/+}$ redox couple, which the colloidal TiO$_2$ particle reaches at pH 5.34. The transmission coefficient α is experimentally 0.52, indicating a symmetrical transition state for the interfacial event. Thus, the interfacial electron transfer step controls the rate at the lower pH's when the overvoltage available to drive the reaction is small. At the higher pH's, mass transfer effects become increasingly important and determine the overall reaction rate.[117]

When the semiconductor is CdS, methylviologen can only be reduced if adsorbed at the particle surface since the lifetime of the e$^-$/h$^+$ pair is $\leq$ 30 ps (k$_{recombination} \geq$ 3×10^{10} s^{-1}).[118] Reduction of adsorbed MV^{2+} on CdS particles is borne out experimentally by the observation that appearance of MV$^{+}\bullet$ occurs within the laser pulse of ~ 15–20 ns.[115] A later study showed that the rate of interfacial electron transfer on MV^{2+}(ads) is k$_{113} \sim 1 \times 10^9$ s^{-1} (Eq. 113).[118]

$$e^-_{CB} + (MV^{2+})_{ads} \xrightarrow{k_{113}} \{(MV^{+\bullet})_{ads} \rightleftharpoons \tfrac{1}{2} (MV^{+\bullet})_2\} \qquad (113)$$

$$\underset{600\ nm}{} \qquad \underset{530\ nm}{}$$

In the presence of an electron donor such as SCN^- $\{E^\circ (SCN\bullet/SCN^-) = + 1.5$ V, NHE$\}$, irradiated TiO_2 particles also show interfacial hole transfer. Valence band hole oxidation of SCN^- to the radical $SCN\bullet$ takes place promptly within the laser pulse indicating that SCN^- must be adsorbed on the particle surface and k of formation of the radical is $> 7 \times 10^7$ s^{-1}; the lifetime of h^+_{VB} is 30 ± 15 ns.[68] In the presence of excess SCN^- anion, formation of the radical dimer occurs (see Figure 32B) the rate of which is 7×10^9 M^{-1}s^{-1} and $\tau_{SCN\bullet}$ is ~ 2 ns (at pH 3, 0.1 M SCN^-, and 0.5 g/L TiO_2).[115] Formation of the thiocyanate radical dimer is favored and experimentally verified[115] at low pH as might be expected from the larger driving force available (see Figure 33). In addition, the greater concentration of $(SCN)_2^{-\bullet}$ at the low pH's is favored by a greater quantity of SCN^- species adsorbed on TiO_2 particles below pH 3.3, which is the point of zero zeta potential for the TiO_2 used.[115] Below pH 3.3, the particles are positively charged, thereby improving the adsorption characteristics of the semiconductor towards anions.

The rate of reduction of MV^{2+} is diffusion-controlled in alkaline solutions (pH > 10), but becomes limited by interfacial electron transfer at lower pH for TiO_2 colloids prepared from titanium(IV) isopropoxide hydrolysis; the transfer coefficient α is 0.52 and $k^\circ_{ct} = 4 \times 10^{-3}$ cm s^{-1}.[115] However, if TiO_2 prepared from hydrolysis of $TiCl_4$ were used, the corresponding values are $\alpha = 0.85 \pm 0.05$ and $k^\circ_{ct} = 1 \pm 0.5 \times 10^{-2}$ cm s^{-1}.[119] Thus, while the heterogeneous electron transfer rate constants for the two preparations are about the same, the α values differ considerably. Note that the Marcus theory predicts α to be 0.5. The relatively larger value of α for the TiO_2 obtained from the $TiCl_4$ hydrolysis is, however, compatible with other free energy relationships such as those derived by Rehm and Weller (see Eq. 27).[50,51] Where α is greater than 0.5, the transition state is unsymmetrical and a large fraction of the overvoltage contributes to decrease the free energy of activation of the reaction.[119] The differences in α for reduction of MV^{2+} by e^-_{CB} between the two TiO_2 preparations must arise from differences in the nature and density of surface hydroxyl groups on the TiO_2 particles.[119]

Contrary to the observations on the reduction of methylviologen,[115] results from the reduction of the surfactant methylviologen $C_{14}MV^{2+}$ species reveals that the log k_{obs} versus pH plots (Figure 34)[119] are linear over a wide pH range of 7 pH units. As well, while k_{obs} increases with $[MV^{2+}]$, the observed rate constant is not affected when the concentration of $C_{14}MV^{2+}$ varies from 2×10^{-4} M to 10^{-3} M. The transfer coefficient α for reduction of the latter viologen is 0.78 ± 0.05 and nearly identical to that for MV^{2+} over the same type of TiO_2.[119] These two viologens differ in two respects: (1) for the surfactant viologen, k_{obs} does not attain a diffusion limited value at pH > 10; and (2) k_{obs} is independent of $[C_{14}MV^{2+}]$. The latter suggests that reduction of $C_{14}MV^{2+}$ by e^-_{CB} involves surface bound acceptor viologen species, because of its amphiphilic nature. Equation 47 or its equivalent,

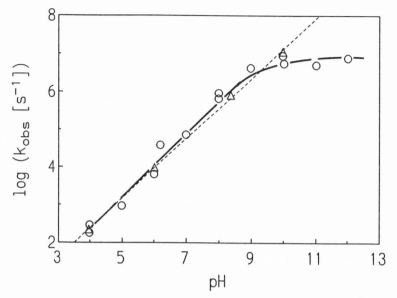

Figure 34. Semilogarithmic plots of k_{obs} versus pH for the reduction of MV^{2+} (O) and $C_{14}MV^{2+}$ (Δ) by $e^-_{CB}(TiO_2)$. The solid line is a computer fit for the MV^{2+} reduction using $\alpha = 0.84$ and $k^o_{ct} = 10^{-2}$ cm s^{-1}; the dashed line was drawn with a slope of 0.78 through the data for the reduction of $C_{14}MV^{2+}$ data points. From ref. 119.

which applies to a bimolecular event, cannot apply when the electron acceptor is surface adsorbed on the semiconductor particle. Rather, simple considerations suggest that k_{obs} is related to the electrochemical rate constant k_{ct} by the relation:[119]

$$k_{ct} = k_{obs}\, d \qquad\qquad (114)$$

where d is the average distance over which electron transfer occurs. The pH at which the overvoltage η is zero is 5.54 and k_{obs} at this pH is 4×10^3 s^{-1} (Figure 34); this yields a value of k^o_{ct} of 10^{-3} cm s^{-1} (distance ~ 25 A°) which is about an order of magnitude smaller than that of MV^{2+}. The smaller value for the surfactant viologen is due to its closer contact with the TiO_2 surface.[119]

The reduction of the dimeric viologen DV^{4+} species (below) has also been examined.[119] It was concluded that reduction occurs via simultaneous two-electron transfer from the excited TiO_2 to surface-adsorbed DV^{4+} to give the DV^{2+} species. The relevant redox potentials for this diviologen are $E^o(DV^{4+/3+}) = 0$ V (NHE) and $E^o(DV^{3+/2+}) = -0.07$ V (NHE); such a small difference in the potentials made this electron acceptor a good candidate for simultaneous two-electron transfer events. Laser pulsed experiments with ns-time resolution showed that for [DV^{4+}] as small as 2×10^{-5} M a major fraction of DV^{2+} is already evident within the 15 to 20 ns laser pulse, following which a slower growing transient in millisecond time forms that is identified as DV^{3+}. Thus the time required for interfacial two-electron transfer (Eq. 115) is $< 10^{-8} s^{-1}$, and considering the average distance for the process

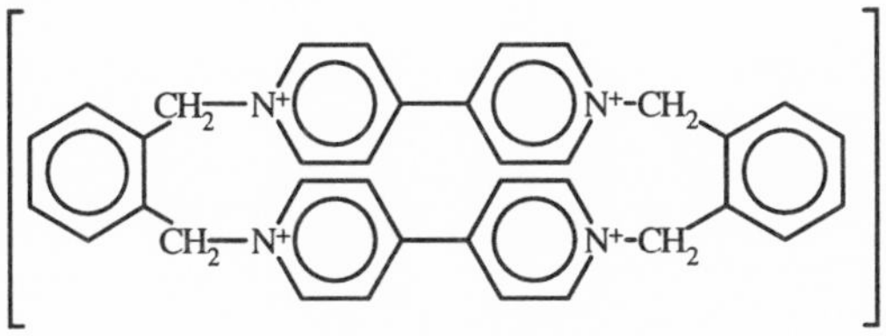

$$DV^{4+}(ads) + 2 e^-_{CB} \rightarrow DV^{2+}(ads) \qquad (115)$$

to be $\sim 5 A°$, where the k_{ct} is > 0.2 cm s^{-1} at pH 7 (from Eq. 114). Kinetic evaluations of the slower transient growth gives for the rate of reduction of DV^{4+} to DV^{3+} (Eq. 116) a value $k_{116} = 35$ s^{-1}; the corresponding reduction rate for DV^{3+} to DV^{2+} was obtained from laser experiments on a pre-irradiated system that consisted of DV^{3+} and TiO_2 particles. Evaluation gives $k_{117} = 20$ s^{-1}.[119] The corresponding second-order constants are respectively, $k'_{116} = 3.5 \times 10^5 M^{-1}s^{-1}$ and $k'_{117} = 2 \times 10^5 M^{-1}s^{-1}$; these are significantly below the diffusion controlled limit, suggesting that electron transfer is controlled by the rate of heterogeneous electron transfer at the semiconductor surface (k_{ct}) because of low overvoltages (0.17 V and 0.1 V, respectively at pH 1) and decreased frequency of encounter between solution DV^{4+} and the positively charged TiO_2 surface. Additional DV^{3+} species also formed by coproportionation of DV^{2+} with DV^{4+} (reaction 118); $k_{118} = 1.5 \pm 0.5 \times 10^7 M^{-1}s^{-1}$.[119]

$$DV^{4+}(ads) + e^-_{CB} \rightarrow DV^{3+}(ads) \qquad (116)$$

$$DV^{3+}(ads) + e^-_{CB} \rightarrow DV^{2+} \qquad (117)$$

$$DV^{2+} + DV^{4+} \rightarrow 2 DV^{3+} \qquad (118)$$

In an early polarographic and coulometric study on the reduction of $Rh(bpy)_3^{3+}$ in alkaline aqueous media (pH 10), it was noted that this complex undergoes a two-electron transfer reduction (reaction 119), $E°(Rh(bpy)_3^{3+/+}) = -0.67$ V.[120] A later cyclovoltammetric study of this reaction concluded that, in fact, reduction occurs via two subsequent one-electron transfer events (reactions 120 and 121).[121] The relevant redox potentials inferred this bipyridine complex also to be an excellent candidate for examination of its reduction by irradiated TiO_2 conduction band electrons.[119] Pulsed laser photolysis data indicate that the conduction band process involves a single electron transfer from excited TiO_2 particles to $Rh(bpy)_3^{3+}$ species in the solution bulk. A Tafel plot of log k_{122} versus pH is linear in the pH range 7 to 12 giving a transfer coefficient (from the slope) α of 0.64, smaller than the MV^{2+} value (0.85) using the same TiO_2 preparation. At the pH where the overvoltage η is zero (pH 10), the observed rate constant k_{122} is 1.7×10^5 s^{-1} and the corresponding electrochemical rate constant k°_{ct} is 0.4 cm s^{-1}. This 40-fold greater value compared to that for the MV^{2+} species implies a lower intrinsic barrier for electron transfer from TiO_2 particles to $Rh(bpy)_3^{3+}$; a small reorganization energy is associated with the Rh^{III} to Rh^{II} electron transition.[119]

$$Rh(bpy)_3^{3+} + 2\ e^- \rightarrow Rh(bpy)_3^+ \tag{119}$$

$$Rh(bpy)_3^{3+} + e^- \rightarrow Rh(bpy)_3^{2+} \tag{120}$$

$$\{E^\circ(Rh(bpy)_3^{3+/2+}) = -0.72\ V\}$$

$$Rh(bpy)_3^{2+} + e^- \rightarrow Rh(bpy)_3^+ \tag{121}$$

$$\{E^\circ(Rh(bpy)_3^{2+/+}) = -0.8\ V\}$$

In alkaline aqueous media, the quantum efficiency of formation of $Rh(bpy)_3^{2+}$ is nearly unity, but is drastically lower at pH < 7.5.[119] At the lower pH's, only a fraction of the initial e^-_{CB} leave the TiO_2 particle; equilibrium between this event and the solution redox couple is established by simultaneous occurrence of reaction 122 (charge ejection) and reaction 123 (charge injection) which takes place in microsecond time scale. Reactions of the type 123 are very important as potential processes for sensitization of wide bandgap metal-oxide semiconductors; this topic is treated later (Section 5.3). Although, reduction of $Rh(bpy)_3^{3+}$ to $Rh(bpy)_3^+$ occurs via one-electron transfer events, the reverse oxidation process via a two-electron transfer event on TiO_2 particles is not excluded.[119]

$$Rh(bpy)_3^{3+} + e^-_{CB} \rightarrow Rh(bpy)_3^{2+} \tag{122}$$

$$Rh(bpy)_3^{2+} + TiO_2 \rightarrow Rh(bpy)_3^{3+} + e^-_{CB}(TiO_2) \tag{123}$$

From the above several examples, it is clear that the transfer coefficient α for MV^{2+} reduction from TiO_2 particles prepared from the hydrolysis of $TiCl_4$ is much greater than α from TiO_2 particles prepared from hydrolysis of $Ti(^iPrO)_4$; this is understandable on the basis that surface states (OH$^-$ groups) on TiO_2 do participate in the electron transfer events. As well, the dramatic pH effects on the rate of reduction of MV^{2+} and other electron acceptors confirm the cathodic shift of the Fermi level of the TiO_2 particles with increasing pH (see Figure 33). Where the electron transfer event takes place with surface-adsorbed substrates, such as the surfactant viologen $C_{14}MV^{2+}$, the conduction band process can take place in the sub-nanosecond time frame.

The scarcity of simultaneous two-electron transfer events led[122] to a re-examination of the reduction of the cofacial dimer DV^{4+} by picosecond laser methods. It was also hoped that we could directly measure the pertinent electron transfer rate constants. The picosecond results precluded a truly simultaneous two-electron transfer event, since the primary transient species was DV^{3+} which was later followed by the appearance of DV^{2+}. Interfacial electron transfer from the conduction band of colloidal TiO_2 to the surface-adsorbed DV^{4+} occurs rapidly (pH 7.8): $k_{124} = 2 \times 10^{10}$ s^{-1}, while that for the formation of the DV^{2+} species is 3 to 10 times slower ($k_{125} \sim 2$ to 7×10^9 s^{-1}). The rate constant for interfacial electron transfer, k_{124}, can be approximated by the expression (Eq. 126):

$$e^-_{CB}(TiO_2) + DV^{4+}(ads) \rightarrow DV^{3+}(ads) \tag{124}$$

$$e^-_{CB}(TiO_2) + DV^{3+}(ads) \rightarrow DV^{2+} \tag{125}$$

$$k_{124} = \nu_o \exp[-\beta(r - r^o)] \exp[-(\Delta G^o + \lambda)^2/4\lambda KT] \tag{126}$$

where $\nu_o \sim 10^{13}$ s^{-1} is the frequency for nuclear reorganization, $\beta \sim 1.2$ A^o is the damping coefficient and $(r - r^o) \sim 5$ A^o is the electron tunneling distance.[122] The value for k_{124} can only be reproduced if the driving force for interfacial electron transfer is optimal; that is, if $-\Delta G^o$ is $\sim \lambda$. The estimated value for k_{124} using Eq. 126 is 2.5×10^{10} s^{-1}.[122] Thus the charge transfer process is extremely rapid under neutral pH conditions, where intimate contact between the acceptor and the semiconductor favors efficient coupling of the $3d^1$ wavefunction manifold of the Ti lattice ions with the wavefunction of the DV^{4+} acceptor orbitals.

Table 9 summarizes the quenching constants for the emission from aqueous ZnO colloidal sols by a variety of metal ions.[123] Many of the bimolecular quenching constants, k_q, appear to exceed the limit for diffusion controlled reactions. A rational explanation obtains if the quenchers are adsorbed onto the semiconductor particle surface at the moment of excitation (static quenching). Variations in pH alters the extent of adsorption and thus k_q, as witnessed by the data in Table 9; changes are seen when the pH is 7.7, where the ZnO surface is positively charged (point of zero charge for these ZnO colloids is 9.3 ± 0.2) and when pH is 12 where the surface is negatively charged. For example, although Ag^+, Mn^{2+}, and MV^{2+} are somewhat poor quenchers at pH 7.7, they become more efficient quenchers at pH 12; this contrasts, as expected, with the anions MnO_4^- and $Cr_2O_7^{2-}$ which are not readily adsorbed on a negative surface at pH 12, but are increasingly adsorbed and are better quenchers at pH 7.7. Cations such as Cr^{3+}, Cu^{2+}, Fe^{3+}, and Fe^{2+} are also very effective quenchers at pH 7.7, suggesting that electrostatic interactions between the cations and the ZnO surface are not the only factors governing adsorption.[123] Luminescence quenching takes place via electron transfer quenching. For the strongly oxidizing adsorbates MnO_4^-, $Cr_2O_7^{2-}$, Fe^{3+} and Ag^+ act as irreversible electron acceptors as back electron transfer to electron traps on the ZnO surface does not occur; the cation Fe^{2+} appears to quench via electron injection into

Table 9. Rate Constants for the
Quenching of the Visible
Luminescence from Excited ZnO
Colloids in Aqueous Media[123]

Quencher	$10^{-11} k_q$, $M^{-1}s^{-1}$
pH = 7.7	
MnO_4^-	6000
$Cr_2O_7^{2-}$	2000
Fe^{2+}	1400
Cu^{2+}	460
Fe^{3+}	250
Cr^{3+}	110
Zn^{2+}	11
Co^{2+}	4.2
Ni^{2+}	1.2
Mn^{2+}	0.7
MV^{2+}	(0.6)
Ag^+	(0.4)
pH = 12	
MnO_4^-	260
Ag^+	200
MV^{2+}	86
Mn^{2+}	26
Co^{2+}	23
$Cr_2O_7^{2-}$	5
Li^+	< 0.07

the valence band (hole scavenger) or into electron traps near the valence band of ZnO.

5.2. Excited Electron Donor to Semiconductor Electron Transfer (Charge Injection)

Electron transfer processes across a semiconductor/solution interface have concentrated on improving the visible light response of wide bandgap metal oxide semiconductors such as ZnO and TiO_2. This is often referred to as photosensitization, which is achieved by adsorption of a dye molecule on the semiconductor surface, and following excitation of the dye leads to injection of an electron into its conduction band.[124] Early studies in this area that have implicated flash photolysis methods have alluded to sensitization of AgI colloids in the presence of dyes,[125] sensitization of TiO_2 by $Ru(bpy)_3^{2+}$ at high temperatures,[126] and erythrosine sensitization of colloidal TiO_2 in acetonitrile, $k_{inj} = 4.2 \times 10^9$ s^{-1}.[127] We consider in some detail the photosensitization of TiO_2 colloids with the dye Eosin-Y (abbreviated EO).[124]

EO:

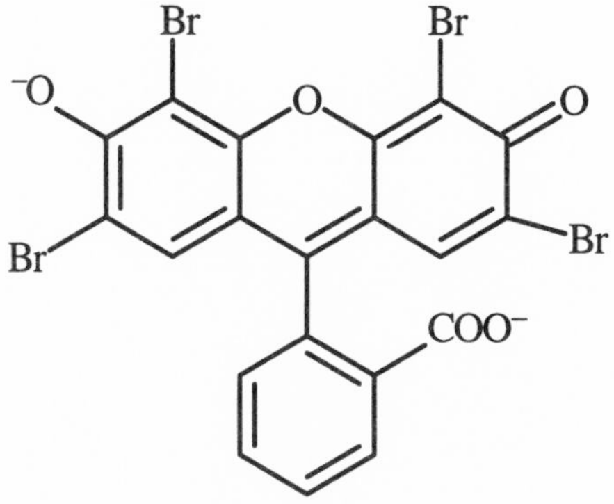

Irradiation of a system comprised of an aqueous solution of colloidal TiO_2 (radius ~ 60 Å) and eosin-Y generates the EO^+ species efficiently and at a high rate, formation of EO^+ being complete in < 15 ns.[124] It is generated by electron injection from the electronically excited singlet state of eosin $[EO(S_1)]$ into the conduction band of the TiO_2 particle (reaction 127), unlike the path in the absence of the semiconductor where the photogeneration of EO^+ in aqueous media is relatively slow and somewhat inefficient. The driving force for reaction 127 is $\Delta G = -1.1 + 0.059$ pH. Although there is sufficient driving force in alkaline media, charge injection is observed only at pH ≤ 6, conditions where eosin is associated (adsorbed) with TiO_2,[124] For charge injection to occur from the S_1 state, it must compete with other decay channels of this state: intersystem crossing into the triplet, $EO(T_1)$, excited state, and by radiative and nonradiative channels back to the ground state $EO(S_o)$. The events are schematized in Figure 35a.[124] Charge injection can, in principle, also originate from the triplet state (reaction 128), but despite a favorable driving force for this reaction (0.5 eV at pH 5) no such reaction seems to occur, under the conditions used, as evidenced by the fact that the lifetime of the triplet state is unchanged whether or not TiO_2 is present.[124] Changes in pH alter the extent of adsorption of EO on the TiO_2 particle surface, and alter the rate of charge injection. All the eosin is associated to TiO_2 in the pH range from 5 to 2, within which the quantum efficiency of formation of EO^+ increases from 0.27 to 0.45, respectively. This is due to an increase in the rate of charge injection, $\phi_{EO+} = k_{inj}\,\tau$, and is consistent with an increase in the driving force from 0.8 eV at pH 5 to 0.98 eV at pH 2. At pH 3, the rate of charge injection is $k_{inj} = 8.5 \times 10^8$ s^{-1}.[124]

$$EO(S_1) + TiO_2 \xrightarrow{k_{inj}} EO^+ + e^-_{CB}(TiO_2) \tag{127}$$

$$EO(T_1) + TiO_2 -x \rightarrow EO^+ + e^-_{CB}\,(TiO_2) \tag{128}$$

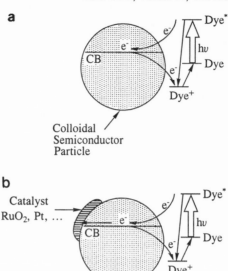

Figure 35. Schematic illustration of charge injection and intraparticle back electron transfer in the photosensitization of a colloidal semiconductor particle: (**a**) without redox catalyst, and (**b**) with a redox catalyst. Adapted from ref. 124.

Once formed on the particle surface, EO^+ decays via biphasic kinetics: a fast decaying component that is complete within 10 μs (intraparticle mechanism; k^1_{decay} = 4×10^5 s^{-1}), and a slower decaying component occurring over several hundred microseconds (interparticle mechanism).[124] Approximately 96% of the eosin is recovered via back electron transfer (Eq. 129 and Figure 35a), which takes place in two steps: (1) about 50% of the eosin undergo the fast decay and is connected with the EO^+ adsorbed on TiO_2; and (2) with the remaining desorbing from the particle surface and which will undergo slow back electron transfer via diffusional encounters between solution EO^+ and TiO_2 particles. The rate of the back electron transfer (fast component) is $k_{-el} = 2 \times 10^5$ s^{-1}, which corresponds to a mean lifetime of ~ 5 μs for a $EO^+...e^-_{CB}$ pair associated with a TiO_2 particle. A more recent determination of k_{-el} gives 2.6×10^5 s^{-1} for back electron transfer from a TiO_2 electrode to EO^+. This rate comprises diffusion of the injected charge to the particle surface and the subsequent interfacial electron transfer to the EO^+ parent ion. Thus, $k_{-el}^{-1} = k_{diff}^{-1} + k_{-inj}^{-1}$ from which k_{diff} is estimated as 1.5×10^9 s^{-1}, a value much greater than k_{-el} suggesting that electron transfer at the interface is rate determining; that is, $k_{-el} \sim k_{-inj}$.[124] The relatively slow rate of back electron transfer is very desirable (k_{-el} about 4000 times slower than k_{inj}) in light energy conversion and storage devices.

$$e^-_{CB}(TiO_2) + EO^+ \rightarrow EO(S_o) \tag{129}$$

Where the TiO_2 particles also contain an additional redox catalyst (electron sinks) such as RuO_2 and Pt islands, the fast decay component of EO^+ in the presence of TiO_2 no longer exists. The role of RuO_2 and Pt is to trap ($k_{trap} = 2.5 \times 10^5$ s^{-1}) the injected electrons and thereby intercept their back reaction with adsorbed EO^+ species; a larger fraction of all the EO^+ formed will thus escape into the solution bulk. It should be noted in this instance that the species reducing EO^+ back to EO is not e^-_{CB}, but rather H• atoms adsorbed at the catalyst surface (Pt), or in the case of RuO_2, it is a ruthenium lattice ion for which the oxidation state is $< 4+$.[124]

The rate of charge injection from the $EO(S_1)$ state to TiO_2 particles is estimated (see above) from the quantum efficiency of formation of EO^+ and from the lifetime of the singlet state ($k_{inj} = 8.5 \times 10^8$ s^{-1} at pH 3). Direct measurement of the rate of interfacial electron transfer by picosecond laser pulsed techniques gave $k_{inj} = 9.5 \pm 1.4 \times 10^8$ s^{-1}, under otherwise identical conditions.[128] The observed risetime for formation of EO^+ in the presence of colloidal TiO_2 is 400 ps. In the present case, theories of nonadiabatic electron transfer reactions[129] relate the rate of interfacial electron transfer, k_{inj}, to the frequency of nuclear reorganization, $v_o \sim 10^{13}$ s^{-1}, the damping coefficient, $\beta \sim 1.2$ A^{-1}, the tunneling distance, r-r$^o \sim 3$ A^o, and to the thermodynamic driving force, $\Delta G^o \sim 0.9$ eV, by Equation 126 from which the reorganizational energy for heterogeneous electron transfer (λ_{hetero}) can be estimated from the experimental value of k_{inj}. This reorganizational energy is related to the analogous one for the homogeneous exchange reaction (Eq. 130):

$$EO(S_1) + EO^+ \rightarrow EO^+ + EO(S_1) \tag{130}$$

by $\lambda_{hetero} > 1/2 \lambda_{homo}$. Although the inner sphere contributions to λ_{homo} are not known, the outer sphere contributions amount to between ~ 0.4 to 1.1 eV; this range calls attention to a lack of knowledge of the activated complex configuration. The two values are estimated for a distance of 6 A^o between the center of $EO(S_1)$ and EO^+ in the activated complex (sandwich configuration) and for the edge-to-edge distance of 14 A^o.[128] From Equation 126, the value of λ_{hetero} turns out to be ~ 0.4 eV. These results confirm the notion that charge injection from an excited dye to a semiconductor particle is favored if the two partners are in intimate contact.

A more recent examination suggests that the triplet state of eosin, $EO(T_1)$, which earlier was noted as inconsequential in the charge injection events between eosin-Y and TiO_2, does in fact also sensitize TiO_2 electrodes.[130] The rate of charge injection from $EO(T_1)$ is $\sim 2 \times 10^5$ s^{-1}, and even though this rate is rather slow compared to k_{inj} from the $EO(S_1)$ state, it appears to be fast enough to produce oxidized EO^+ species from $EO(T_1)$ states with a yield of $\sim 50\%$.

Rose bengal (RB) photosensitizes colloidal TiO_2 particles via a monophotonic charge injection process from its excited singlet state (Eq. 131) which is at -1.3 V (vs NHE) against the conduction band of TiO_2 which is at -0.5 V (vs. NHE).[131] Only about 10% of $RB^{+\bullet}$ undergoes back electron transfer with the injected charge carrier; the remainder survives for milliseconds because the injected charge is trapped in surface traps within the forbidden bandgap region of the semiconductor. Eventually, in the absence of other regenerative channels for rose bengal, the dye undergoes photodegradation on TiO_2, but not when the heterogeneous phase is comprised of SiO_2 or Al_2O_3.[131] The trapped electrons have lifetimes from milliseconds to seconds.[132,133]

Electron injection from anthracene-9-carboxylic acid (9AC) to TiO_2 takes place from the first excited singlet state (reaction 132), subsequent to which $\sim$ 90% of $9AC^{+\bullet}$ undergoes back electron transfer in less than 30 ns (reaction 133). The remainder of the injected charge can be used to reduce N,N,N',N'-tetraethyloxonine (OX) via reaction 134 to produce the corresponding anion radical.[134,135]

$$RB(S_1) + TiO_2 \rightarrow RB^{+\bullet} + e^-_{CB}(TiO_2) \qquad (131)$$

$$9AC(S_1) + TiO_2 \rightarrow 9AC^{+\bullet} + e^-_{CB}(TiO_2) \qquad (132)$$

$$9AC^{+\bullet} + e^-_{CB}(TiO_2) \rightarrow TiO_2 + 9AC(S_0) \qquad (133)$$

$$e^-_{CB}(TiO_2) + OX \rightarrow TiO_2 + OX^-\bullet \qquad (134)$$

Colloidal SnO_2 quenches the luminescence from $^*Ru(bpy)_3^{2+}$ with $k_q = 3.2 \times 10^{11}$ $M^{-1}s^{-1}$ (in terms of particle concentration) compared to an estimated diffusion constant, k_{diff} of 2×10^{10} $M^{-1}s^{-1}$.[136] Taking into account the electrostatic contributions to the mass transfer limit gives a collision frequency $\leq 8 \times 10^{10}$ $M^{-1}s^{-1}$, which is smaller than k_q. As noted earlier for quenching of ZnO emission by metal ions, only the $^*Ru(bpy)_3^{2+}$ adsorbed onto SnO_2 must undergo emission quenching; this has been verified by addition of salts which increases the ionic strength and decreases adsorption of the dye onto SnO_2, as evidenced by enhanced emission in the presence of $NaClO_4$.[136]

5.3. Photosensitization of Wide Band Metal Oxide Semiconductors

It is evident from the above examples that adsorption of an excited dye, via either physisorption or chemisorption, onto a semiconductor particle can photosensitize the latter which in turn can be used to carry out a variety of tasks that the semiconductor would normally do if it were directly excited with suitable light energy. In this regard, one of the early demonstrations of this originates from derivatizing the TiO_2 particle surface with a ruthenium(II)-based dye (Figure 36),[137] in which the dye was actually chemically bound (chelated) to surface titanium ions via Ti-O-Ru bonds. Irradiation of the derivatized TiO_2 system at wavelengths longer than 400 nm, where TiO_2 is transparent, produced dihydrogen

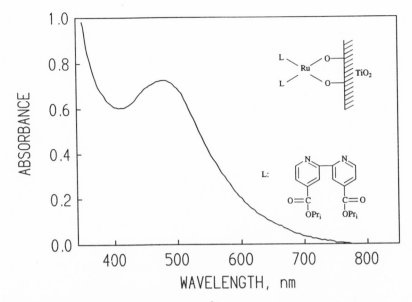

Figure 36. Reflectance spectrum of RuL_2^{2+}-derivatized TiO_2 particles loaded simultaneously with 0.5 wt/% Pt and 0.5 wt.% RuO_2. The absorption maximum in the visible region is at 480 nm. From ref. 137.

from reduction of water. Since that early report,[137] other examples have been documented (see ref. 42). One of these is sensitization of TiO_2 colloids with the inorganic dye $Fe(CN)_6^{4-}$ (reactions 135 and 136).[138] Charge injection from the excited dye takes place within the 20-ns duration of the laser pulse, from which k_{inj} $\geq 5 \times 10^7$ s^{-1}. This event is followed by back electron transfer (Eq. 136). Here k_{-el} $\sim 2 \times 10^5$ s^{-1}, which seems to be rather typical of back electron transfer rate constants from sensitized TiO_2 back to the oxidized dyes (see above).

$$\{Fe(CN)_6^{4-}\}_{TiO_2} \xrightarrow[k_{inj}]{h\nu} e^-_{CB}(TiO_2) + \{Fe(CN)_6^{3-}\}_{TiO_2} \qquad (135)$$

$$e^-_{CB}(TiO_2) + \{Fe(CN)_6^{3-}\}_{TiO_2} \xrightarrow{k_{-el}} \{Fe(CN)_6^{4-}\}_{TiO_2} \qquad (136)$$

The effect of surface chelation of metal oxide semiconductors on the interfacial electron transfer process has recently been examined,[139] in which the dye phenylfluorone (PF) was used as the chelating agent to derivatize TiO_2 colloidal particles. Electronic excitation in the visible absorption band of the chelated particles is followed by an extremely rapid and efficient ($k_{inj} > 10^8$ s^{-1}) electron injection in the conduction band of TiO_2. The recapture of the injected electron by the chelated oxidized dye takes place in microsecond time ($k_{-el} = 2.8 \times 10^5$ s^{-1}). An additional advantage of chelated PF is that it drastically accelerates interfacial

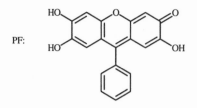

PF:

A

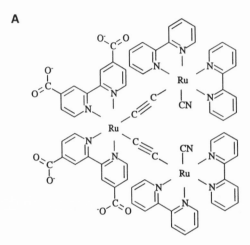

B

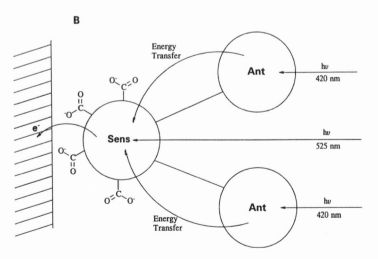

Figure 37. (**A**) Graph showing the supermolecule based on ruthenium(II) fragments. (**B**) Block diagram showing the function of the trinuclear complex as a sensitizer/antennae molecular device. From ref. 141

112

electron transfer from the TiO_2 to electron acceptors such as MV^{2+}.[139] The electrochemical rate constant for charge ejection from the PF^+-TiO_2 system to MV^{2+} is 0.25 cm s^{-1}, some 142 times faster (see the earlier discussion above) than when the chelated dye is absent. This dramatic enhancement of interfacial electron transfer arises from the removal of surface states on TiO_2 particles by the chelated dye PF. Normally, the surface titanium ions which are coordinated to water molecules (surface states) act as electron traps (reaction 137), whence they play a key role in heterogeneous charge transfer processes. In dye-free single crystal rutile TiO_2 electrodes, the concentration of these trap sites is 1.8×10^{13} cm^{-2} and lie about 0.8 eV below the conduction band[140] In TiO_2 colloids, the traps should be distributed over a whole range of energies.[139]

$$(Ti^{IV})_{surf} + e^-_{CB} \rightarrow (Ti^{III})_{surf} \tag{137}$$

A final example is worth mentioning. It embodies much of the advantages of supramolecular systems with organized assemblies such as semiconductors. The ruthenium based sensitizer/antennae system (Figure 37A) is adsorbed onto TiO_2 via electrostatic attraction between the positive surface and the carboxylate groups of the sensitizer.[141] As illustrated in Figure 37B, the two antennae fragments of the supermolecule absorb light at 420 nm following which intramolecular energy transfer, $k \geq 10^9$ s^{-1}, shifts the energy onto the sensitizer fragment; alternatively, 525-nm radiation can also excite this fragment. Photocurrents were measured but rates of interfacial electron transfer were not determined.[141] Additional dyes that have also been explored as potential photosensitzers of TiO_2 particles include those depicted below. A detailed discussion on the merits of these dyes and others noted above may be found in references 42 and 142.

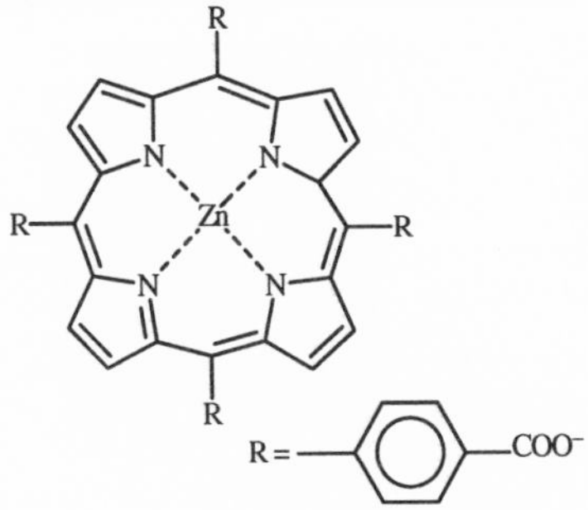

Zinc tetra(4-carboxaphenyl)porphyrin

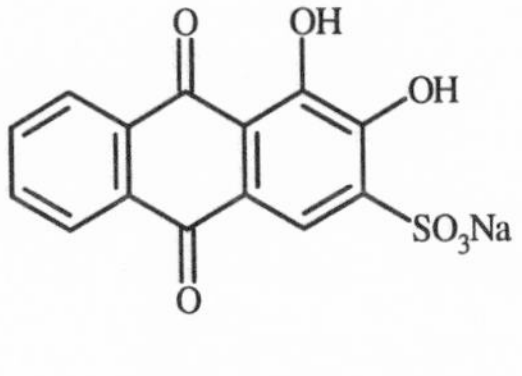

3-alizarinsulfonic acid

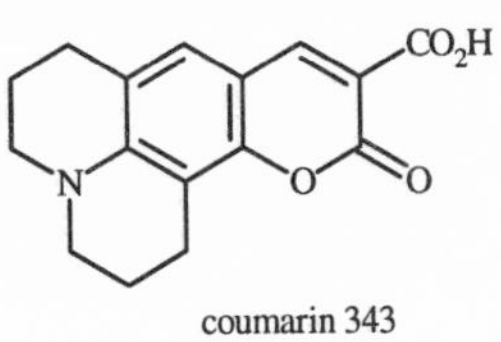

coumarin 343

6. ELECTRON TRANSFER IN HETEROGENEOUS PHOTOCATALYTIC SYSTEMS

6.1. Electron Transfer between a Photosensitive Surface and an Adsorbate (Photoionosorption)

Since band-gap absorption of light induces the formation of electrical charges, the presence of a gas can lead to electron transfer from (to) the solid to (from) the adsorbed molecules. This phenomenon is known as photoadsorption or photoionosorption. Several excellent articles have appeared on the subject.[143–147] In this section, we treat photoinduced gas/solid effects, since electron transfers are best highlighted as elementary steps without the interference often associated with solvent molecules.

Photoadsorption of Oxygen on Semiconductor Oxides

Under UV illumination, various oxides such as titania can adsorb a certain quantity of oxygen in addition to that previously adsorbed in the dark. The amount of photoadsorbed oxygen strongly depends on oxygen pressure and pretreatment of the solid. At oxygen pressures of a few torr, a maximum amount of $\sim 3 \times 10^{17}$ molecules m^{-2} has been reported.[148–151]

The nature of the photoadsorbed oxygen species has been tentatively identified in several studies, and depends on the technique employed and on the experimental conditions required. For instance, electron spin resonance (ESR) spectroscopy on TiO_2 at 77 K reveals the existence of several oxygen adspecies: $O_2^{-}\bullet$, O_3^{-}, O_3^{3-} and O_2^{2-}.[152–155] Only the superoxide radical, $O_2^{-}\bullet$, is stable at ambient temperature.[153] It is the least reactive species.[155] All these oxygen species are formed by different mechanisms. For example, the O_3^{-} species is formed by the interaction of a molecule of O_2 with an O^{-} ionosorbed species. Many of these oxygen species are not present at ambient temperatures under illumination. Electron transfer in photoadsorption can be inferred from thermoelectronic work function measurements and photoelectrical conductivity.

The first technique, based on the Kelvin-Zismann method (vibrating capacitor), examines the electronic interactions between gaseous oxygen and powder TiO_2 under illumination.[156] The vibrating capacitor enables measuring the difference of the contact potential ϕ between TiO_2 and a reference gold electrode, which is electronically inert under UV light and oxygen atmosphere.

The contact potential ϕ is related to the work function ω of titania by the relation:

$$\omega_{TiO_2} - \omega_{Au} = e\phi \tag{138}$$

where e is the electron charge. The work function of gold, ω_{Au}, is constant, whereas ω_{TiO_2} is given by the relationship:

$$\omega_{TiO_2} = \chi + \varepsilon_F + eV_s \tag{139}$$

where χ is the electron affinity of TiO_2, ε_F is the difference in energy between the Fermi level and the lower limit of the conduction band and V_s is the surface potential. χ is constant and the Fermi level in the bulk does not vary under illumination. Consequently, variations of the contact potential $\Delta\phi$ equal those of the surface potential ΔV_S. An accumulation of free electrons decreases $\Delta\phi$. A depletion of photoelectrons by electron transfer to chemisorbed species increases $\Delta\phi$. A detailed study in vacuum and in oxygen[157] has concluded: (1) that simultaneous photodesorption and photoadsorption of oxygen occur and that the photodesorption prevails in the initial period of illumination; (2) that oxygen photoadsorption follows a kinetic model of the Roguinskii-Zeldovitch-Elovitch type; and (3) that oxygen is ionosorbed as two distinct species. Unfortunately, the technique is unable to provide a definite identification of the two species, tentatively described as $O_2^-\bullet$ and O^-.[157] The exact nature of these two species was later confirmed by photoconductivity measurements. These will now be described further.

As mentioned earlier, oxygen photoadsorption depends on the amount of photoelectrons available and on the state of the surface of the metal oxide solid. In particular, the quantity of photoadsorbed oxygen depends directly on the presence of water and hydroxyl groups.[148,149,158,159] Thermal removal of water and hydroxyl groups decreases the amount of photoadsorbed oxygen. The beneficial effect of water and OH^- groups is attributed to hole trapping by OH^- surface species which allow the electron/hole pair to dissociate and oxygen to be photoadsorbed.[148,149,154,158-160]

$$TiO_2 + h\nu \rightarrow e^-/h^+ \tag{140}$$

$$e^-/h^+ + OH^- \rightarrow OH\bullet + e^- \tag{141}$$

$$O_2(g) + e^- \rightarrow O_2^- \text{ (ads)} \tag{142}$$

At ambient temperature, the proposed ionosorbed oxygen species is O_2^- since it is the only stable species detected by electron spin resonance (ESR) techniques.[153] The concomitant production of $OH\bullet$ radicals and the subsequent formation of hydrogen peroxide must be noted. Munuera and co-workers[159] observed no photodesorption of oxygen at oxygen pressures below ~ 1 torr. Other radicals or ions are involved during oxygen photoadsorption: HO_2^- and $HO_2\bullet$; their existence was proposed in a study of the catalyzed photodecomposition of H_2O_2 by TiO_2.[153,160]

$$2\, OH\bullet \rightarrow H_2O_2 \tag{143}$$

$$H_2O_2 \xrightarrow{TiO_2} H_2O + O_2 \tag{144}$$

$$O_2^- + H^+ \rightarrow HO_2\bullet \tag{145}$$

$$OH\bullet + H_2O_2 \rightarrow HO_2\bullet + H_2O \tag{146}$$

Oxygen photoadsorption on TiO_2 follows several kinetic laws. For instance, photoadsorption of oxygen on hydroxylated surfaces follows first order kinetics, whereas on dehydroxylated surfaces photoadsorption is diffusion-controlled.[159] Several workers have noted a Roguinski-Zeldovich-Elovich behavior for the photoadsorption of O_2 on TiO_2,[157,161] ZnO,[161] and SnO_2.[162] This kinetic behavior is consistent with a progressive inhibition of electron transfer to oxygen by a Schottky barrier, which is induced by pre-adsorbed negatively charged species. This point has been demonstrated by surface potential measurements under UV light.[157]

A theoretical approach of the kinetics of oxygen adsorption on titania in aqueous media has recently been reported.[163] The kinetics of the reaction between dioxygen and an electron on a semiconductor particle was analyzed by two models: (1) electrons move freely in the particle and react anywhere on the particle surface, and (2) electrons are transferred from surface states or from subsurface traps. The second model appears the most realistic.

Photoelectron Transfer Between a Solid and an Adsorbed Phase by Electrical Conductivity

Photoconductivity is one of the best means to examine photoinduced electron transfer between a solid and a gaseous reactant through an adsorbed phase. The technique can be used *in situ* and under conditions that prevail in photocatalysis. The general expression for the electrical photoconductivity, σ, in a single crystal[163–165] is given by:

$$\sigma = [e^-]\, q\, \mu_e + [h^+]\, q\, \mu_h \tag{147}$$

where $[e^-]$ and $[h^+]$ are the electron and hole concentrations, respectively, q is the absolute value of the elementary charge of the electron and μ_e and μ_h are the respective mobilities of electrons and holes. The mobilities, which vary with temperature, can be considered as constant at or near ambient temperatures. For n-type semiconductors, the ones most often examined, the $[e^-]$, is much greater than $[h^+]$, so that σ is directly proportional to $[e^-]$.[166] For a polycrystalline or powder sample, this relationship is still valid,[166,167] with the proportionality factor C (Eq. 148) including various parameters such as compression, particle surface texture, and the number of contact points between particles, among others. The relative variations in σ for a given sample as a function of a chosen parameter yields important quantitative information.[167] In what follows, photoinduced electron transfer between titania and various gases will be presented first. Then, the case of other semiconductor oxides will be examined and photoconductivity rules shall be established for determining the potential photoactivity of a solid.

$$\sigma = C \, [e^-] \tag{148}$$

Electrical Photoconductivity of Titania[166,168,169]

Influence of Vacuum. A reference state needs first to be defined and subsequently to follow photoconductivity variations and to establish the direction and the intensity of photoinduced electron transfer during chemisorption. This reference state has been determined under vacuum.

In the dark, the electrical conductivity is too low to be measured at room temperature ($\sigma < 10^{-14}$ ohm^{-1}). As soon as the sample is illuminated with UV-light, the photoconductivity increases sharply and then tends to a plateau. The suppression of illumination has no effect on the (dark) conductivity. This has been interpreted as follows. UV light creates electron/hole pairs which after separation contribute to the photoconductivity according to Eq. 147. The charge carrier separation is favored by the reaction of positive holes with negatively charged ionosorbates whose heat of adsorption is large enough to let them resist dark thermal desorption by dynamic vacuum (~ 10^{-6} torr). Most frequently, A$^-$ species are the oxygen species O_x^- (x = 1 or 2) and OH$^-$ groups (Eqs. 141 and 142). Each hole that is consumed corresponds to one free electron accumulated. When light is switched off, the concentration of accumulated electrons [e_o^-] is much greater than the concentration of the instantaneously photoproduced electrons, [e_i^-] (Eq. 151). The same behavior was witnessed with the work function measurements.[156]

$$A^-_{(ads)} + h^+ \rightarrow A^*_{(ads)} \rightarrow A(g) \tag{149}$$

$$O_x^- + h^+ \rightarrow x/2 \, O_2(g) \tag{150}$$

$$\sigma = C \, ([e_o^-] + [e_i^-]) \approx C \, [e_o^-] = \text{constant} \tag{151}$$

Influence of Oxygen Pressure. After the photodesorption treatment described above, introduction of oxygen produces a steep initial decrease of σ until it reaches a steady-state value within a few hours. With increasing oxygen pressures, a plot of the photoconductivity isotherm, log $\sigma = f$(log P_{O_2}), gives a straight line whose slope is equal to -1 (Figure 38). This value is accounted for by an associative chemisorption of oxygen molecules followed by electron transfer to O_2:

$$O_2(g) \rightleftharpoons O_2(ads) \tag{152}$$

$$O_2(ads) + e^- \rightleftharpoons O_2^- \bullet (ads) \tag{153}$$

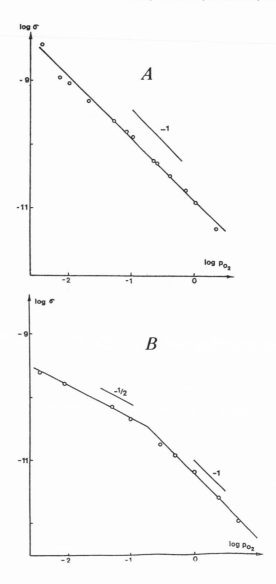

Figure 38. Log-log plot of the photoconductivity σ (in ohm^{-1}) of TiO$_2$ as a function of increasing (**A**) or decreasing (**B**) oxygen pressure (in Torr). From ref. 168.

When the oxygen pressure is progressively decreased from a few torr to 10^{-3} torr, the isotherm log σ = f(log P$_{O_2}$) gives two straight lines whose slopes are, respectively, −1 and −1/2. The slope of −1 is due to the presence of O$_2^-$ (ads) species which photodesorb by interacting with the holes:

$$O_2^- \bullet (ads) + h^+ \rightarrow O_2(g) \tag{154}$$

and is also due to the displacement of the equilibria in Eqs. 152 and 153. The slope of $-1/2$ obtained at low pressures ($P_{O_2} < 0.1$ torr), is due to the existence of a dissociatively ionosorbed oxygen species, Eq. 155:

$$O_2 (g) \rightleftharpoons 2 O (ads) \tag{155}$$

and

$$O (ads) + e^- \rightleftharpoons O^-(ads) \tag{156}$$

These O^- species are more tightly bound to the surface owing to their greater electron affinity (-1.47 eV) versus -0.45 eV for O_2. At higher pressures, the adsorption sites of O^- are immediately saturated, while the excess electrons accumulated during UV-illumination under vacuum permit the concomitant ionosorption of O_2^- species. These two species correspond to those detected by work function measurements under the same conditions of illumination.[157] As mentioned earlier, $O_2^- \bullet$ species are readily identified by ESR methods,[152–155] but detection of O^- species by the same technique did not materialize.[170] It must be noted that O^- is formed by dissociative adsorption of O_2 *before electron transfer*. A photoinduced two-electron transfer to a chemisorbed oxygen molecule prior to its own dissociation to O species is precluded by the observation that σ varies linearly with the light flux instead of the square root dependence expected otherwise.[166]

Influence of Other Gases. As demonstrated for oxygen, variations in the photoconductivity with changes in gas pressure can point to the direction and to the extent of electron transfer during photoadsorption. The method has been applied to other oxygenated and nonoxygenated gases.

Carbon oxides. Measurements carried out under a CO atmosphere[171] showed that the photoconductivity isotherm $\sigma = f(P_{co})$ exhibits no variations with changes in P_{CO} ($\partial \sigma / \partial P_{CO} = 0$), indicating that no photoinduced electron transfer takes place between TiO_2 and CO. The possible existence of CO^- or of CO^+ species eluded verification. CO seems to be neither an electron donor nor an acceptor at room temperature towards UV-illuminated titania. It merely chemisorbs associatively and is photocatalytically oxidized to CO_2.[172]

Carbon dioxide, CO_2, exhibits the same behavior.[171] There is no photo-induced electron transfer to CO_2. As well, CO_2 has no donor character. The consequence of such a behavior is important. It shows that carbon dioxide is rather inert under conditions similar to those of photocatalytic oxidation. This explains why CO_2 is a final product in many oxidation reactions. It desorbs easily and does not compete with oxygen for photoadsorption. CO_2 has never been implicated as an inhibitor in kinetic models of gas phase reactions.

Nitrogen oxides. Nitrous oxide (N_2O) in contact with UV-illuminated titania displays no variations of photoconductivity at room temperature ($\partial\sigma/\partial[N_2O] = 0$).[171] Conclusions analogous to those for the carbon oxides can be inferred. In particular, the absence of N_2O^- species is confirmed by the disappearance of the ESR spectrum when the temperature is increased from 77 to 300 K.[173] Moreover, nitrous oxide cannot constitute a source of active oxygen for oxidation reactions since it cannot be decomposed by neutralization with a positive photohole. This is experimentally verified by the negative results obtained in the photocatalytic oxidation of some alkanes with N_2O as the potential source of oxygen.[171,174] Rather, the stability of N_2O in contact with UV-illuminated TiO_2 explains why this molecule is one of the final products in the gas-phase photocatalyzed oxidation of ammonia.[175]

$$\overset{TiO_2}{10\ NH_3 + 8\ O_2 \rightarrow 4\ N_2 + N_2O + 15\ N_2} \tag{157}$$

Contrary to N_2O, nitric oxide (NO) is not inert. The photoconductivity isotherm $\log \sigma = f(\log P_{NO})$ gives a straight line (Figure 39) whose slope is -1: $\sigma \propto P_{O_2}^{-1}$. This dependence can be accounted for by the associative adsorption:

$$NO\ (g) \rightleftharpoons NO\ (ads) \tag{158}$$

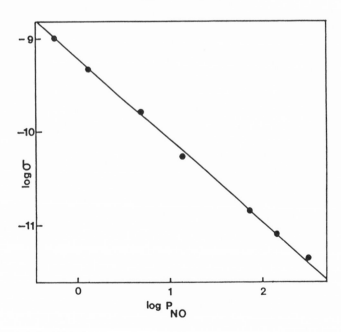

Figure 39. Log-log plot of the photoconductivity σ (in ohm^{-1}) of TiO_2 as a function of NO pressure (in Pa). From ref. 169.

followed by the transfer of one photoelectron to one NO molecule in accord with its high electron affinity (–0.90 eV). However, the formation of negatively charged ionosorbed species cannot leave the photoholes insensitive. These progressively react with NO^- species for electrostatic reasons. The heat of reaction drives the dissociation of the activated intermediate into nitrogen and oxygen atoms.[176]

$$NO\ (ads) + e^- \to NO^-\ (ads) \qquad (159)$$

$$NO^-(ads) + h^+ \to (NO)^* \to N\bullet + O\bullet \qquad (160)$$

At low NO pressures, nitrogen atoms recombine to yield dinitrogen, and the oxygen atoms remain on the semiconductor particle surface, probably as ionosorbed species. At higher NO pressures, the probability of reaction of $N\bullet$ with NO increases and N_2O is formed (Eq. 161):[176]

$$N\bullet \begin{cases} + N\bullet \longrightarrow N_2(g) & \text{(at low } P_{NO}) \\ + NO \longrightarrow N_2O(g) & \text{(at high } P_{NO}) \end{cases} \qquad (161)$$

NO is a source of atomic oxygen species which are active in photocatalysis. In the presence of an oxidizable molecule such as butanol, NO can be photoactivated by illuminated titania to selectively oxidize the butanol to butanal or butanone.[169,176] Similarly, in the presence of gaseous labeled $N^{18}O$, there is an isotope exchange with ^{16}O coming from the surface of titania.[177] In the presence of 2-butanol, the nitric oxide isotopic exchange is suppressed, while the butanone formed is labelled principally with ^{16}O. This suggests that the same oxygen species intervenes in the exchange and in the oxidation, and that 2-butanol conserves its original ^{16}O atom.[177]

Other gases. The photoconductivity test has also been applied to nonoxygenated molecules. Nitrogen (N_2) displays no changes in photoconductivity ($\partial\sigma/\partial P_{N_2} = 0$).[171] Analogous to CO_2, N_2 is photocatalytically inert. This explains why N_2 can neither be photocatalytically oxidized nor reduced in the gas phase (absence of NO_x and/or NH_3), and why it is found as a final (stable) product in the oxidation of ammonia.[175] As well, alkanes (and principally isobutane) induce no variation of the photoconductivity of titania. It appears that alkanes chemisorb nondissociatively at the TiO_2 surface without electron transfer.[171] This is corroborated by a photogravimetric study of isobutane adsorption under UV light; the temperature was 294 K.[178] Adsorption of isobutane is completely reversible displaying no differences in σ between dark and illumination conditions. The adsorption isotherm accords with a Langmuirian model and with a BET isotherm with a single monolayer. From the maximum number of isobutane molecules adsorbed, and with a molecule area of 47.4 A^2,[179] the surface area of titania is estimated as 66 $m^2\ g^{-1}$ at 294 K. This is in good agreement with the conventional

BET surface area for the titania sample used, determined by N_2 adsorption at 77 K (70 m^2 g^{-1}). This result suggests that aliphatic hydrocarbons are more likely physisorbed and constitute a bidimensional gas on the whole surface of the photocatalyst without localized adsorption sites.

The absence of photoinduced electron transfer during hydrocarbon chemisorption under UV light persists during photocatalysis. This was demonstrated by *in situ* photoconductivity measurements carried out during the photocatalytic oxidation of isobutane[180] in a cell reactor described by Pichat, Herrmann and co-workers.[169] Under the conditions used, two kinetic laws were concurrently determined for the photoconductivity σ and for the reaction rate r (Eqs. 162 and 163).

$$\sigma = k \, P_{O_2}^{-1} P^o_{iso} \tag{162}$$

$$r = k_r \, P_{O_2}{}^o P_{iso}{}^{0.35} \tag{163}$$

The independence of σ on isobutane pressure, P_{iso}, indicates that the hydrocarbon is not in electronic interaction with the illuminated solid. The fractional kinetic order of 0.35 suggests that isobutane reacts in the adsorbed phase since this value corresponds to the apparent order of adsorption (0.30), determined by thermogravimetry in the pressure range investigated for Langmuir-type adsorption without dissociation : $\theta = KP/(1 + KP) \approx K'P_{iso}{}^{0.30}$. With regard to the effect of oxygen on σ, the apparent order of -1 in the expression for σ (Eq. 162) is indicative of the fact that $O_2^-\bullet$ species control the adsorption equilibrium at the pressures chosen. The absence of any effect of oxygen pressure on the activity implies that $O_2^-\bullet$ species cannot be the precursor of the photoactive species. By contrast, O^- precursors can be at the origin of the active species, since their adsorption sites are saturated under the pressures used according to photoconductance measurements performed at lower pressures.

Electrical Photoconductivity of Various Oxides

In addition to TiO_2, several other metal oxides can photosorb oxygen; for example, ZnO[161] and SnO_2.[162] The electron transfer event during oxygen photoadsorption on various oxides has been examined systematically by photoconductivity measurements on ZnO, SnO_2, ZrO_2, Sb_2O_4, CeO_2, WO_3, and V_2O_5 , with titania taken as the reference material.[166] The photoconductivity isotherms are presented in Figure 40. To achieve stability in the electrical measurements, the solids are first equilibrated in UV light under the highest oxygen pressure used. The case of titania has been discussed above. Taking into account the various oxygen adsorption equilibria (Eqs. 152,155), the electron transfer reactions (Eqs. 153,156), and the electron/hole recombination events,

$$e^- + h^+ \rightarrow h\nu' \text{ (or heat) } (\nu' \leq \nu) \tag{164}$$

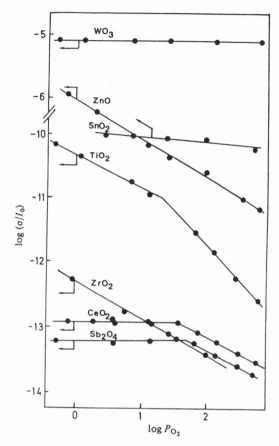

Figure 40. Log-log plot of the photoconductivity σ (in ohm⁻¹) per radiant light flux unit as a function of oxygen pressure (in Pa) for various semiconductor oxides. From ref. 166.

The steady-state photoconductivity can be written:

$$d\sigma/dt = C\,\{d[e^-]/dt\} = 0 \tag{165}$$

That is,

$$d[e^-]/dt = \alpha\varphi - k_{153}K_{152}P_{O_2}[e^-] - k_{156}K_{155}{}^{1/2}P_{O_2}{}^{1/2}[e^-] - k_{164}[e^-][h^+] = 0 \tag{166}$$

where C is the proportionality constant between $[e^-]$ and σ, φ is the light flux, α is the light absorption coefficient of the solid, and k_i and K_i are the respective rate and equilibrium constants. The resulting general expression for σ is consequently:

$$\sigma = \frac{C \, \alpha \, \varphi}{k_{153}K_{152}P_{O_2} + k_{156}K_{155}^{\frac{1}{2}}P_{O_2}^{\frac{1}{2}} + k_{164}[h^+]} \tag{167}$$

Equation 167 can account for all the curves presented in Figure 40.

TiO₂. The two slopes (-1 and $-1/2$) are indicative of the simultaneous existence of O_2^- and O^- species, with the latter detectable only at low oxygen pressures when the second term in the denominator of Eq. 167 predominates: that is, when $k_{156}K_{155}^{\frac{1}{2}}P_{O_2}^{\frac{1}{2}} \gg \{k_{164}[h^+] + k_{153}K_{152}P_{O_2}\}$.

ZrO₂ and ZnO. In the whole pressure range examined, a single slope ($-1/2$) is obtained. This is in line with the existence of O^- (ads) species on these two oxides.

Sb₂O₄ and CeO₂. At pressures $P_{O2} \geq 0.5$ torr, O^- species determine the equilibrium between photogenerated electrons and gaseous oxygen (slope, $-1/2$ in Figure 40). At lower oxygen pressures, the second term in the denominator of Eq. 167 becomes smaller than the third, $k_{164}[h^+] > k_{156}K_{155}^{\frac{1}{2}}P_{O_2}^{\frac{1}{2}}$, and no more photoinduced electron transfer event is detected.

SnO₂ and WO₃. The slopes $\partial \log \sigma / \partial \log P_{O_2}$ are nearly zero. In fact, these oxides are photosensitive and an important electron transfer to chemisorbed oxygen occurs during the first introduction of oxygen. However, these oxides display a non-insignificant electrical conductivity in the dark, σ_d. The measured apparent photoconductivity, σ_{app}, is the sum of σ_d and the true photoconductivity which as a result of electron transfer is a function of P_{O_2}. Log σ_{app} is actually the logarithm of a sum which, when transposed in the log-log plot of Figure 40, yields an apparent slope that is practically independent of P_{O2}. Calculations based on Eq. 168 yield the relationships $\sigma = f(P_{O_2})$ which are better fitted with the square root of P_{O_2}, thus indicating the preferential presence of O^- (ads) species.[166]

$$\sigma_{app} = \sigma_d + \sigma(P_{O_2}) \tag{168}$$

V₂O₅. This oxide is completely inert under UV illumination exhibiting no photoconductivity variation, no photoadsorption, and no electron transfer. It seems that the photoproduced electrical charges spontaneously recombine, possibly with the help of vanadium cations which can easily change their oxidation state and thereby become recombination centers.[166]

Consequences for Oxidation Photocatalysis in the Gas Phase

In the preceding discussion, we indicated that electron transfer occurs from the illuminated solid to photoadsorbed oxygen or nitric oxide. From the various systems examined by photoconductivity with different solids and gases, some rules can now be formulated and some criteria can be defined to predict whether or not a solid is photocatalytically active in oxidation reactions.[167]

Criterion 1: The photoconductivity of a solid must increase under dynamic vacuum (ability to: (1) create electron/hole pairs; and (2) to photodesorb ionosorbed species A^- via their neutralization by photoholes).

Criterion 2: The photoconductivity has to decrease with increasing oxygen pressures (ability to ionosorb oxygen).

Criterion 3: The photoconductivity must vary with $P_{O_2}^{-\frac{1}{2}}$ (ability to photoadsorb oxygen as O^-, precursor of the dissociated atomic oxygen species which is the active agent in photocatalytic reactions in the gas phase and in nonaqueous media).[177,180,181]

6.2. Photoinduced Electron Transfer between a Semiconductor and a Metal Deposit

The beneficial effect of group VIII metal deposits in certain reactions carried out over illuminated semiconductors was initially reported in the 1980s.[182–207] Generally, these reactions involve hydrogen either as a product (dehydrogenation reactions), or as a reactant (alkane-deuterium isotopic exchange), or as intermediate protons. When the photosensitive phase acts alone (naked particles) under illumination, the reaction ceases after a while owing to exhaustion of the active sites.[206,207] The reaction yield corresponds to a "stoichiometric threshold" and the reaction is not catalytic. This stoichiometric threshold can be overcome by group VIII metal deposits on the semiconductor with the consequence that the reaction is then made catalytic. The metal deposit functions as a cocatalyst necessary to: (1) dissociate or recombine dihydrogen; and (2) to play a favorable role in photoinduced electron transfer events. These events have been examined in inorganic heterogeneous systems by photoconductivity techniques.[208]

To get semiquantitative or quantitative results, the solids have to be well characterized and have to exhibit a regular texture, close to a simple geometric model. The preparation of such solids can be achieved by the impregnation and reduction method,[209] which has been applied to the preparation of platinum,[206] rhodium, and nickel deposits.[210] The metal deposits are small, homodispersed crystallites, regularly distributed on all the grains of the support. The mean size, <d>, of these metal deposits is determined by transmission electron microscopy (TEM) and/or by hydrogen chemisorption.<d> varies with the nature of the metal and is equal to ~ 1.5 nm for platinum and ~ 3 nm for Rh. The texture is stable and the extent of reduction is 100% in contrast with the *in situ* photoreduction for which the total reduction is often incomplete as noted by Teratani and co-workers.[211]

In Metal-Support Interactions Under Vacuum

On well-defined Pt/TiO₂ catalysts with different loadings, the kinetics of the variations in the photoconductance σ has been followed as a function of the illumination time (Figure 41). Two main features are worth noting: the higher the metal content (in wt %) is, the smaller is the photoconductivity under steady state,

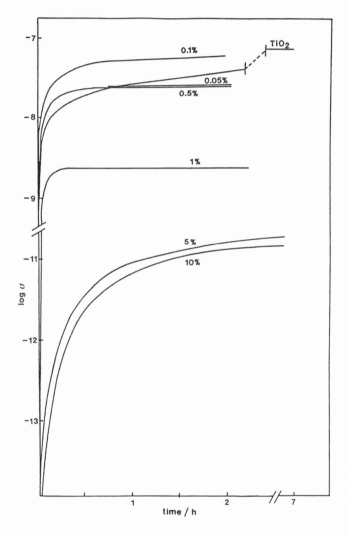

Figure 41. Variations of the photoconductivity (in ohm⁻¹) under vacuum as a function of illumination time for naked TiO₂ and Pt/TiO₂ samples with different platinum wt.%. From ref. 208.

and the longer is the time needed to reach steady state. This can be interpreted by a reversible electron transfer between the photosensitive support and the metal. Photons of suitable energy ($h\nu \geq E_{BG} \sim 3.0$ eV) create photoinduced electron/hole pairs (Eq. 169).

$$TiO_2 + h\nu \rightarrow e^-/h^+ \rightarrow e^- + h^+ \tag{169}$$

Electrons and holes contribute to the photoconduction process; the steady state is reached when the rate of generation of the charge carriers G (Eq. 169) equals the rate of their recombination (Eq. 170). Thus,

$$e^- + h^+ \rightarrow N + Energy \qquad (N = neutral\ center) \tag{170}$$

In the presence of platinum, some photogenerated electrons can be transferred reversibly to the metal (Eq. 171).

$$e^- + Pt \rightleftharpoons e_{Pt}^- \tag{171}$$

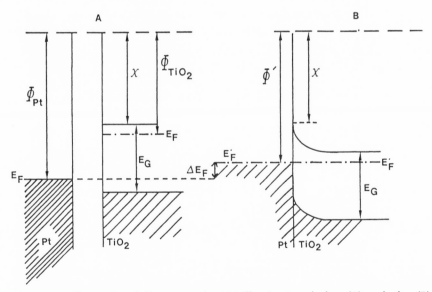

Figure 42. Energy band diagram under UV illumination: before (**A**) and after (**B**) contact between Pt and TiO_2; χ and E_{BG}: electron affinity (~ 4 eV) and energy bandgap (~ 3 eV) of TiO_2, respectively; ϕ_{TiO_2}: work function of illuminated TiO_2, estimated close to that of the reduced state (~ 4.6 eV); ϕ_{Pt}: work function of Pt (~ 5.36 eV); ΔE_F = $\phi_{Pt} - \phi'_{Pt}$: increase in the Fermi level of Pt after contact with illuminated TiO_2. From ref. 208.

This is thermodynamically possible when one considers the energy band diagram of the system under illumination (Figure 42). The steady state requires the condition $d\sigma/dt = 0$. Consequently,

$$d\sigma/dt = 0 = d[e^-]/dt = R_{169} - R_{170} - R_{171} + R_{-171}$$

$$= G - k_{170}[e^-][h^+] - k_{171}[e^-][Pt] + k_{-171}[e_{Pt}^-] \tag{172}$$

where the rate G is proportional to the light flux.

The free electron concentration $[e_o]$ at steady state is given by Eq. 173:

$$[e_o^-] = \frac{G + k_{-171}[e_{Pt}^-]}{k_{170}[h^+] + k_{171}[Pt]} \tag{173}$$

Therefore, the photoconductivity appears as a hyperbolic function of the metal content as illustrated in Figure 43, deduced from the data of Figure 41. This confirms the reversible electron transfer process given by Eq. 171. The slower kinetics for σ to reach the steady state at higher metal (percent) loadings is demonstrated by the negative sign of the derivative of the photoconductivity rate with respect to [Pt] calculated from Eq. 174:

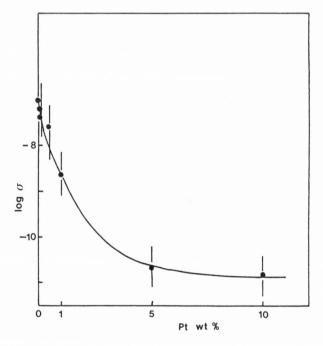

Figure 43. Variations of the photoconductivity σ at steady state under vacuum of various Pt/TiO$_2$ samples as a function of Pt wt.%. From ref. 208.

$$\frac{\partial}{\partial[Pt]} \frac{d[e^-]}{dt} = -k_{171}[e^-] < 0 \qquad (174)$$

In Metal-Support Interactions Under Hydrogen

In the dark and at room temperature, all group VIII metals chemisorb hydrogen dissociatively with a stoichiometry of one H atom adsorbed per surface metal atom:

$$\tfrac{1}{2} H_2 + Pt_s \rightleftharpoons Pt_s - H \qquad (175)$$

However, the (dark) electrical conductivity increases with the hydrogen pressure at steady state. This has been attributed to a migration of dissociated hydrogen from the metal onto the support,[212,213] a phenomenon denoted as *hydrogen spillover* in heterogeneous catalysis.[214]

$$Pt_s - H + O^{2-}\{TiO_2\} \rightarrow Pt_s + 0H^-\{TiO\} + e^- \qquad (176)$$

Application of the steady-state principle to the electrical conductivity σ gives a linear variation of σ with the square root of the hydrogen pressure P_{H_2}.

$$\sigma = a + b\, P_{H_2}^{1/2} \qquad (177)$$

This relationship is quite general and has been observed for various supported metal catalysts: Pt, Rh, and Ni/TiO$_2$;[210] Pd/ZnO;[171] and Ni/CeO$_2$.[215]

Under illumination, the concentration in free electrons originates from the two sources described by Eqs. 169 and 176. Consequently, considering Eqs. 169–171, 175 and 176 yields Eq. 178:

$$[e^-] = \frac{G + k_{171}[e_{Pt}^-] + k_{176}K_{175}[O^{2-}]Pt_s]P_{H_2}^{1/2}}{k_{170}[h^+] + k_{171}[Pt]} \qquad (178)$$

Equation 178 is formally identical to that obtained for the electrical conductivity of nonilluminated catalysts.

$$\sigma = a' + b'P_{H_2}^{1/2} \qquad (179)$$

Since the coefficient b' does not contain the term G indicates that hydrogen spillover occurs under illumination but is not a photoactivated phenomenon. This will be confirmed and illustrated by the different photocatalytic examples given below.

During Photocatalytic Reactions Involving Hydrogen

These photoinduced electron transfer reactions require the total absence of oxygen (or air), since noble metals easily catalyze the oxidation of dihydrogen into water. The different examples presently given will only refer to nonsacrificial reactions to highlight the electron transfer process.

Alcohol Dehydrogenation. Alcohols, either in liquid or gas phase, can easily be oxidized into aldehydes or ketones either by oxygen (or air) on a naked metal oxide photocatalyst or by dehydrogenation in contact with a metal-deposited photocatalyst.[194,206,211,216–229] Under anaerobic conditions, a naked oxide catalyst, such as TiO_2, can dehydrogenate alcohols; after a certain period, the solid progressively turns blue and hydrogen production ceases.[206] This color is typical of a slightly reduced oxide sample, or of a solid which, at room temperature, contains Ti^{3+} ions. An accumulation of free electrons results owing to an electron transfer inhibition which leads to a corresponding inhibition of the reaction.

In the presence of a metal deposit, site blocking is avoided and the reaction runs catalytically. This can easily be understood by the reaction mechanism determined for various metals deposited on titania. The mechanism is broken down into the various following steps:

1. The alcohol chemisorbs dissociatively at the surface of titania on acidic and basic sites provided by the OH^- groups, known for their amphoteric character.

$$R–OH \rightleftharpoons RO^- + H^+ \tag{180}$$

The dissociative nature of the adsorption of the reactive alcoholic species is determined by kinetic studies. The reaction rate r follows a dissociative Langmuir–Hinshelwood mechanism with a dependence on the square root of the concentration of alcohol:

$$r = k \bullet \frac{K^{\frac{1}{2}}C^{\frac{1}{2}}}{(1 + K^{\frac{1}{2}}C^{\frac{1}{2}})} \tag{181}$$

where k is the rate constant and K is the adsorption coefficient.

2. The UV light creates photoinduced electron/hole pairs which easily dissociate because of the spontaneous electron transfer from the photosensitive support to the metal.

$$TiO_2 + h\nu \rightarrow e^-/h^+ \tag{182}$$

$$e^-/h^+ + Pt \rightleftharpoons e_{Pt}^- + h^+ \tag{183}$$

3. The photoholes remain free and can react with the negative alkoxide ions.

$$RO^- + h^+ \rightarrow RO\bullet^* \tag{184}$$

4. The alkoxide radical, which contains excess energy due to the heat of neutralization, deactivates into a stable molecule (aldehyde or ketone) by loosing a second hydrogen atom.

$$-CH(O\bullet)- \rightarrow \diagdown C = O + H\bullet \tag{185}$$

5. Simultaneously, the proton created in step 1 (Eq. 180) can migrate to the surface of titania by a *reverse spillover* mechanism where it is attracted by

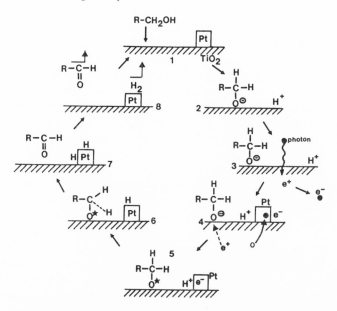

Figure 44. Cyclic mechanism of alcohol dehydrogenation on Pt/TiO$_2$. From ref. 210.

the excess negative charges originating from electron transfer to platinum crystallites at the contact of which it is cathodically neutralized.

$$H^+ + e_{Pt}^- \rightleftharpoons H\bullet_{ads} \tag{186}$$

This electron transfer, which is not photoassisted, enables the reduction of the proton and, when the irreversible adsorption sites of hydrogen at the surface of platinum are saturated, gaseous hydrogen evolution is initiated.

$$2\,Pt_s–Pt_s\text{-}H \rightleftharpoons H_2(g) + 2\,Pt_s \tag{187}$$

6. The carbonylated molecule formed in step 4 is desorbed in the liquid phase and the hydrogen atom can migrate to the metal particles as a proton.

$$H\bullet \rightleftharpoons H^+ + e^- \tag{188}$$

This reaction also involves an electron transfer from a hydrogen atom to the oxide, which is permanently in electronic interaction with the metal. This concerted mechanism can be illustrated by the reaction cycle illustrated in Figure 44.

The photoinduced electron transfer emerges clearly between TiO$_2$ and the peripheral metal island. It has the beneficial role of: (1) separating the electron/hole pair; (2) avoiding electron/hole recombination; and (3) of reducing the two protons into hydrogen atoms before desorption as dihydrogen.

If one considers that one cycle described in Figure 44 exhausts one active site, comparison of the hydrogen yield given by naked titania at its maximum conversion to that of a Pt/TiO$_2$ catalyst yields an estimate of the number of catalytic cycles

sion to that of a Pt/TiO_2 catalyst yields an estimate of the number of catalytic cycles accomplished by the bifunctional catalyst. With a 5 wt% Pt/TiO_2 catalyst, more than 800 cycles can be executed on the same sites without detecting any inhibition, provided that gaseous hydrogen is pumped out regularly to avoid the reverse, nonphotoassisted hydrogenation of carbonylated molecules.[217] Such an example illustrates the demonstration of a true catalytic character of a photoinduced reaction.

Deuterium-Alkane Isotopic Exchange. The deuterium-alkane isotopic exchange reaction is of pure academic interest. It is generally used in heterogeneous catalysis to identify the different pools of active sites on metals.[230] For instance, in cyclopentane-deuterium isotopic exchange (CDIE) on a series of Pt/SiO_2, several metal active sites were highlighted from the selectivity patterns in the various deuterated molecules formed.[231] This reaction was transposed to heterogeneous photocatalysis on various titania-deposited metals to investigate the (electronic) metal-support interactions under photoassisted working conditions.[207,232] The thermal activity is eliminated by decreasing the temperature to below $-10\ °C$ ($T \leq 263$ K).

The CDIE reaction exhibits the following characteristics: The initial selectivity is 100% in monodeuterocyclopentane (C_5H_9D) formation, whereas the thermally activated reaction produces all the 10 possible deuterated molecules with a double maximum in $C_5H_5D_5$ and in C_5D_{10} formation.[231] Naked titania is photoactive for a certain time (~ 1 hr) before its activity ceases definitively, whereas the photoactivity of M/TiO_2 remains constant without inhibition. There is no photoactivity on dehydroxylated samples; the hydroxyl groups present on titania are deuterated.[207,232] The deuteration of the OH^- groups of naked titania is obtained by heating TiO_2 in gaseous deuterium at 150 °C. By contrast, the hydroxyl groups present on Pt/TiO_2 are spontaneously deuterated at room temperature. This can be ascertained by infrared spectroscopy and interpreted by a *reversible hydrogen (deuterium) spillover*.[210] All these considerations suggest the following mechanistic steps:

1. creation of electron/hole pairs by band-gap illumination and electron transfer to the metal;

$$TiO_2 + h\nu \rightarrow e^-/h^+ \tag{189}$$

$$e^-/h^+ + M \rightleftharpoons e_M^- + h^+ \tag{190}$$

2. dissociative adsorption of deuterium on the metal and direct spillover of D atoms onto titania;

$$D_2 + M_s \rightarrow 2\ M_s\text{–}D \tag{191}$$

$$M_s\text{-}D + O^{2-} \rightleftharpoons OD^- + M_s \tag{192}$$

$$O'D^- + OH^- \rightleftharpoons O'H^- + OD^- \tag{193}$$

3. weak associative adsorption of cyclopentane on titania. (This occurs in agreement with what was described earlier in Section 6.1.);

$$C_5H_{10} \text{ (g)} \rightleftharpoons C_5H_{10} \text{ (ads)} \qquad (194)$$

4. photoactivation of OD^- groups by neutralization with photoproduced holes;

$$OD_s^- + h^+ \rightarrow OD_s\bullet \qquad (195)$$

5. isotopic reaction through an associative complex;

$$C_5H_{10} \text{ (ads)} + OD_s\bullet \rightarrow C_5H_9D \text{ (ads)} + OH_s\bullet \qquad (196)$$

6. and regeneration of the initial OD_s^- active sites.

Step 6 involves: (a) the spillover of deuterium atoms (probably as deuterons) from the metal particles to the surface of TiO_2; (b) the reverse migration of H atoms (probably as protons) to the metal particles where they are recombined as HD molecules subsequently desorbing into the gas phase; and (c) the capture of photoproduced electrons by deuteroxyl groups.

All the steps 1 through 6 are summarized in the cycles depicted in Figure 45. The photoassisted electron transfer between the photosensitive support and the metal is clearly demonstrated. This electron transfer enables the reaction to be catalytic; this cycle was repeated more than 100 times on the same sites at the same rate, with no inhibition.[207,232] This reaction mechanism is quite general and can be applied to other bifunctional photocatalysts and to other hydrocarbons. In particular, the isotopic exchange of deuterium with propane produced monodeuteropropane with an initial selectivity of 100%.[233]

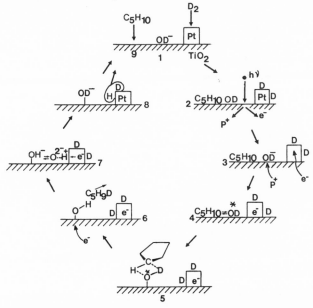

Figure 45. Cyclic mechanism of cyclopentane-deuterium isotopic exchange (CDIE) on Pt/TiO$_2$. Note that p$^+$ refers to the positively charged hole denoted h$^+$ in the text. From ref. 210.

The 100% initial selectivity in monodeuterocyclopentane and in monodeutero-propane[233] can be explained by the exothermicity of the activation process (see steps 4 and 5) in the cyclic mechanism of Figure 45.

$$OD_s^- + h^+ \rightarrow \underset{-OH_s\bullet}{\overset{C_3H_8}{OD_s\bullet}} \rightarrow C_3H_7D \text{ (ads)} \rightarrow C_3H_7D \text{ (g)} \tag{197}$$

The excess energy created by the neutralization of OD^- groups by the photoholes h^+ is dissipated through the endothermic desorption of the monodeuterated molecule. Consequently, each alkane molecule can only react once during one period of adsorption. An important consequence can be deduced from reactions of deuterium-alkane isotopic exchange concerning the heat transfer in deposited metal bifunctional photocatalysts. It is difficult to measure or estimate the actual temperature of the small metal crystallites under UV illumination. Deposited metal catalysts are generally dark and it could be imagined that part of the light is directly converted to heat. The two isotopic exchange reactions[207,232,233] permit the rejection of such thermal effects since the initial selectivities of the reaction would otherwise be identical to those obtained on thermally activated metals,[230,231] that is with the simultaneous formation of all the various deuterated molecules.

Optimum Metal Content.   When the quantity of a deposited metal is varied, an optimum metal content has often been observed in various photocatalytic reactions, for example in alcohol dehydrogenation,[217,223,234] deuterium-alkane isotopic exchange,[207,232] oxygen isotopic exchange,[235] and inter- or intramolecular hydrogen transfer.[227] The maximum depends on the nature of the metal and on its texture. It is 0.5 to 1 wt% for Pt,[207,217,220,232,235] 5 wt% for Ni,[223] and 1% for Ag.[234] The optimum metal content is the result of detrimental and beneficial effects; these are collected in Table 10.

The first two detrimental effects are rather limited. For instance, a calculation based on electron microscopy examination shows that a deposit of 10 wt% Pt covers only 6% of the total surface area of titania.[217] Perusal of Table 10 shows that, in addition to its catalytic properties towards hydrogen, the metal plays a key role in bifunctional photocatalysis associated to an electronic effect. Electron transfer to the metal is indeed a general phenomenon which has been observed in many cases with different supports, such as CdS (see Section 6.3), MoS_2 and $MoSe_2$,[236] and with different techniques, such as photoacoustic spectroscopy,[237] time-resolved photoconductivity,[236] and time-resolved microwave conductivity.[238]

In Deposited Metal Photocatalysts Under "SMSI" Conditions

Definition of the "SMSI" Effect.   The "Strong Metal-Support Interaction" (SMSI) effect was historically discovered at Exxon Corporation in 1978.[239,240] Initially discovered on Pt/TiO_2, this effect was later extended to other metals and

Table 10. Influence of the Loading of a Deposited Metal (M = Pt, Ni, Ag) in Photocatalytic Processes

Beneficial Effects of M	*Detrimental Effects of M*
1. M favors electron/hole pair dissociation $TiO_2 + h\nu \rightarrow (e^-/h^+)$ $(e^-/h^+) + M \rightleftharpoons e_M^- + h^+$	1. M may catalyze reverse or side-reactions e.g., $-CO- + H_2 \rightarrow -CHOH-$
2. M catalyzes the cathodic-like reduction of protons $H^+ + e_M^- \rightarrow M-H$	2. M masks a fraction of the photosensitive support area
3. M catalyzes the evolution of dihydrogen in the gas phase recombination $2\,M-H \rightleftharpoons H_2(g) + 2\,M$	3. Once enriched in photoelectrons, M attracts holes and becomes an ensemble of centers N. $h^+ + e_M^- \rightarrow N$
4. M initiates direct and reverse hydrogen spillover	
5. M favors the regeneration of photoactive sites on the support[207,217,232]	

to other semiconductor oxides.[241] When a metal catalyst is deposited on a reducible metal oxide support, and is reduced (or pretreated) in hydrogen at "low temperature" (T = 473 K), its capacity for hydrogen (or for CO) chemisorption at room temperature is maximum, while its corresponding catalytic activity in various reactions that involve hydrogen (hydrogenation, hydrogenolysis, methanation, Fisher–Tropsch, among others) is *normal;* that is, it is considered as the reference state. If the reduction temperature T_R of the catalyst is increased from 473 to 773 K, the hydrogen (or CO) chemisorption at room temperature is strongly decreased, whereas the catalytic activity is strongly altered or modified with important increase or decrease according to the type of reaction. This phenomenon occurs without any apparent changes in the metal texture. This is why the pioneers called this the SMSI effect.[239–241] Additionally, the early workers established that a subsequent treatment in oxygen at various temperatures restores the initial adsorptive and catalytic activities.

Attempts to explain the SMSI effect gave birth to a large number of publications. Useful references can be found in several articles[242–246] and in the proceedings of two symposia devoted to this subject.[247,248]

Various hypotheses have been proposed to account for the origin of SMSI effect in titania: (1) formation of intermetallic compounds; (2) encapsulation of the metal particles; (3) effect of hydrogen spillover; (4) occurrence of an electronic effect with electron transfer from the support to the metal; and (5) *decoration* of the metal particles by TiO_x mobile suboxide species. This last explanation is essentially based on results obtained on model catalysts, although some evidence of decoration by TiO_x species by high resolution electron microscopy (HREM) has been demon-

strated on real, well-dispersed Pt/TiO_2 catalysts.[249,250] Independent of the presence
or absence of decoration by TiO_x species, the permanent existence of an electronic
effect in SMSI was clearly demonstrated by *in situ* electrical conductivity measurements,[212,213,251,252] and by replacing the high temperature reduction of Pt/TiO_2
by a n-type doping of TiO_2 by pentavalent or hexavalent heterocations.[253,254]

Comparison with previous results from microcalorimetry[229] establishes that
high temperature reduction by hydrogen creates a large number of anion vacancies
in the reducible oxide support.[252] (See also Eqs. 6 and 7).

$$O^{2-} + H_2 \text{ (g)} \rightleftharpoons H_2O \text{ (g)} + V_O \qquad (198)$$

where V_O represents an anion vacancy with two electrons trapped at the oxygen
site. This site is neutral with respect to the lattice of the supports. However, anion
vacancies V_O easily lose a first electron and become singly ionized ($V_O^{\bullet}$); that is,
it is positively charged with respect to the lattice.

$$V_O \rightleftharpoons V_O^{\bullet} + e^- \qquad (199)$$

The quasi-free electrons are promoted into the conduction band and increase the
energy of the Fermi level of the metal oxide support with respect to the Fermi level
of the metal, thus inducing electron transfer, even if limited, from the support to
the metal. Under these conditions, if molecules such as H_2 or CO chemisorb at the
surface of the metal deposit by forming dipoles with a donor character (M^{d-}–
H^{d+}),[255–257] it is easily seen that excess electrons in the metal particles will
counteract this chemisorption.[252] This explanation is supported by: (1) the decrease
in the initial heat of hydrogen chemisorption, from 22 kcal mol^{-1} in the normal state
to 18.5 kcal mol^{-1} in the SMSI state;[229] and (2) the observation of an artificial SMSI
effect obtained by the n-type doping of titania.[253,254]

The electron transfer interpretation explains why the SMSI state is suppressed
by oxygen treatment. As soon as gaseous oxygen comes in contact with a titania-
deposited metal catalyst in the SMSI state, the anion vacancies get repopulated
with oxygen atoms which consume free electrons, some of which are returned to
the oxide by the metal deposit:

$$V_O^{\bullet} + 1/2 \ O_2 \text{ (g)} + e^- \rightleftharpoons O^{2-} \qquad (200)$$

This reaction is witnessed and confirmed by the variations in electrical conductivity
with changes in the partial pressure of oxygen.[252]

$$\sigma = kP_{O_2}^{-1/4} \qquad (201)$$

The process is highly exothermic (-81 kcal mol^{-1} of V_O), which explains why it is
spontaneous at ambient temperatures. In fact, partial reoxidation of the support and
the consecutive suppression of the SMSI state can be realized not only by gaseous
oxygen but also by oxygen-containing molecules such as water or alcohols.[252,258]

$$V_O^{\bullet} + H_2O + e^- \rightarrow O^{2-} + H_2 \text{ (g)} \qquad (202)$$

SMSI Effect in Heterogeneous Photocatalysis. Because of the mode of activation of photocatalysts and the ensuing photoinduced electron transfer involved in heterogeneous photocatalysis, it is instructive to examine a photocatalytic reaction taking place on a metal catalyst in the SMSI state. From the conditions described earlier, neither oxygen nor oxygen-containing reactants have proven useful. This explains why only the deuterium-alkane isotope exchange is worth considering. The photocatalytic rate of cyclopentane-deuterium isotope exchange (CDIE) has been determined under otherwise identical conditions on a series of Pt/TiO$_2$ catalysts either in the *normal* state (after reduction at T$_R$ ≤ 300 °C) or in the SMSI state (after reduction at T$_R$ = 500 °C).[207,232,258] The results are depicted in Figure 46. It is clear that the SMSI effect: (1) strongly decreases the photocatalytic activity of the catalysts; and (2) inhibits the reaction to an extent which depends on the quantity of platinum (weight percent). The lower the metal content is, the greater is the *inhibition* factor, defined as the ratio of the reaction rate in the normal state to the rate in the SMSI state. Several explanations are possible: (1) partial depletion in deuterated hydroxyl groups OD$^-$ due to the high temperature treatment; (2) strong decrease in chemisorbed deuterium, that is a low density in Pt$_s$–D species; and (3) increased electron transfer from the support to the metal. The partial depletion in OD$^-$ can be ruled out since the deuteroxyl groups are spontaneously formed by deuterium spillover from Pt particles. The strong decrease in chemisorbed deuterium due to SMSI does not necessarily imply that it is the only responsible factor for CDIE inhibition. Adsorption of deuterium is not the rate-limiting step of the process illustrated in Figure 45. The hydrogen remaining adsorbed in the SMSI state is the more weakly bound; that is, it is the more reversibly adsorbed.[229] It corresponds to the hydrogen species involved in the reaction mechanism. This hydrogen is located close to the metal-support perimeter and can easily spillover to the surface of titania. Electron transfer certainly remains the major factor to account for the inhibition observed in Figure 46.

In the SMSI state, there are several anion vacancies, each one liberating one electron. The Fermi level of titania in the SMSI state is positioned at higher energy than is the Fermi level in the normal state. Consequently, photoinduced electron transfer from the oxide support to the metal is strongly augmented. The metal particles being more negatively charged will now attract more efficiently the photogenerated positive holes and consequently will become more efficient recombination centers.

Since the metal crystallites have the same size, there is no textural effect to explain the variation of the inhibition factor as a function of metal content (Figure 47). With the number of crystallites being proportional to the platinum content, the high inhibition factor at low loadings corresponds to a relatively more efficient electron transfer to metal particles which are present in smaller concentration. By contrast, at high loadings (10 wt%), the density of metal crystallites per support grain is higher (~ 20) and the relative electron enrichment per particle is less

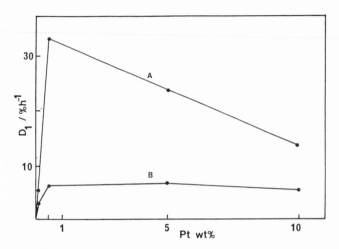

Figure 46. Cyclopentane-deuterium isotopic exchange. Conversion (in %) in monodeuterocyclopentane as a function of Pt wt.%. Curve **A**: Pt/TiO$_2$ in the normal state; Curve **B**: Pt/TiO$_2$ in the SMSI state. From ref. 232.

pronounced. This is the semiquantitative explanation of the SMSI effect observed on a photoassisted catalytic reaction.

The correlation between photoinduced and nonphotoinduced electron transfer in this type of reaction is an indirect proof of the occurrence of an electronic factor in the origin of the SMSI effect.

In Pt/TiO$_2$ Photocatalysts Under Oxygen

According to previous studies of electrical photoconductivity carried out either under vacuum or under a hydrogen atmosphere, a reversible electron transfer between a metal catalyst and its support can easily be ascertained. A few questions are worth addressing. Does this electron transfer persist under an atmosphere of oxygen? What is the role of the electrophilic character of oxygen?

Independent of light excitation, the introduction of oxygen on a previously desorbed metal surface produces a spontaneous dissociative oxygen chemisorption on two surface metal atoms.

$$2 \, M_s + O_2 \, (g) \rightarrow 2 \, M_s - O \tag{203}$$

Like hydrogen, oxygen chemisorption can also be used to determine the metal dispersion (D), the number of surface metal atoms per total number of metal atoms from which the mean particle size <d> can be calculated using the simple relationship: <d> (nm) ~ 1/D.

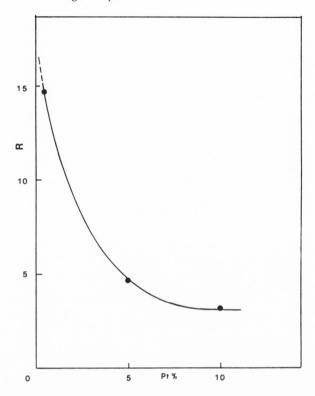

Figure 47. Variations of the inhibition factor R as a function of Pt wt.%. R is defined as the ratio of the photoactivity of Pt/TiO$_2$ in the normal state to that in the SMSI state. From ref. 258.

It must be noted, however, that there exist some specific problems that are inherent to the nature of the transition metal considered. For instance, the adsorption of oxygen on nickel produces a nonreducible nickel oxide (NiO) layer. The chemisorption of oxygen on highly dispersed rhodium gives a Rh$_2$O$_3$ surface layer[251] with a stoichiometry O/M$_s$ of 1.5, different from the value of 1 expected from Eq. 203. The simplest case is obtained when M is platinum for which the hydrogen stoichiometry is close to unity, giving in most cases a <d> from dispersion measurements which accords with the mean particle size <d> determined by transmission electron microscopy (TEM).

Photoinduced electron transfer under oxygen has been detected by dc electrical photoconductivity[235] on a series of texturally well defined Pt/TiO$_2$ catalysts, prepared by impregnation of TiO$_2$ in a H$_2$PtCl$_6$ solution and followed by thermal reduction with hydrogen. The metal crystallites are homodispersed, about 1.5 nm in size. Whether *in vacuo* or in a hydrogen atmosphere, the photoconductivity at steady state decreases monotonically when the platinum loading or the surface density of the metal particles is increased. This means that in spite of the formation of a Pt-O surface layer, photoinduced electron transfer from the support to the metal still occurs.

The consequences of this photoinduced electron transfer under oxygen are manifold. The higher the metal loading is, the more significant is the number of electrons transferred to the metal, and the smaller is the free photoelectron density at the surface of titania. The depletion in photoelectrons decreases the photoadsorption of ionosorbed oxygen species that form according to the following reactions:

$$TiO_2 + h\nu \rightarrow e^-/h^+ \tag{204a}$$

$$e^-/h^+ + Pt \rightleftharpoons e_{Pt}^- + h^+ \tag{204b}$$

$$h^+ + e_{Pt}^- \rightarrow N \text{ (neutral center)} \tag{204c}$$

$$O_2\,(g) \rightleftharpoons O_2\,(ads) \tag{204d}$$

$$O_2\,(ads) + e^- \rightleftharpoons O_2^-\,(ads) \tag{204e}$$

$$O_2\,(g) \rightleftharpoons 2\,O\,(ads) \tag{204f}$$

$$O\,(ads) + e^- \rightleftharpoons O^-\,(ads) \tag{204g}$$

In the absence of platinum (pure naked titania), photoadsorption and photodesorption studies show that dissociated oxygen O^- species photoadsorb first (tenaciously) and saturate their adsorption sites. The remaining free photoinduced electrons are then in equilibrium with gaseous oxygen O_2 molecules to form reversibly labile $O_2^-\bullet$ species.[166,168] The existence of O_2^- species (Eqs. 204e,f) makes the photoconductivity vary as the reciprocal of the oxygen pressure ($\sigma = k\,P_{O_2}^{-1}$), whereas σ of O^- species varies as the reciprocal of the square root of P_{O_2} ($\sigma = k\,P_{O_2}^{-\frac{1}{2}}$).[166,168] The log-log plots of σ versus f (P_{O_2}) give straight lines whose slopes, $\partial \log \sigma/\partial \log P_{O_2}$, are correspondingly -1 and $-1/2$ (e.g., see Figure 38). In the present case, as the number of platinum crystallite deposits on titania increase, the quantity of free photogenerated electrons available at the surface of titania decrease and only the more strongly bound oxygen species, O^- (ads), form. This is clearly indicated by the progressive increase of the slope $\partial \log \sigma/\partial \log P_{O_2}$, which varies from a value of -1 for naked titania and for the 0.05 wt% Pt/TiO$_2$ sample to a value of $-1/2$ for the 5 wt% and 10 wt% Pt/TiO$_2$ samples (Table 11). Consequently, under an oxygen atmosphere, photoinduced electron transfer from the photosensi-

Table 11. Relative Electron Enrichment of Titania-Deposited Platinum in Oxygen Atmosphere and Under UV-Illumination

Pt loading (wt %)	Slope $\partial \log \sigma / \partial \log P_{O_2}$	Main oxygen species photoadsorbed	Number of photoadsorbed O_2 per m^2	Electron enrichment per Pt atom e^-/Pt
0	−1	O_2^-	3.66×10^{17}	—
0.5	−1	O_2^-	1.34×10^{17}	0.7
5	−1/2	O^-	2.5×10^{16}	0.1
10	−1/2	O^-	$.8 \times 10^{16}$	0.05

tive support to the metal not only decreases the amount of photoadsorbed oxygen but also modifies the nature of the ionosorbed species.

By combining the photoconductivity values with the quantitative oxygen chemisorption data, it is possible to estimate the relative electron enrichment per Pt atom as a function of the metal loading. It is generally assumed that the number of free electrons transferred from TiO_2 to Pt corresponds to the difference between the number of oxygen molecules photoadsorbed on naked TiO_2 and the number of oxygen species ionosorbed on Pt/TiO_2. The nature of these ionosorbed species (O_2^- and O^-) varies with the concentration of Pt (wt%; see Table 11, 3rd column) and has to be taken into account in the calculation. Such an electron enrichment is given in Table 11 (last column) and is illustrated in Figure 48. The decrease in the ratio e^-/Pt as a function of wt% Pt is of the hyperbolic type, similar to the variations of $\log \sigma = f(wt\% \, Pt)$. This means that for samples with a low Pt loading (about 1 Pt crystallite per TiO_2 particle), although the overall photoinduced electron transfer is limited, the relative electron transfer per Pt atom is extensive (about 0.7 e^-/Pt atom). By contrast, at high Pt loadings (from 5 to 10 wt% Pt), although the overall electron transfer is significant, the electron transfer per Pt atom is rather limited (about 0.1 to 0.05 e^-/Pt atom). This kind of compensation is related to the constant rate of electron/hole pair formation under constant light flux illuminating the same fraction of the surface of titania particles. The extent of electron transfer (up to 0.7 e^-/Pt atom) is of the same order of magnitude as that found in the literature on model catalysts in the dark: 0.6 e^-/Pt atom for Pt/$SrTiO_3$ (100)[259] and 0.1 e^-/Ni atom in Ni/TiO_2 (110).[260] These values were estimated from binding energy shifts in metal-core levels.

Consequences for Photocatalytic Oxidation Reactions on Pt/TiO_2. The Pt/TiO_2 photocatalysts described above have been tested in various reactions: (1) the oxygen isotope exchange (OIE);[235] (2) the selective oxidation of tertiary butyltoluene into tertiary butylbenzaldehyde;[261] and (3) the mild oxidation of cyclohexane into cyclohexanone.[262] In the first case, the OIE proceeds basically via a mechanism where surface oxygen atoms exchange one at a time.[263,264]

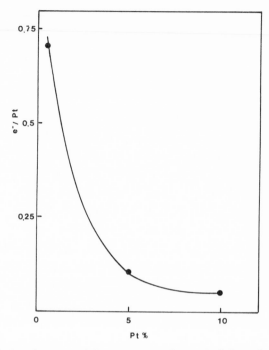

Figure 48. Relative electron enrichment per metal atom (e⁻/Pt) as a function of weight percent Pt.

$$^{18}O_2 \text{ (g)} + {}^{16}O \text{ (ads)} \rightarrow {}^{16}O-{}^{18}O \text{ (g)} + {}^{18}O \text{ (ads)} \tag{205}$$

The influence of Pt deposits changes with the nature of the pretreatment. After a preoxidizing treatment, the initial rate of OIE decreases slightly as a function of Pt content, while the samples evacuated at 423 K exhibit a maximum at 0.5 wt% Pt,[217,223,234] analogous to findings in reactions involving hydrogen (alcohol dehydrogenation and alkane-deuterium isotopic exchange).[207,232] This underlines the influence of the pretreatment on the electron density and on the nature of the ionosorbed oxygen species. No improvement is observed in the oxidation of liquid 4-tertiary-butyltoluene by gaseous oxygen when replacing TiO₂ with 0.5 wt% Pt/TiO₂.[261] By contrast, in the mild oxidation of isobutane and cyclohexane,[262] a regular decrease of the initial rate as a function of Pt wt% is observed (Figure 49). The detrimental role of the metal can be rationalized by electron transfer to the metal: (1) the oxide is depleted in free electrons, thereby decreasing the ionosorption of precursor active oxygen species; and (2) once negatively charged, the metal particles attract positive photoholes and become recombination centers as witnessed in photocatalytic reactions involving hydrogen.

It is clear from the above discussion that the presence of oxygen in contact with a deposited metal photocatalyst does not interfere with photoinduced electron

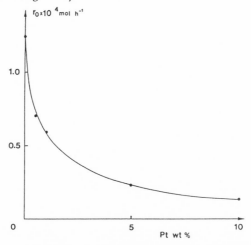

Figure 49. Initial rate variations as a function of Pt wt.% in the mild photocatalytic oxidation of cyclohexane in contact with Pt/TiO_2. From ref. 262.

transfer to the metal. Except in some favorable cases in photodegradation reactions in aqueous media,[186,265] a metal deposit has a detrimental effect in the oxidation of organic compounds in nonaqueous media.[262] The depletion of free electrons in the support yields only the formation of adsorbed O^- species, which control the electron transfer process between the solid and the gas phase. Since Pt/TiO_2 samples are active in photocatalytic oxidation reactions, it demonstrates that O^- species are the precursors of active oxidizing species in both gas and liquid organic phases. This is different from the oxidative degradation of organic compounds, which preferentially involves OH• radicals resulting from the activation of hydroxyl groups by photoholes.

In Metal Photodeposition

When a photosensitive semiconductor is irradiated in the presence of a noble metal salt or complex with photons of suitable energy ($h\nu \geq E_{BG}$), the metal can be deposited as small crystallites. This phenomenon has been known for over two decades,[266,267] but systematic investigations have been carried out only recently. The metals used have been Pt,[268–275] Pd,[265,273–282] Rh,[273,274] Au,[274,283,284] Ir,[275] and Ag,[266,267,285–291] deposited on various metal oxide semiconductors: ZnO, WO_3, but predominantly on TiO_2. Photosensitive metal sulfides such as CdS can also be used.[275,292–294] The photocatalytic deposition has been employed as a means of preparing metal supported-catalysts[265,270,271,275,278] or photocatalysts,[270] and as a potential method of metal recovery from aqueous effluents [273,274,283,284,291,295] or of metal separation.[274,283,291]

The true catalytic character of the photodeposition of metals has been clearly established by examining the influence of various parameters.[273,291] The initial step is adsorption of the cation or of the metal complex at the surface of the semiconductor. The adsorption of silver ions occurs on OH^- groups[291] since it leads to a decrease in pH according to the equilibrium:

$$TiOH + Ag^+ \rightleftharpoons TiO–Ag + H^+ \tag{206}$$

Hexachloroplatinate ions, as H_2PtCl_6 or as Na_2PtCl_6, adsorb on TiO_2 in the dark with a maximum coverage of 2×10^{18} ions m^{-2}.[273] This value corresponds to about half the maximum coverage of titania by OH groups (5×10^{18} m^{-2}).[296] This indicates the necessity of two hydroxyl groups for the adsorption of one chloroplatinate complex ion.

The adsorption of metal ions or complexes during the catalytic process is confirmed by the Langmuir–Hinshelwood mechanism of the reaction determined by the influence of the initial concentration on the initial rate of reaction r_0:

$$r_0 = k \ K \ C_0 / (1 + K \ C_0) \tag{207}$$

This reaction rate tends to a plateau at $C_0 \geq 5 \times 10^{-3}$ M. The influence of temperature is revealed by a small apparent activation energy of 6 kJ mol^{-1},[273] which demonstrates that the reaction is a true photoactivated process, practically independent of temperature. In the case of silver photodeposition, the rate decreases above 60 °C. This is understandable if at higher temperatures the reaction becomes limited by adsorption.[291]

The initial rate of deposition increases proportionately with the light flux. Such variations afford an estimate of the photochemical efficiency of the process, defined here as the ratio of deposited metal atoms per second to the number of photons reaching the reactor walls per second. The value estimated for Pt is 0.05;[273] values from 0.1 to 0.16 have been reported for Ag.[286,287,291]

Various studies have explored the stoichiometry of the overall reaction, given by the reaction:

$$M^{n+} + n/2 \ H_2O \rightarrow M^\circ + n \ H^+ + n/4 \ O_2 \ (g) \tag{208}$$

The deposition produces n protons per n-valent metal atoms deposited on the semiconductor particle surface; the kinetics of the evolution of gaseous oxygen does not follow the expected stoichiometry.[273,291] This is explained by the concomitant occurrence of, (1) oxygen photoadsorption by TiO_2, and (2) by a dissociative oxygen adsorption on the recently formed metal crystallites.

Metal deposition is inferred to occur via the mechanism proposed below on the basis of the examination of the various parameters noted earlier:

1. adsorption of the dissolved metal;

$$M^{n+} \ (aq) \rightleftharpoons M^{n+} \ (ads) \tag{209}$$

2. creation of electron/hole pairs by suitable photons ($h\nu \geq E_{BG}$) in the illuminated semiconductor SC;

$$(SC) + h\nu \rightarrow e^-/h^+ \rightarrow e^- + h^+ \tag{210}$$

3. oxidation of water by photogenerated holes;

$$H_2O \rightleftharpoons H^+ + OH^- \tag{211}$$

$$OH^- + h^+ \rightarrow OH\bullet \rightarrow 1/2\ H_2O + 1/4\ O_2\ (g) \tag{212}$$

4. stepwise reduction of the metal cation by photogenerated electrons;

$$M^{n+}\ (ads) + n\ e^- \rightarrow M^o\ (ads) \tag{213}$$

This basic mechanism of photoassisted deposition of noble metals has a true catalytic character despite the fact that one of the reaction products is a solid which deposits at the surface of the photocatalyst.

Photoinduced electron transfer is the key element in all the processes. First, electron transfer from the semiconductor to the metal ions or complexes only occurs in favorable cases, when the redox potential of the metal is more anodic (more positive) than the flatband potential of the conduction band of the semiconductor. This is one reason why only noble metals deposit on titania, whereas such common metals as nickel, copper (as the NO_3- salt), and iron do not, at least as M^{n+} aqueous ions. Second, photoinduced electron transfer governs the relative reactivity of the different metals. Under identical experimental conditions, the following relative reactivity pattern is established:[275]

$$Ag^I > Pd^{II} > Au^{III} > Pt^{IV} >> Rh^{III} >> Ir^{IV} > Cu^{II} \sim Ni^{II} \sim Fe^{II} = O \tag{214}$$

The corresponding metal compounds or complexes used to establish this series were: $AgNO_3$, $PdCl_2$ or $Pd(NO_3)_2$; $AuCl_3$; H_2PtCl_6 or Na_2PtCl_6; $RhCl_3$; H_2IrCl_6; $Cu(NO_3)_2$; $Ni(NO_3)_2$; and $Fe(NO_3)_3$.

For the first four metal ions, the reactivity appears to vary inversely with the oxidation state of the metal; that is, to the number of photoinduced electrons to be transferred to the metal ion. This is not a general rule since, for instance, Pt^{IV} is more reactive than Rh^{III}, which requires one electron less to be reduced. Pt^{IV} is also much more reactive than its isoelectronic homolog Ir^{IV}. Photoinduced electron transfer also depends on the nature of the metal ion. Additionally, for a given metal, it also depends on the nature of the metal complex. For instance, Pt^{II} in $Pt^{II}(NO_2)_2(NH_3)_2$ is more difficult to photoreduce than Pt^{IV} in $Pt^{IV}Cl_6{}^{2-}$. although it needs two photogenerated electrons less to be reduced.[273]

Photoinduced electron transfer is an important process since it governs the extent and the texture of the metal deposits. The reduction reaction is not the final step of the overall photocatalytic process. Once produced, zerovalent metal atoms agglomerate spontaneously into small crystallites, as witnessed by transmission electron microscopy (TEM).[271,273,279,280,291,292] The mean diameter seems to depend on the nature of the metal and on the experimental conditions. This agglomeration may occur by two possible ways, either via clustering of free mobile metal atoms:

$$m \, M^{o} \, (ads) \rightarrow M_m \, (supported) \tag{215}$$

or via a cathodic-like reduction at small primary metallic nuclei:

$$M^{o} + M^{n+} \rightarrow (M_2)^{n+} \xrightarrow{+ \, ne^-} M_2{}^{o} \xrightarrow{M^{n+}} (M_3)^{n+} \xrightarrow{+ \, ne^-} M_3{}^{o} \rightarrow ... \rightarrow M_m \tag{216}$$

Since it has been shown in the preceding sections that there is photoinduced electron transfer from the support to the metal, the second pathway (Eq. 216) seems more likely. Electron transfer is also responsible for the size of the metal particles and for their distribution on the support.

The initial deposition of silver provides particles of 3- to 8-nm diameter,[291] whereas the average size of platinum crystallites is 1 to 1.5 nm,[273] surprisingly close to those prepared by thermal deposition. When a metal, such as platinum, is well dispersed and distributed on all the particles of the support, it exhibits an excellent texture for constituting a highly divided metal catalyst. However, the amount of deposited metal is limited to a few wt%.[275] This is understandable if electron transfer to the numerous metal particles attracted positive photogenerated holes and became recombination centers. In contrast to the metal deposits, which show poor metal dispersion and a poor particle distribution on the support, the deposition can reach high values. For instance, in a concentrated (10^{-1} M) solution of $AgNO_3$, large deposits of silver are obtained with a "photodeposition capacity" C, defined as the mass of deposited metal per unit of mass of support, reaching a value of 2.4 after 100 hr of illumination.[291] This means that metal deposits agglomerated into large particles do not markedly mask the photosensitive support from the light source, and that the initial crystallites, acting as electron/hole recombination centers, may have little consequence in overall metal deposition.

The photoinduced electron transfer process also governs the formation of alloys in the preparation of bimetallic catalysts. This has been clearly demonstrated for various couples (Pt-Rh, Ag-Rh and Pt-Pd/TiO_2) that electron transfer to both types of metal ions (or complexes) must proceed at similar rates to get alloy particles, as evidenced by scanning transmission electron microscopy (STEM).[297]

Insofar as non-noble metals are concerned, a photoassisted reduction of ions into metal atoms has been observed for mercury[298–301] which has a favorable redox potential for reduction. For heavier common metals, after Hg in the periodic table (Tl^+ and Pb^{2+}), electron transfer occurs in the opposite direction to that witnessed in noble metals. Here, the deposits consist of metal oxides supported on titania, and there is an increase in the oxidation state of the metal.[302]

$$Pb^{2+} + 1/2 \, O_2 + H_2O \rightarrow PbO_2 \, (deposited) + 2 \, H^+ \tag{217}$$

$$Tl^+ + 1/2) \, O_2 + H_2O \rightarrow Tl_2O_3 \text{ (deposited)} + 2 \, H^+ \tag{218}$$

Other workers have coupled the photoassisted oxidation of lead into PbO_2 with the photoassisted reduction of palladium on a TiO_2 single crystal, thus realizing a concerted photoinduced electron transfer between two metal ions:[292]

$$Pb^{2+} + Pd^{2+} + 2 \, H_2O \rightarrow PbO_2(dep) + Pd^0(dep) + 4H^+ \tag{219}$$

In this case, no gaseous oxygen (or air) is needed.

6.3. Photoinduced Electron Transfer between Two Semiconductors

Vectorial displacement of photogenerated charge carriers between two different semiconductors, first reported by Serpone and co-workers[64a] for electron transfer from CdS to TiO_2 loaded with ruthenium dioxide, RuO_2/TiO_2, has now been demonstrated for various systems: from CdS to TiO_2[60,63,67,303] or Pt/TiO_2;[65] CdS to ZnO,[60] Ag_2S,[304] $K_4Nb_6O_{17}$,[305] or AgI;[63] ZnS to ZnO;[306] ZnS to AgI and Ag_2S;[307] Cd_3P_2 to TiO_2 and ZnO;[303] and ZnS to CuS and mixed sulfides $Zn_x \, Cd_{(n-x)}S$.[308]

The interest in vectorial electron transfer between two phases is manifold. It enhances the charge separation and increases the lifetime of charge carriers. It can improve the photocatalytic rates of various reactions and enable a catalyst to be sensitized by another semiconductor using wavelengths appropriate to the sensitizer.[59]

The CdS-TiO_2 coupled system is of particular interest since it illustrates all the points mentioned above, and therefore will be described in some detail.

Evidence for Electron Transfer between CdS and TiO$_2$ In Vacuo and In Oxygen

Photoconductivity Under Vacuum. Photoinduced electron transfer between CdS and TiO_2 has been demonstrated by photoconductivity measurements;[59,67] however, because different conditions prevail in aqueous dispersions *versus* solid phases treated under vacuum, and because of ionosorption processes (Section 6.1) which involve electron exchange, we first examine electron transfer between two solids *in vacuo* taken as a reference state.

TiO$_2$. Under vacuum, UV-illumination increases the photoconductivity of titania, reaching a plateau within a few hours owing to an accumulation of free electrons as a result of the photogeneration of electron/hole pairs (Eq. 220) and from the depletion of holes by the photodesorption of oxygen (Eq. 221) from pre-adsorbed ionosorbed species.[166,168]

$$TiO_2 + h\nu \rightarrow e^-/h^+ \tag{220}$$

$$e^-/h^+ + O_x^- \rightarrow x/2\ O_2\ (g) + e^- \qquad (221)$$

The excess number of electrons is ascertained by the constant value of the conductivity when the light is switched off. With a bandgap energy of ~ 3.0 eV, titania is only sensitive to near-UV excitation. However, photoconductivity is a very sensitive technique and progressive accumulation of free electrons can be observed even with visible illumination ($\lambda > 400$ nm), albeit the kinetics are slower and the photoconductivity is two orders of magnitude smaller.

CdS. Illumination of naked CdS in the visible region increases its photoconductivity as expected from the value of its bandgap energy (2.4 eV). However, when irradiation is terminated (dark), the photoconductivity drops instantaneously to nonmeasurable values. This suggests that electron/hole recombination in the dark is complete for equal concentrations of electrons and holes:

$$CdS + h\nu \rightarrow e^-/h^+ \rightarrow e^- + h^+\ (h\nu \geq 2.4\ eV) \qquad (222)$$

$$e^- + h^+ \rightarrow N + h\nu'\ (\text{or heat}) \qquad\qquad (N = \text{neutral center}) \quad (223)$$

This indicates that, in contrast with TiO_2, no accumulation of photogenerated electrons occurs during illumination because of an absence of photodesorption of negatively charged pre-ionosorbed species. The variations in photoconductivity of CdS are reproducible between dark and illumination periods.

TiO₂ - CdS mechanical mixture. An intimate mixture of CdS and 5 wt.% TiO_2, prepared by vigorous prolonged agitation, contained 300 TiO_2 grains per CdS crystallite because of the respective textures of the two semiconductors (TiO_2 - Degussa P25, 50 $m^2\ g^{-1}$; $<d> = 31$ nm; Fluka CdS, 2.4 $m^2\ g^{-1}$ with different particle mean sizes of 40, 150, and 200–700 nm). Owing to the size and concentration (wt%) of CdS, this phase is expected to impose its conductivity to the whole sample according to the percolation threshold theory[309] and to the direct observations from transmission electron microscopy (TEM).

Illumination of this mixture at appropriate wavelengths induced photoconductivity variations (see Figure 4 of ref. 67) which can be summarized as follows: (1) the photoconductivity with visible irradiation is higher than that of pure CdS alone; and (2) after illumination is terminated, the dark conductivity decreases abruptly but to measurable values, in contrast to observations on pure CdS. These findings are understandable if electron transfer occurs from TiO_2 to CdS, under both illumination and in the dark. From a theoretical point of view, this can be explained by an alignment of the Fermi levels of both phases. The accumulation of quasi-free electrons generated in titania by such methods as surface reduction[251,252] and prolonged illumination in dynamic vacuum increases the Fermi level of titania. If the Fermi level of a second phase in contact with titania does not vary, as is the case for deposited metals such as Pt, Rh, or Ni,[251,252] or a semiconductor such as

CdS[59,67] a progressive electron transfer occurs from titania to this second phase. This electron transfer to CdS, under illumination and in the dark, has been confirmed experimentally by replacing TiO_2 by a photo-inert solid; for example, silica possessing the same texture. Qualitatively the CdS-SiO$_2$ mixture behaved as the single-phase CdS.

Pt/TiO$_2$-CdS mechanical mixture. Analogous photoconductivity experiments have been carried out with a triphasic system in which titania supports a deposited metal (platinum; metal loadings, 0.5 wt% and 5 wt%). Since the metal particles are homodispersed, their number is proportional to the concentration of the metal. Under vacuum, the photoconductivity of Pt/TiO$_2$- CdS is lowered by the presence of the metal; increasing the metal content decreases the photoconductivity (Figure 50). This can easily be understood if the photoelectronic interactions

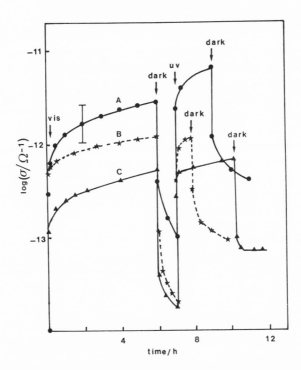

Figure 50. Variation of the photoconductivity σ under vacuum in visible and UV light for various mixtures: (**A**), CdS + 5 wt.% TiO$_2$; (**B**), CdS + 5 wt.% of 0.5 wt.% Pt/TiO$_2$; (**C**), CdS + 5 wt.% of 5 wt.% Pt/TiO$_2$. From ref. 321.

between Pt and TiO_2 are known (see Section 6.2). Because of the respective positions of the Fermi levels under illumination, there is a spontaneous electron transfer from the semiconductor to the metal.[208,212] This means that in the present case, photogenerated electrons are preferentially transferred to platinum, which is detrimental to electron transfer from TiO_2 to CdS. When the Pt wt% is increased, the number of electrons trapped in Pt crystallites is increased; this decreases the electron transfer to CdS and consequently diminishes the resulting photoconductivity of the whole sample.

When the visible illumination is interrupted (Figure 50), there is an abrupt decrease of the dark conductivity for solids containing platinum. Presumably, electron withdrawal by Pt is important in the dark and fewer electrons are transferred to CdS. Changing the illumination from the visible to the UV region (Figure 50) increases the photoconductivity of CdS-TiO_2 mixtures because of a higher efficiency of photoinduced electron formation in TiO_2 and a more favorable electron transfer to CdS. In contrast, the photoconductivity σ levels of Pt/TiO_2-CdS mixtures under UV illumination remain similar to those obtained with visible light. This shows that the electron transfer from TiO_2 to Pt is detrimental to the electron transfer from TiO_2 to CdS, and that it is more important as Pt particles can trap a greater number of electrons.

Photoconductivity in Oxygen.   Since catalysts are conserved in air and are often used without special pretreatment, it is useful to examine the photoelectronic behavior of a set of samples under an oxygen atmosphere.

TiO_2.   In agreement with results described above, the introduction of oxygen abruptly decreases the photoconductivity σ of TiO_2 [67] and variations of σ as a function of oxygen pressure at steady state follow the law:

$$\sigma = k \, P_{O_2}^{-1} \qquad (224)$$

This infers the existence of $O_2^-\bullet$ ionosorbed species in equilibrium with the illuminated surface (Eqs. 204d,e).

CdS.   In contrast to TiO_2, the photoconductivity of CdS varies as $P_{O_2}^{-1/4}$.[67] This is consistent with the formation of O^{2-} surface species:

$$O_2 \, (g) \rightarrow 2 \, O \, (ads) \qquad (225)$$

$$O \, (ads) + 2 \, e^- \rightarrow O^{2-}_{surface} \qquad (226)$$

A species such as $O^{2-}_{surface}$ has eluded photoconductivity measurements on various semiconductor oxides.[310] The fixation of anionic O^{2-} species could correspond to a photoinduced oxidation of CdS. Photocorrosion of CdS has been studied extensively in aqueous suspensions;[311] however, in the present case the system is water-free. Although photoconductivity measurements are silent on the nature of the resulting surface compounds between CdS and O^{2-} species, formation of the latter species should favor the formation of a sulfate layer at the surface of CdS.

TiO₂–CdS mechanical mixture. The introduction of oxygen on a pre-illuminated TiO_2–CdS mechanical mixture under vacuum strongly decreases the photoconductivity of the sample.[67,309] The steady-state photoconductivity decreases as a function of increasing oxygen pressures according to the kinetic expression:

$$\sigma = k \, P_{O_2}^{-0.04} \tag{227}$$

The negative exponent indicates that the mixture is still an n-type photoconductor but the photoconductivity is substantially lower than that of TiO_2 and CdS taken separately. This means that photoinduced electron transfers occur even in oxygen. Titania is a strong photoadsorbent of oxygen and captures free electrons to form O^- and O_2^- adsorbed species, with the former being adsorbed at saturation.[310] CdS photoadsorbs O^{2-} species but this phenomenon remains limited relative to the small variations of σ_{CdS} as a function of P_{O_2}.

The low value of the photoconductivity of TiO_2-CdS mixtures in oxygen, which is smaller than that of each constituent separately, clearly indicates that CdS, which ensures the overall conductivity of the samples between the two electrodes, is depleted of electrons, the main charge carriers for an n-type semiconductor. The chemisorption of oxygen on TiO_2 inhibits the electron transfer event from TiO_2 to CdS observed under vacuum, but more remarkably it produces an electron transfer in the opposite direction from CdS to TiO_2. Because of the strong photoionosorption of oxygen species on TiO_2, the Fermi level of titania is now located at lower energy, whereas that of CdS is maintained at a constant position because of the photoelectron generation in the visible and the limited electrophilic adsorption of oxygen on it.

Pt/TiO₂-CdS mechanical mixture. The photoconductivity of 0.5 wt% and 5 wt% Pt/TiO_2-CdS mechanical mixtures behave similarly to the σ of naked CdS with the same kinetic expression as a function of oxygen pressure:

$$\sigma = k \, P_{O_2}^{-\frac{1}{4}} \tag{228}$$

and under identical oxygen pressure the photoconductivities followed the sequence:

$$\sigma(CdS) > \sigma(CdS\text{-}0.5wt\% \; Pt/TiO_2) > \sigma(CdS\text{-}5wt\% \; Pt/TiO_2) \tag{229}$$

These two observations indicate that the presence of Pt/TiO_2 does not modify the type of photoadsorption of oxygen as O^{2-} species from the gas phase onto CdS, but strongly decreases the number of photoelectrons of conduction. The photoconductivity in oxygen for Pt/TiO_2–CdS samples is close to that of TiO_2–CdS mixtures, but with a different law for variations in σ as a function of P_{O_2}. This observation can be explained by a double electron transfer process. Electrons photogenerated by visible light in CdS are transferred to TiO_2 from which they are transferred to O_2 adsorbed species to give O_2^- (ads) on TiO_2. When TiO_2 is replaced by Pt/TiO_2, the same electron transfer from CdS to TiO_2 occurs, but then is followed by a second transfer from TiO_2 to Pt. The higher the metal content, the greater is the number of photoelectrons transferred to the metal and the lower is the photoconductivity of the samples. The metal is thus enriched in electrons photogenerated in a semiconductor that is different from its own support.

$$CdS + h\nu \rightarrow e^-/h^+\{CdS\} \tag{230a}$$

$$e^-/h^+\{CdS\} + Pt/TiO_2 \rightarrow e^-\{Pt/TiO_2\} + h^+\{CdS\} \tag{230b}$$

$$e^-\{Pt/TiO_2\} \rightarrow e_{Pt}^-\{Pt/TiO_2\} \tag{230c}$$

Conclusion. Interparticle electron exchange occurs between two different photosensitive solids. The direction of the transfer depends on the relative positions of the conduction bands of both solids and on their Fermi levels whose positions are linked to the nature and concentrations of ionosorbates. In the particular case of CdS–TiO_2 mixtures under vacuum, electrons are vectorially displaced from TiO_2 to CdS, whereas under oxygen (or air) they migrate from CdS to TiO_2. In the presence of a metal on TiO_2, electrons are transferred to the metal via the surface of titania.

Consequences for Heterogeneous Photocatalysis

Light at wavelengths above 400 nm is absorbed by CdS particles whose band gap energy (2.4 eV) corresponds to an absorption threshold of ~ 520 nm. The electron/hole pairs formed dissociate into e^-_{CB} and h^+_{VB}. The redox reactions occur at the conduction band (reduction) and at the valence band (oxidation) in competition with electron/hole pair recombination. For CdS particles of colloidal size (5–10 nm), this recombination occurs in $\leq$ 30 ps.[312] Since the presence of a second semiconductor induces a vectorial interparticle electron transfer to the other phase, a more efficient charge separation is favored. Different reactions have been carried out with such coupled photocatalytic systems. We now examine a selection of some of these.

Photocleavage of Hydrogen Sulfide[64a,c,313–316]. Photogenerated holes in the valence band of CdS react with HS^- and/or S^{2-} anion, depending on the basicity of the pH, to produce elemental sulfur which dissolves in sulfhydric media as polysulfides:

$$CdS + h\nu \rightarrow e^-/h^+ \rightarrow e^- + h^+ \tag{231}$$

$$2 e^-/h^+ + S^{2-} \rightarrow S + 2 e^- \tag{232}$$

$$n S + S^{2-} \rightarrow S^{2-}_{n+1} \ (n = 1,2,3,4...) \tag{233}$$

In the absence of TiO_2, direct reduction operates in naked CdS suspensions:

$$2 e^- + 2 H_2O \rightarrow H_2 + 2 OH^- \tag{234}$$

In the presence of TiO_2, a photoinduced interparticle electron transfer occurs from CdS to TiO_2:

$$e^-(CdS) + TiO_2 \rightarrow e^-(TiO_2) + CdS \tag{235}$$

and enables the generation of hydrogen from water:

$$2 e^-(TiO_2) + 2 H_2O \rightarrow H_2 + 2 OH^- \tag{236}$$

The same reaction occurs if titania, added as a cocatalyst, is replaced by deposited platinum or RuO_2. This illustrates that an interparticle electron transfer process can lead to a photocatalytic reaction as efficient as those in which the catalyst is constituted by a photodeposited phase with an intimate interface. Hydrogen evolution from sodium sulfite solutions has been demonstrated by electron transfer from CdS to $K_4Nb_6O_{17}$.[305]

Photodehydrogenation of Alcohols[65,317]. Visible light illumination of platinized CdS particles generate hydrogen from primary and secondary alcohols.[65] Coupling two semiconductor catalysts in the same suspension improves the production rate of H_2. For instance, the addition of Pt/TiO_2 to a slurry of CdS in methanol yields a near twofold increase in $r(H_2)$ relative to a Pt/CdS suspension.[65]

The details of the overall process are:

1. creation of electron/hole pair by visible light in CdS (Eqs. 222 and 233);
2. dissociative adsorption of the methanol;

$$CH_3OH \rightarrow CH_3O^- + H^+ \tag{239}$$

3. neutralization of the alkoxide ions by the holes;

$$CH_3O^- + h^+(CdS) \rightarrow HCHO + H\bullet \tag{240}$$

4. migration of conduction band electrons to titania (Eqs. 230b and 235), and subsequently to the metal deposits on titania (Eq. 230c);
5. migration of protons and $H\bullet$ atoms to platinum particles by reverse spillover.[318] $H\bullet$ atoms can migrate as protons at the surface of titania by releasing an electron to the solid and use OH groups as migration sites;[318]

$$H\bullet \rightarrow H^+ + e^- \tag{241}$$

$$H^+ + Ti–OH \rightarrow TiOH_2^+ \tag{242}$$

$$TiOH_2^+ + TiOH \rightarrow TiOH + TiOH_2^+ \rightarrow ... \tag{243}$$

6. neutralization of protons by excess electrons accumulated on platinum;

$$H^+ + e^-(Pt) \rightarrow H(Pt) \tag{244}$$

7. desorption of hydrogen when all the irreversible hydrogen adsorption sites are saturated:

$$2 \, Pt–H \, (rev) \rightarrow 2 \, Pt + H_2 \, (g) \tag{245}$$

The sites of reversible dissociative adsorption of hydrogen are characterized by a low heat of adsorption ($\Delta H_{(ads)} \sim 9$ to $10 \, kcal \, mol^{-1}$) in contrast with the strong adsorption sites on which hydrogen is irreversibly chemisorbed (at least at room temperature) with an initial heat of adsorption of $22 \, kcal \, mol^{-1}$.[229]

Photosensitization of Titania by CdS in Visible (Solar) Light. Titania has been recognized universally as the best photocatalyst since it combines two important complementary qualities for a photocatalyst: good absorption efficiency for the light harvesting process and good adsorption capacities, due in particular to the density of OH^- groups of amphoteric character. However, the bandgap energy (3–3.1 eV) requires that near-UV light be used to photoactivate this very attractive photocatalyst. This is confirmed by the experiments described in Figure 51.[319] Illumination at wavelengths longer than 400 nm, a Pt/TiO₂ sample is totally inefficient in producing hydrogen from methanol, even if the optimum platinum concentration (in wt%: 0.5 wt% Pt) is chosen. Addition of a small amount of deaerated CdS without any perturbation of the catalytic system immediately leads to the dehydrogenation of methanol. This means that photoinduced electron transfer occurs readily from photoactivated CdS to platinum via the TiO₂ support (or possibly directly). The charge separation allows the photoholes to neutralize the alkoxide anions. The protons follow the same path as the electrons. They are neutralized in a cathodic-like process at Pt particles at which dihydrogen is evolved. Small quantities of H₂ are formed with naked CdS, but at several orders of magnitude smaller than the quantity of dihydrogen obtained with CdS-Pt/TiO₂. Identical results are obtained when the experiment is initiated with CdS to which Pt/TiO₂ is subsequently added.

Photosensitization of titania in the visible (solar) region by addition of CdS has also been attempted in oxidation reactions. Oxidation of propanol into acetone was chosen as a test reaction.[171] Using wavelengths longer than 400 nm to irradiate titania, the reaction is relatively slow ($r = 2.53 \, \mu mol \, h^{-1}$), about two orders of magnitude smaller than under conditions of UV-light illumination [$r(UV) = 338 \, \mu mol \, h^{-1}$], but is not insignificant in contrast with methanol dehydrogenation by

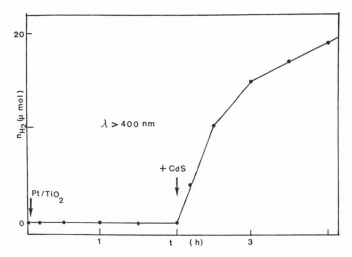

Figure 51. Photosensitization of Pt/TiO$_2$ (50 mg) in the visible (solar) light by addition of 10 mg CdS in the photocatalytic dehydrogenation of methanol. From ref. 319.

Pt/CdS in visible light. This is probably due to: (1) the presence of platinum in the latter case which constitutes a sink for photogenerated electrons; and (2) the presence of surface states or active sites which can be photoactivated by visible photons of energy close to the absorption threshold. The addition of CdS to naked titania increases the reaction rate sixfold (r(TiO$_2$ + CdS) = 15.3 μmol h^{-1}); this reaction rate roughly corresponds to the rate obtained with naked CdS [r(CdS) = 16.5 μmol h^{-1}]. Clearly, there is no sensitization of TiO$_2$ from visible excitation of CdS in this oxidation reaction. This can be explained by several factors. Photocatalytic oxidation reactions involve ionosorbed oxygen species which are tightly bound to titania. It was noted earlier that photoadsorption of oxygen created O$_2^-$ species on TiO$_2$ and anionic O^{2-} species at the surface of CdS,[67] with the possible formation of sulfate ions on the surface. The negatively charged oxygen entities adsorbed on both solids consume photoformed electrons and inhibit the electron transfer process between particles since electrons have to cross Schottky barriers induced by the negative ionosorbates. This quasi-absence of photoinduced electron transfer explains the absence of synergy between CdS and TiO$_2$. By contrast, in the absence of oxygen, that is under a hydrogen reducing atmosphere, the electron transfer is facilitated with a driving force induced by the presence of a metal.

The absence of photosensitization of TiO$_2$ in the visible wavelengths by CdS under oxidizing conditions has also been witnessed in the mild oxidation of cyclohexane; no products are detected that might have originated from the mild oxidation process.

Photoinduced Electron and Hydrogen Atom Transfer in Bronze Formation

Photoinduced interparticle electron transfer gives birth to a new solid phase when it is associated with hydrogen transfer. This is illustrated by the photoassisted synthesis of hydrogen-molybdenum bronzes H_xMoO_3.[320] By illuminating alcoholic suspensions of MoO_3 (in methanol or 2-propanol), the solid turns rapidly to a dark blue color. X-ray analysis indicates the formation of the $H_{0.34}MoO_3$ bronze. The addition of an equivalent mass of titania (Degussa P-25) produces a darker solid constituted by a mixture of TiO_2 and $H_{0.93}MoO_3$ hydrogen-molybdenum bronze.[320] The presence of Pt deposited either on MoO_3 or on TiO_2 completely inhibits the formation of a bronze. The results are presented in Table 12.

The results are understandable on the basis that illumination of an alcoholic suspension of a photosensitive naked metal oxide produces hydrogen; in the absence of a metal deposit on this metal oxide the reaction ceases after a short period (~ 1hr).[320,321] The reaction is not catalytic.[166] For naked MoO_3, rather than desorb, hydrogen incorporates into the oxide and forms the $H_{0.34}MoO_3$ bronze, in which the hydrogen content is somewhat low since x can vary from 0 to 2. Addition of naked TiO_2 to the MoO_3 system substantially increases the amount of photo-generated atomic hydrogen; the presence of MoO_3 permits H• to migrate to MoO_3, thereby effectively inducing vectorial charge displacement.

$$H\bullet(TiO_2) + MoO_3 \rightarrow TiO_2 + H\bullet(MoO_3) \qquad (246)$$

$$x\ H\bullet(MoO_3) + MoO_3 \rightarrow H_xMoO_3\ (x = 0.93) \qquad (247)$$

Since titania is more photoactive than MoO_3, the production of atomic hydrogen, which is a prerequisite for bronze formation,[322] is increased and the hydrogen stoichiometric mole fraction is about 3 times greater (x = 0.93). However, in the MoO_3 + TiO_2 mixtures, photoinduced interparticle electron transfer is not a necessary step for bronze formation, since transfer of atomic hydrogen can *per se* lead to an increase in the H content in the bronze.

The presence of platinum deposits on MoO_3 induces electron transfer from TiO_2 to the metal, and the attraction of protons by the excess of electrons accumulated

Table 12. Photoassisted Hydrogen-Molybdenum Bronze Synthesis[320]

Semiconductor illuminated	Stoichiometry H_xMoO_3 of the bronze formed
MoO_3	$H_{0.34}MoO_3$
Pt/MoO_3	—
MoO_3 + TiO_2	$H_{0.93}MoO_3$
MoO_3 + 0.5 wt% Pt/TiO_2	—

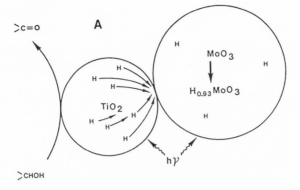

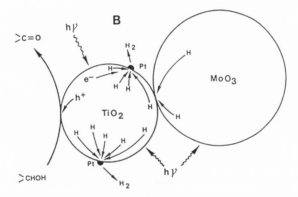

Figure 52. (**A**): $H_{0.93}MoO_3$ bronze formation by hydrogen transfer from TiO_2 to MoO_3 during the photocatalytic dehydrogenation of an alcohol. (**B**): Inhibition of the bronze formation due to preferential migration of H atoms or protons to Pt particles before evolution of dihydrogen. From ref. 321.

on the metal constitutes a driving force superior to the incorporation of these protons into MoO_3 to form a bronze. Similarly, the presence of platinum deposits on titania induces not only electron transfer from TiO_2 to Pt but an additional transfer from MoO_3 to Pt via TiO_2.[320,321] This is illustrated by the scheme of Figure 52. As a consequence, the protons issued from the dissociative adsorption of the alcohol are readily attracted by the metal, at the surface of which they are reduced before evolution into the gas phase, thus explaining the absence of bronze formation.

ACKNOWLEDGMENTS

Our work in Montreal is supported by the Natural Sciences and Engineering Research Council of Canada, and in Lyon by the Centre National de la Recherche Scientifique (Paris). We are also particularly grateful to the North Atlantic Treaty Organization (Brussels) for a Collaborative Exchange Grant (No. CRG 890746) between our respective laboratories in Montreal (Serpone), Lyon (Pichat), Torino (Pelizzetti), and Austin, Texas (Fox). We are most thankful to Journal Editors, Copyright Administrators, and the following Publishers and Chemical Societies for their kind responses for permission to reproduce certain figures from the respective publications: Verlag Helvetica Chimica Acta, Elsevier Science Publishers, the CRC Press, the American Chemical Society, The Royal Society of Chemistry, Elsevier Sequoia, Springer-Verlag, the Electrochemical Society, J. C. Baltzer Scientific Publications, the Canadian Society for Chemical Engineering, the Chemical Society of Japan, John Wiley & Sons, and Academic Press.

REFERENCES

1. Fox, M. A.; Chanon, M., Eds. *Photoinduced Electron Transfer*, Part C. Elsevier, Amsterdam, 1988.
2. Fox, M. A.; Chanon, M., Eds. *Photoinduced Electron Transfer*, Part D. Elsevier, Amsterdam, 1988.
3. Serpone, N. In: *Photoinduced Electron Transfer*, Part D. (Fox, M. A.; Chanon, M., Eds.). Elsevier, Amsterdam, 1988, Chapter 5.3, pp. 47–147.
4. Balzani V.; Scandola, F. In: *Photoinduced Electron Transfer*, Part D (Fox, M. A.; Chanon, M., Eds.). Elsevier, Amsterdam, 1988, Chapter 5.4, pp. 148–178.
5. Vogler, A. In: *Photoinduced Electron Transfer*, Part D (Fox, M. A.; Chanon, M., Eds.). Elsevier, Amsterdam, Chapter 5.5, pp. 179–199.
6. Pichat, P.; Fox, M. A. In *Photoinduced Electron Transfer*, Part D (Fox, M. A.; Chanon, M., Eds.). Elsevier, Amsterdam, 1988, Chapter 6.1, pp. 241–302.
7. Juris, A.; Balzani, V.; Barigeletti, F.; Campagna, S.; Belser, P.; von Zelewsky, A. *Coord. Chem. Rev.* **1988**, *84*, 85.
8. Balzani, V., Ed. *Supramolecular Photochemistry*. Reidel, Dordrecht, The Netherlands, 1987.
9. Balzani, V.; Scandola, F. *Supramolecular Photochemistry*. Ellis Horwood, London, 1991.
10. Meyer, T. J. *Acc. Chem. Res.* **1989**, *22*, 163.
11. Scandola, F.; Indelli, M. T. In: *Photochemical Conversion and Storage of Solar Energy* (Pelizzetti, E.; Schiavello, M., Eds.). Kluwer Academic, Dordrecht, The Netherlands, 1991, p.1.
12. Denti, G.; Campagna, S.; Sabatino, L.; Serroni, S.; Ciano, M.; Balzani, V. In: *Photochemical Conversion and Storage of Solar Energy* (Pelizzetti, E.; Schiavello, M., Eds.). Kluwer Academic, Dordrecht, The Netherlands, 1991, p. 27.
13. Balzani, V. *J. Int. Rec. Mater.,* **1989**, *17*, 339.
14. Balzani, V. *J. Photochem. Photobiol., A: Chem.*, **1990**, *51*, 55.
15. Balzani, V.; Moggi, L. *Coord. Chem. Rev.*, **1990**, *97*, 313.
16. Scandola, F. In: *Photochemical Energy Conversion* (Norris, J. R.; Meisel, D., Eds.). Elsevier, Amsterdam, 1989, p. 60.
17. Kalyanasundaram, K.; Gratzel, M.; Pelizzetti, E. *Coord. Chem. Rev.* **1986**, *69*, 57.
18. Roundhill, D. M.; Gray, H. B.; Che, C.-M. *Acc. Chem. Res.* **1989**, *22*, 55.
19. Hill, C. L.; Bouchard, D. A.; Kadkhodayan, M.; Williamston, M. M.; Schmidt, J. A.; Hilinski, E. F. *J. Am. Chem. Soc.* **1988**, *110*, 5471.
20. Rodman, G. S.; Daws, C. A.; Mann, K. R. *Inorg. Chem.*, **1988**, *27*, 3347; Rodman, G. S.; Mann, K. R. *Inorg. Chem.* **1985**, *24*, 3507.

21. Partigianoni, C. M.; Chang, I-JY.; Nocera, D. G. *Coord. Chem. Rev.* **1990**, *97*, 105.
22. Somorjai, G. A. *Chemistry in Two Dimensions: Surfaces.* Cornell University Press, Ithaca, N.Y., 1981, Chapter 1.
23. Gratzel, M., Ed. *Energy Resources through Photochemistry and Photocatalysis. Academic Press, London, 1983.*
24. Willner, I.; Degani, Y. *J. Am. Chem. Soc.* **1983**, *105*, 6228.
25. Wheeler, J.; Thomas, J. K. *J. Phys. Chem.* **1982**, *86*, 4540.
26. Willner, I.; Otvos, J. W.; Calvin, M. J. *J. Am. Chem. Soc.* **1981**, *103*, 3203.
27. Thomas, J. K. *Chem. Rev.* **1980**, *80*, 283; *J. Phys. Chem.* **1987**, *91*, 267.
28. Matsuo, T.; Takuma, K.; Nishizima, T.; Tsutsui, Y. *J. Coord. Chem.* **1980**, *10*, 195.
29. Milosavijevic, B. H.; Thomas, J. K. *J. Phys. Chem.* **1983**, *87*, 616.
30. Gratzel, M. *Acc. Chem. Res.* **1981**, *14*, 376.
31. Vlachopoulos, N.; Liska, P.; Augustynski, J.; Gratzel, M. *J. Am. Chem. Soc.* **1988**, *110*, 1216.
32. Lehn, J. M.; Sauvage, J. P.; Ziessel, R. *Nouv. J. Chim.* **1980**, *4*, 623.
33. Clark, W. D.; Sutin, N. *J. Am. Chem. Soc.* **1977**, *99*, 4676.
34. Novak, A. J. In: *Photochemical Conversion and Storage of Solar Energy* (Connolly, J. S., Ed.). Academic Press, New York, 1981, p. 271; *Appl.Phys.Lett.* **1977**, *30*, 567.
35. Krueger, J. S.; Mayer, J. E.; Mallouk, T. E. *J. Am. Chem. Soc.* **1988**, *110*, 8232.
36. Li, Z.; Lai, C.; Mallouk, T. E. *Inorg. Chem.* **1989**, *28*, 178.
37. Gafney, H. D. *Coord. Chem. Rev.* **1990**, *104*, 113.
38. Scandola, F.; Indelli, M. T.; Chiorboli, C.; Bignozzi, C. A. *Top. Curr. Chem.* **1990**, *158*, 73.
39. Billing, R.; Rehorek, D.; Hennig, H. *Top. Curr. Chem.* **1990**, *158*, 151.
40. Vogler, A.; Kunkely, H. *Top. Curr. Chem.* **1990**, *158*, 1.
41. Balzani, V.; Barigeletti, F.; DeCola, L. *Top. Curr. Chem.* **1990**, *158*, 31.
42. Gratzel, M. *Heterogeneous Photochemical Electron Transfer.* CRC Press, Boca Raton, Florida, 1989.
43. Rillema, D. P.; Jones, D. S.; Levy, H. A. *J. Chem. Soc. Chem. Commun.* **1979**, 849.
44. (a) Serpone, N.; Pelizzetti, E., Eds. *Photocatalysis—Fundamentals and Applications*, Wiley, New York, 1989. (b) Pelizzetti, E.; Serpone, N., Eds. *Homogeneous and Heterogeneous Photocatalysis.* Reidel Publ., Dordrecht, The Netherlands, 1986.
45. Pankove, J. I. *Optical Processes in Semiconductors.* Dover Publ., New York, 1971.
46. Gratzel, M. in ref. 44a, Chapter 5, p. 123.
47. Lewis, N. S.; Rosenbluth, M. L. in ref. 44a, Chapter 3, p. 45.
48. Albery, W. J.; Bartlett, P. N. *J. Electrochem. Soc.* **1984**, *131*, 315.
49. Sutin, N. *Prog. Inorg. Chem.* **1983**, *30*, 441.
50. Rehm, D.; Weller, A. *Israel J. Chem.* **1970**, *8*, 259.
51. Rehm, D.; Weller, A. *Ber. Bunsenges. Phys. Chem.* **1969**, *73*, 834.
52. Marcus, R. A. *Annu. Rev. Phys. Chem.* **1964**, *15*, 155.
53. Hush, N. S. *Electrochim. Acta* **1968**, *13*, 1005.
54. Sutin, N. In *Inorganic Biochemistry* (Eichorn, G. L., Ed.). Elsevier, New York, 1979, p. 611.
55. Marcus, R. A.; Sutin, N. *Biochim. Biophys. Acta* **1985**, *811*, 265.
56. Balzani, V.; Bolletta, F.; Gandolfi, M. T.; Maestri, M. *Top. Curr. Chem.* **1978**, *75*, 1.
57. Chiorboli, C.; Indelli, M. T.; Rampi Scandola, M. A.; Scandola, F. *J. Phys. Chem.* **1988**, *92*, 156.
58. Scandola, F. In *Photocatalysis and Environment* (Schiavello, M., Ed.). Kluwer Academic, Dordrecht, The Netherlands, 1988, p. 3.
59. Serpone, N.; Pichat, P.; Herrmann, J.-M.; Pelizzetti, E. in ref. 8, p. 415.
60. Spanhel, L.; Weller, H.; Henglein, A. *J. Am. Chem. Soc.* **1987**, *109*, 6632.
61. Gerischer, H.; Luebke, M. *J. Electroanal. Chem. Interfacial Electrochem.* **1986**, *204*, 225.
62. Sobczynski, A.; White, J. M.; Bard, A. J.; Campion, A.; Fox, M. A.; Mallouk, T.; Webber, S. E. *J. Phys. Chem.* **1987**, *91*, 3316.
63. Gopidas, K. R.; Bohorquez, M.; Kamat, P. V. *J. Phys. Chem.* **1990**, *94*, 6435.

64. (a) Serpone, N.; Borgarello, E.; Gratzel, M. *J. Chem. Soc. Chem. Commun.* **1984**, 342; (b) Barbeni, M.; Pelizzetti, E.; Borgarello, E.; Serpone, N.; Gratzel, M.; Balducci, L.; Visca, M. *Int. J. Hydrogen Energy* **1985**, *10*, 249; (c) Borgarello, E.; Serpone, N.; Gratzel, M.; Pelizzetti, E. *Inorg. Chim. Acta* **1986**, *112*, 197.

65. Serpone, N.; Borgarello, E.; Pelizzetti, E.; Barbeni, M. *Chim. Ind. (Milano)* **1985**, *67*, 318.

66. Serpone, N.; Borgarello, E.; Pelizzetti, E. *J. Electrochem. Soc.* **1988**, *135*, 2760.

67. Pichat, P.; Borgarello, E.; Disdier, J.; Herrmann, J.-M.; Pelizzetti, E.; Serpone, N. *J. Chem. Soc. Faraday Trans. 1* **1988**, *84*, 261.

68. Rothenberger, G.; Moser, J.; Gratzel, M.; Serpone, N.; Sharma, D. K. *J. Am. Chem. Soc.* **1985**, *107*, 8054.

69. Moser, J.; Gallay, R.; Gratzel, M. *Helv. Chim. Acta* **1987**, *70*, 1596.

70. Albery, W. J.; Bartlett, P. N. *J. Electroanal. Chem.* **1982**, *57*, 139.

71. Chiorboli, C.; Indelli, M. T.; Rampi-Scandola, M. A.; Scandola, F. *J. Phys. Chem.* **1988**, *92*, 156.

72. Rampi-Scandola, M. A.; Scandola, F.; Indelli, A. *J. Chem. Soc. Faraday Trans. 1* **1985**, *81*, 2967.

73. Olson, A. R.; Simonson, T. R. *J. Phys. Chem.* **1949**, *17*, 1167.

74. Ohno, T.; Yoshimura, A.; Mataga, N. *J. Phys. Chem.* **1990**, *94*, 4871.

75. Ohno, T.; Yoshimura, A.; Mataga, N. *J. Phys. Chem.* **1086**, *90*, 3295.

76. Eigen, M. *Z. Phys. Chem. (Munich)* **1954**, *1*, 176.

77. Greiner, G.; Pasquini, P.; Weiland, R.; Orthwein, H.; Rau, H. *J. Photochem. Photobiol., A: Chem.* **1990**, *51*, 179.

78. Rau, H.; Frank, R.; Greiner, G. *J. Phys. Chem.* **1986**, *90*, 2476.

79. Rehm, D.; Weller, A. *Ber. Bunsenges. Phys. Chem.* **1969**, *73*, 834; *Israel J. Chem.* **1970**, *8*, 259.

80. Katal'nikov, I. V.; Kuz'min, V. A.; Levin, L. P.; Shafirovich, V. Ya. *Izvestiya Akademii Nauk.SSSR*, Seriya Khimicheskaya, 1989, No.1, pp. 176–178.

81. Kavarnos, G. J.; Turro, N. J. *Chem. Rev.* **1986**, *86*, 401.

82. (a) Kitamura, N.; Okano, S.; Tazuke, S. *Chem. Phys. Lett.* **1982**, *90*, 13; (b) Tazuke, S.; Kitamura, N. *Pure Appl. Chem.* **1984**, *56*, 1269; (c)Tazuke, S.; Kitamura, N.; Kawanishi, Y. *J. Photochem.* **1985**, *29*, 123.

83. Bock, C. R.; Connor, J. A.; Gutierrez, A. R.; Meyer, T. J.; Whitten, D. G.; Sullivan, B. P.; Nagle, J. K. *J. Am. Chem. Soc.* **1979**, *101*, 4815.

84. Kitamura, N.; Obata, R.; Kim, H.-B.; Tazuke, S. *J. Phys. Chem.* **1987**, *91*, 2033.

85. Bolletta, F.; Juris, A.; Maestri, M.; Sandrini, D. *Inorg. Chim. Acta* **1980**, *44*, L175.

86. Gorner, H.; Kuhn, H. J.; Schulte-Frohlinde, D. *EPA Newsletter* **1987**, *31*, 13.

87. Horvath, A.; Bako, Z.; Papp, S.; Keszei, C. S. *J. Photochem. Photobiol. A:Chem.* **1990**, *52*, 271.

88. Ferraudi, G.; Arguello, G. A. *J. Phys. Chem.* **1988**, *92*, 1846.

89. Yamazaki-Nishida, S.; Kimura, M. *Inorg. Chim. Acta* **1990**, *174*, 231.

90. Kimura, M.; Nakamura, T.; Nakamura, H.; Nishida, S. *Polyhedron* **1987**, *6*, 1571.

91. Kurimura, Y.; Hiraizumi, K-I.; Harakawa, T.; Yamashita, M.; Osada, Y.; Shigehara, K.; Yamada, A. *J. Chem. Soc. Faraday Trans., 1* **1990**, *86*, 609.

92. Amatore, C.; Saveant, J. M. *J. Am. Chem. Soc.* **1981**, *103*, 5021.

93. Lehn, J.-M.; Ziessel, R. *J. Organometal. Chem.* **1990**, *382*, 157.

94. Fukuzumi, S.; Mochizuki, S.; Tanaka, T. *J. Phys. Chem.* **1990**, *94*, 722.

95. Ohkubo, K.; Ishida, H.; Hamada, T.; Inaoka, T. *Chem. Lett.* **1989**, 1545.

96. Willner, I.; Tsfania, T.; Eichen, Y. *J. Org. Chem.* **1990**, *55*, 2656.

97. Gsponer, H. E. *J. Photochem. Photobiol. A: Chem.* **1990**, *55*, 233.

98. Rybak, W.; Haim, A.; Netzel, T. J.; Sutin, N. *J. Phys. Chem.* **1981**, *85*, 2856.

99. Ballardini, R.; Gandolfi, M. T.; Balzani, V. *Chem. Phys. Lett.* **1985**, *119*, 459.

100. Ballardini, R.; Gandolfi, M. T.; Balzani, V. *Inorg. Chem.*

101. Simic, M.; Lilie, J. *J. Am. Chem. Soc.* **1974**, *96*, 291.

102. Vogler, A.; Osman, A. H.; Kunkley, H. *Coord. Chem. Rev.* **1985**, *64*, 159.

103. Osman, A. H. *Inter- und intramolekulare Photoredoxreaktionen von Ubergangs-metallkomplexen*, Thesis, Universitat Regensburg, Regensburg, Germany, 1987.
104. Lilie, J.; Shinohara, N.; Simic, M. *J. Am. Chem. Soc.* **1976**, *98*, 6516.
105. Shinohara, N.; Lilie, J.; Simic, M. *Inorg. Chem.* **1977**, *16*, 2809.
106. Danielson, E.; Elliott, C. M.; Merkert, J. W.; Meyer, T. J. *J. Am. Chem. Soc.* **1987**, *109*, 2519.
107. Moore, T. A.; Gust, D.; Mathis, P.; Mialocq, J. C.; Chachaty, C.; Bensasson, R. V.; Land, E. J.; Doizi, D.; Lidell, P. A.; Lehman, W. R.; Nemeth, G. A.; Moore, A. L. *Nature (London)*, **1984**, *307*, 630.
108. Meyer, T. J. *Pure & Appl.Chem.* **1990**, *62*, 1003.
109. Meyer, T. J. In *Supramolecular Photochemistry* (Balzani, V., Ed.). Reidel Publ., Dordrecht, The Netherlands, 1987, pp. 103–120.
110. Chen, P.; Danielson, E.; Meyer, T. J. *J. Phys. Chem.* **1988**, *92*, 3708.
111. Chen, P.; Curry, M.; Meyer, T. J. *Inorg. Chem.* **1989**, *28*, 2271.
112. Chen, P.; Westmoreland, T. D.; Danielson, E.; Shanze, K. S.; Anthon, D.; Neveux, P. E., Jr.; Meyer, T. J. *Inorg. Chem.* **1987**, *26*, 1116.
113. Deusing, R.; Tapolsky, G.; Meyer, T. J. *J. Am. Chem. Soc.* **1990**, *112*, 5378.
114. Schmehl, R. H.; Ryu, C. K.; Elliott, C. M.; Headford, C. L. E.; Ferrere, S. *Adv. Chem. Series* **1990**, *226*, 211.
115. Duonghong, D.; Ramsden, J. J.; Gratzel, M. *J. Am. Chem. Soc.* **1982**, *104*, 2977.
116. Trudinger, P. A. *Anal. Biochem.* **1970**, *36*, 222.
117. Gratzel, M.; Frank, A. J. *J. Phys. Chem.* **1982**, *86*, 2964.
118. Serpone, N.; Sharma, D. K.; Jamieson, M. A.; Gratzel, M.; Ramsden, J. J. *Chem. Phys. Lett.* **1985**, *115*, 473.
119. Moser, J.; Gratzel, M. *J. Am. Chem. Soc.* **1983**, *105*, 6547.
120. Kirch, M.; Lehn, J. -M.; Sauvage, J. P. *Helv. Chim. Acta* **1979**, *62*, 1345.
121. Chan, S. F.; Chu, M.; Creutz, C.; Matsubara, T.; Sutin, N. *J. Am. Chem. Soc.* **1981**, *103*, 369.
122. Serpone, N.; Sharma, D. K.; Moser, J.; Gratzel, M. *Chem. Phys. Lett.* **1987**, *136*, 47.
123. Bahnemann, D. W.; Kormann, C.; Hoffmann, M. R. *J. Phys. Chem.* **1987**, *91*, 3789.
124. Moser, J.; Gratzel, M. *J. Am. Chem. Soc.* **1984**, *106*, 6557.
125. (a) Terenin, A.; Akimov, I. A. *J. Phys. Chem.* **1964**, *69*, 730. (b) Terenin, A.; Akimov, I. A. *Z. Phys. Chem.* **1961**, *217*, 307.
126. Kiwi, J. *Chem. Phys. Lett.* **1981**, *83*, 594.
127. Kamat, P. V.; Fox, M. A. *Chem. Phys. Lett.* **1983**, *102*, 379.
128. Moser, J.; Gratzel, M.; Sharma, D. K.; Serpone, N. *Helv. Chim. Acta* **1985**, *68*, 1686.
129. (a) Sutin, N. *Acc. Chem. Res.* **1982**, *15*, 275; (b) Marcus, R. A.; Siders, P. *J. Phys. Chem.* **1982**, *86*, 622.
130. Ryan, M. A.; Fitzgerald, E. C.; Spitler, M. T. *J. Phys. Chem.* **1989**, *93*, 6150.
131. Gopidas, K. R.; Kamat, P. V. *J. Phys. Chem.* **1989**, *93*, 6428.
132. Fessenden, R. W.; Kamat, P. V. *Chem. Phys. Lett.* **1986**, *123*, 233.
133. Highfield, J. G.; Gratzel, M. *J. Phys. Chem.* **1988**, *92*, 464.
134. Kamat, P. V. *J. Phys. Chem.* **1989**, *93*, 859.
135. Kamat, P. V.; Dimitrijevic, N. M. *Solar Energy* **1990**, *44*, 83.
136. Mulvaney, P.; Grieser, F.; Meisel, D. *Langmuir* **1990**, *6*, 567.
137. Duonghong, D.; Serpone, N.; Gratzel, M. *Helv. Chim. Acta* **1984**, *67*, 1012.
138. Vrachnou, E.; Vlachopoulos, N.; Gratzel, M. *J. Chem. Soc. Chem. Commun.* **1987**, 868.
139. Frei, H.; Fitzmaurice, D. J.; Gratzel, M. *Langmuir* **1990**, *6*, 198.
140. Siripala, W.; Tomkiewicz, M. *J. Electrochem. Soc.* **1982**, *129*, 1240.
141. Amadelli, R.; Argazzi, R.; Bignozzi, C. A.; Scandola, F. *J. Am. Chem. Soc.* **1990**, *112*, 7099.
142. Gratzel, M. In *Photochemical Energy Conversion* (Norris, J.; Meisel, D., Eds.). Elsevier, New York, 1989, pp. 284–296.

143. Morrison S. R. In *The Chemical Physics of Surfaces*. Plenum Press, New York, 1977, Chapter 9, p. 301.

144. Cunningham J. In *Comprehensive Chemical Kinetics* (Bamford, C. H.; Tipper, C. F. H., Eds.). Elsevier, Amsterdam, 1984, Vol. 19, Chapter 3, p. 291.

145. Bickley R. I. In *Photoelectrochemistry, Photocatalysis and Photoreactors* (Schiavello, M., Ed.). Reidel, Dordrecht, The Netherlands, 1985, pp. 379 and 491.

146. Pichat, P. In *Homogeneous and Heterogeneous Photocatalysis* (Pelizzetti, E.; Serpone, N., Eds.). Reidel, Dordrecht, The Netherlands, 1986, p. 533.

147. Pichat, P.; Herrmann, J. M. In *Photocatalysis, Fundamentals and Applications* (Serpone, N.; Pelizzetti, E., Eds.). Wiley, New York, 1989, Chapter 8, p. 217.

148. Bickley, R. I.; Stone, F. S. *J. Catal.* **1973**, *31*, 389.

149. Boonstra, A. H.; Mutsaers, C. A. H. A. *J. Phys. Chem.* **1975**, *79*, 1694.

150. Munuera, G.; Gonzalez, F. *Rev. Chim. Min.* **1967**, *4*, 207.

151. Courbon, H.; Herrmann, J. M.; Pichat, P. *J. Phys. Chem.* **1984**, *88*, 5210.

152. Mériaudeau, P.; Védrine, J. C. *J. Chem. Soc. Faraday Trans. 1* **1976**, *72*, 472.

153. Gonzalez-Elipe, A. R.; Munuera, G.; Soria, J. *J. Chem. Soc., Faraday Trans. 1* **1979**, *75*, 748.

154. Che, M.; Tench, A. J. *Adv. Catal.* **1982**, *31*, 77.

155. Che, M.; Tench, A. J. *Adv. Catal.* **1983**, *32*, 1.

156. Bourasseau, S.; Martin, J. R.; Juillet, F.; Teichner, S. J. *J. Chim. Phys.* **1973**, *70*, 1472.

157. Bourasseau, S.; Martin, J. R.; Juillet, F.; Teichner, S. J. *J. Chim. Phys.* **1974**, *71*, 122; **1974**, *71*, 1017.

158. Bickley, R. I.; Jayanty, R. K. M. *Disc. Faraday Soc.* **1974**, *58*, 194.

159. Munuera, G.; Rives-Arnau. V.; Saucedo, A. *J. Chem. Soc. Faraday Trans. 1* **1979**, *75*, 736.

160. Gonzalez-Elipe, A. R.; Munuera, G.; Sanz, J.; Soria, J. *J. Chem. Soc. Faraday Trans. 1* **1980**, *76*, 1535.

161. Solonitsyn, Yu. P. *Kinet. Katal.* **1966**, *7*, 424.

162. Volodin, A. M.; Zakharenko, V. S.; Cherkashin, A. E. *React. Kinet. Catal. Lett.* **1981**, *18*, 321.

163. Bube, R. H. In: *Photoconductivity of Solids*. Wiley, New York, 1960.

164. Rose, A. In *Concepts in Photoconductivity and Allied Problems*. Wiley, New York, 1963.

165. Heijne, L. *Philips Techn. Rev.* **1965**, *27*, 47.

166. Herrmann, J. M.; Disdier, J.; Pichat, P. *J. Chem. Soc., Faraday Trans. 1* **1981**, *77*, 2815.

167. Herrmann, J. M. In *Les Techniques Physiques d'étude des catalyseurs* (Imelik, B.; Védrine, J. C., Eds.). Editions Technip (Paris) 1988, Chapter 22, p. 753.

168. Herrmann, J. M.; Disdier, J.; Pichat, P. *Proc. 5th Intern. Vacuum Congr. and 3rd Intern. Conf. Solid Surfaces* (Dobrozenski, R., et al., Eds.). F. Berger und Söhne, Horn, Austria, 1977, Vol. II, p. 951.

169. Pichat, P.; Herrmann, J. M.; Courbon, H.; Disdier, J.; Mozzanega, M. N. *Canad. J. Chem. Eng.* **1982**, *60*, 27.

170. Mikheikin, D.; Mashenko, A. I.; Kazanskii, V. B. *Kinet. Katal.* **1967**, *8*, 1363.

171. Herrmann, J. M. (unpublished results).

172. Thévenet, A.; Juillet, F.; Teichner, S. J. *Jpn. J. Appl. Phys.* (suppl. 2) **1974**, *2*, 529.

173. Anpo, M.; Aikawa, N.; Kubokawa, Y. *J. Chem. Soc. Chem. Commun.* **1984**, 644.

174. Juillet, F.; Leconte, F.; Mozzanega, H.; Teichner, S. J.; Thévenet, A.; Vergnon, P. *Faraday Synp. Chem. Soc.* **1973**, *7*, 57.

175. Mozzanega, H.; Herrmann, J. M.; Pichat, P. *J. Phys. Chem.* **1979**, *83*, 2251.

176. Pichat, P.; Courbon, H.; Disdier, J.; Mozzanega, M. N.; Herrmann, J. M. In *New Horizons in Catalysis Proc. 7th Intern. Congr. Catal. Tokyo*, Studies in Surf. Sci. Catal. (Seiyama, T.; Tanabe, K., Eds.), 1981, Vol. 5A, part B, p. 1498.

177. Courbon, H.; Pichat, P. *J. Chem. Soc., Faraday Trans. 1* **1984**, *80*, 3175.

178. Herrmann, J. M.; Perrichon, V. unpublished results.

179. Nay, M. A.; Morrison, J. L. *Comm. J. Res.* **1949**, *27*, 205.

180. Herrmann, J. M.; Disdier, J.; Mozzanega, M. N.; Pichat, P. *J. Catal.* **1979**, *60*, 369.

181. Herrmann, J. M.; Courbon, H.; Disdier, J.; Mozzanega, M. N.; Pichat, P. In *New Developments in Selective Oxidation* (Centi, G.; Trifiro, F., Eds.). Studies Surf. Sci. Catal., 1990, Vol. 55, Elsevier, Amsterdam, p. 675.

182. Kraeutler, B.; Bard, A. J. *J. Am. Chem. Soc.* **1978**, *100*, 5985.

183. Kraeutler, B.; Jaeger, C. D.; Bard, A. J. *J. Am. Chem. Soc.* **1978**, *100*, 4903.

184. Kraeutler, B.; Bard, A. J. *Nouv. J. Chim.* **1979**, *3*, 31.

185. Izumi, I.; Fan, F. F.; Bard, A. J. *J. Phys. Chem.* **1981**, *85*, 218.

186. Izumi, I.; Dunn, W. W.; Wilbourn, K. O.; Bard, A. J. *J. Phys. Chem.* **1980**, *84*, 3207.

187. Dunn, W. W.; Aikawa, Y.; Bard, A. J. *J. Am. Chem. Soc.* **1981**, *103*, 6893.

188. Wrighton, M. S.; Wolczanski, P. T.; Ellis, A. B. *J. Solid State Chem.* **1977**, *22*, 17.

189. Hemminger, J. C.; Carr, R.; Somorjai, G. A. *Chem. Phys. Lett.* **1978**, *57*, 100.

190. Wagner, F. T.; Ferrer, S.; Somorjai, G. A. *Surf. Sci.* **1981**, *101*, 462.

191. Wagner, F. T.; Somorjai, G. A. *J. Am. Chem. Soc.* **1980**, *102*, 5944.

192. Carr, R. G.; Somorjai, G. A. *Nature (London)* **1981**, *290*, 576.

193. Lehn, J. M.; Sauvage, J. P.; Ziessel, R. *Nouv. J. Chim.* **1980**, *4*, 623.

194. Kawai, T.; Sakata, T. *J. Chem. Soc., Chem. Commun.* **1980**, 694; *Nature (London)* **1980**, *286*, 474; *Chem. Lett.* **1981**, 81.

195. Sakata, T.; Kawai, T. *Chem. Phys. Lett.* **1981**, *80*, 341; *Nouv. J. Chim.* **1980**, *5*, 279.

196. Kiwi, J.; Borgarello, E.; Pelizzetti, E.; Visca, M.; Grätzel, . *Angew. Chem., Int. Ed. Engl.* **1980**, *19*, 646.

197. Kalyanasundaram, K.; Borgarello, E.; Duonghong, D.; Grätzel, M. *Angew. Chem., Int. Ed. Engl.* **1981**, *20*, 987.

198. Borgarello, E.; Kiwi, J.; Pelizzetti, E.; Visca, M.; Grätzel, M. *J. Am. Chem. Soc.* **1981**, *103*, 6324.

199. Sato, S.; White, J. M. *Chem. Phys. Lett.* **1980**, *70*, 131 and **1980**, *72*, 83; *Ind. Eng. Chem., Process. Res. Dev.* **1980**, *19*, 542; *J. Am. Chem. Soc.* **1980**, *102*, 7206; *J. Phys. Chem.* **1981**, *85*, 592; *J. Catal.* **1981**, *69*, 128.

200. Kautek, W.; Gobrecht, J.; Gerischer, H. *Ber. Bunsenges. Phys. Chem.* **1980**, *84*, 1034.

201. Kogo, K.; Yoneyama, H.; Tamura, H. *J. Phys. Chem.* **1980**, *84*, 1705.

202. Darwent, J. R.; Porter, G. *J. Chem. Soc., Chem. Commun.* **1981**, 145.

203. Darwent, J. R. *J. Chem. Soc., Faraday Trans. 2* **1981**, *77*, 1703.

204. Reichman, B.; Byvik, C. E. *J. Phys. Chem.* **1981**, *85*, 2255.

205. Harbour, J. R.; Wolkow, R.; Hair, M. L. *J. Phys. Chem.* **1981**, *85*, 4026.

206. Pichat, P.; Herrmann, J. -M.; Disdier, J.; Courbon, H.; Mozzanega, M. -N. *Nouv. J. Chim.* **1981**, *5*, 627.

207. Courbon, H.; Herrmann, J. -M.; Pichat, P. *J. Catal.* **1981**, *72*, 129.

208. Disdier, J.; Herrmann, J. -M.; Pichat, P. *J. Chem. Soc., Faraday Trans., 1* **1983**, *79*, 651.

209. Dorling, T. A.; Lynch, B. W. J.; Moss, R. L. *J. Catal.* **1971**, *20*, 190.

210. Herrmann, J. -M.; Pichat, P. In: *Spillover of Adsorbed Species* (Pajonk, G,. M.; Teichner, S. J.; Germain, J. E., Eds.). Studies Surf. Sci. Catal. vol. 17, Elsevier, Amsterdam, 1983, p. 77.

211. Teratani, S.; Nakamiski, J.; Taya, K.; Tanaka, K. *Bull. Chem. Soc. Jpn.* **1982**, *55*, 1688.

212. Herrmann, J. -M.; Disdier, J.; Pichat, P. In: *Metal-Support and Metal-Additive Effects in Catalysis* (Imelik, B.; et al., Eds.). Studies Surf. Sci. Catal. vol. 11, Elsevier, Amsterdam, 1981, p. 27.

213. Herrmann, J. -M.; Pichat, P. *J. Catal.* **1982**, *78*, 425.

214. Sermon, P. A.; Bond, G. C. *Catal. Rev.* **1973**, *8*, 211.

215. Herrmann, J. -M.; Romaroson, E.; Guilleux, M. F.; Tempere, F. *Appl. Catal.* **1989**, *53*, 117.

216. Benderskii, V. A.; Zolovitskii, Ya. M.; Kogan, Ya. L.; Khidekel, M. L.; Shub, D. M. *Dokl. Akad. Nauk SSSR* **1975**, *222*, 606.

217. Pichat, P.; Mozzanega, M. N.; Disdier, J.; Herrmann, J. -M. *Nouv. J. Chim.* **1982**, *6*, 559.

218. Domen, K.; Naito, S.; Onishi, T.; Tamaru, K. *Chem. Lett.* **1982**, 555.

219. Muradov, N. Z.; Buzhutin, V. Y.; Bezugalya, A. G.; Rustamov, M. I. *J. Phys. Chem.* **1982**, *86*, 1082.

220. Borgarello, E.; Pelizzetti, E. *Chim. Ind. (Milano)* **1983**, *65*, 474.

221. Oosawa, Y. *Chem. Lett.* **1983**, 577.

222. Taniguchi, Y.; Yoneyama, H.; Tamura, H. *Chem. Lett.* **1983**, 269.

223. Prahov, L. T.; Disdier, J.; Herrmann, J. -M.; Pichat, P. *Int. J. Hydrogen Energy* **1984**, *9*, 397.

224. Matsumura, M.; Hiramoto, M.; Iehara, T,; Tsubomura, H. *J. Phys. Chem.* **1984**, *88*, 248.

225. Ait Ichou, I.; Formenti, M.; Teichner, S. J. In *Spillover of Adsorbed Species* (Pajonk, G. M.; Teichner, S. J.; Germain, J. E., Eds.). Elsevier, Amsterdam, 1983, p. 63; and in *Catalysis on the Energy Scene* (Kaliaguine, S.; Mahay, A., Eds.). Elsevier, Amsterdam, 1984, p. 297.

226. Ait Ichou, I.; Formenti, M.; Pommier, B.; Teichner, S. J. *J. Catal.* **1985**, *91*, 293.

227. Pichat, P.; Disdier, J.; Mozzanega, M. N.; Herrmann, J. -M. In *Proceedings, 8th International Congress on Catalysis*, Vol. III, Verlag Chemie, Weinheim, Germany, 1984, pp. 487–498.

228. Hussein, F. H.; Pattenden, G.; Rudham, R.; Russel, J. J. *Tetrahedron Lett.* **1984**, *25*, 3363.

229. Herrmann, J. -M.; Rumeau-Maillot-Gravelle, M.; Gravelle, P. C. *J. Catal.* **1987**, *104*, 136.

230. Burwell, Jr., R. L. *Acc. Chem. Res.* **1969**, *2*, 289.

231. Inoue, Y.; Herrmann, J. -M.; Schmidt, H.; Burwell, Jr., R. L.; Butt, J. B.; Cohen, J. B. *J. Catal.* **1978**, *53*, 401.

232. Courbon, H.; Herrmann, J. -M.; Pichat, P. *J. Catal.* **1985**, *95*, 539.

233. Herrmann, J. -M.; Courbon, H.; Pichat, P. *J. Catal.* **1987**, *108*, 426.

234. Sclafani, A.; Mozzanega, M. N.; Pichat, P. *J. Photochem. Photobiol. A. Chem.* **1991**, *59*, 181.

235. Courbon, H.; Herrmann, J. -M.; Pichat, P. *J. Phys. Chem.* **1984**, *88*, 5210.

236. Schindler, K. M.; Birkholz, M.; Kunst, M. *Chem. Phys. Lett.* **1990**, *173*, 513.

237. Highfield, J. G.; Pichat, P. *New J. Chem.* **1989**, *13*, 61.

238. Warman, J. -M.; de Haas, M. P.; Pichat, P.; Koster, T. P. M.; Van der Zouwen-Assink, E. A.; Mackor, A.; Cooper, R. *Radiat. Phys. Chem.* **1991**, *37*, 433.

239. Tauster, S. J.; Fung, S. C.; Garten, R. L. *J. Am. Chem. Soc.* **1978**, *100*, 170.

240. Tauster, S. J.; Fung, S. C. *J. Catal.* **1978**, *56*, 29.

241. Tauster, S. J.; Fung, S. C.; Baker, R. T. K.; Horsley, J. A. *Science* **1981**, *211*, 1121 and refs. therein.

242. Bond, G. C.; Burch, R. In: *Catalysis* (Bond, G. C.; Webb, G., Eds.). The Royal Society of Chemistry, London, 1983, Vol. 6, p. 27.

243. Tauster, S. J. In *Strong Metal-Support Interactions* (Baker, R. T. K.; Tauster, S. J.; Dumesic, J. A., Eds.), Chapter 1, American Chemical Society, Washington, D.C., 1986, p. 1.

244. Wanke, S. E.; Academiec, J.; Fiedorow, R. M. J. *Proc. 10th Ibero-American Symp. Catal.* **1986**, *1*, 107.

245. Burch, R. In *Hydrogen Effects in Catalysis* (Paal, Z.; Menon, P. G., Eds.). Marcel Dekker, New York, 1988, p. 347.

246. Haller, G. L.; Resasco, D. E. In *Advances in Catalysis* (Eley, D. D.; Selwood, P. W.; Weisz, P. B., Eds.). Academic Press, New York, **1989**, *36*, 173.

247. *Metal-Support and Metal-Additive Effects in Catalysis* (Imelik, B. et al., Eds.). Elsevier, Amsterdam, 1982.

248. *Strong Metal-Support Interactions* (Baker, R. T. K.; Tauster, S. J.; Dumesic, J. A., Eds.). ACS Symposium series, Vol. 298, American Chemical Society, Washington, D.C., 1986.

249. Wang, L.; Qias, G. W.; Ye, H. Q.; Kus, K. H.; Chu, Y. X. *Proc. 9th Int. Congr. Catal.*, Calgary, Canada, Vol. 3, 1988, p. 1253.

250. Logan, A. D.; Braunschweig, E. J.; Datye, A. K. *Langmuir* **1988**, *4*, 827.

251. Herrmann, J. -M. *J. Catal.* **1984**, *89*, 404.

252. Herrmann, J. -M. *J. Catal.* **1989**, *118*, 43.

253. Solymosi, F.; Tombacz, I.; Koszta, J. *J. Catal.* **1985**, *95*, 578.

254. Akubuiro, E. C.; Verykios, X. E. *J. Catal.* **1987**, *103*, 320 and **1988**, *113*, 106.

255. Yamamoto, N.; Tomomura, S.; Matsuoka, T.; Tsubomura, H. *J. Appl. Phys.* **1981**, *52*, 6227.

256. Aspnes, D. E.; Heller, A. *J. Phys. Chem.* **1983**, *87*, 4919.

257. Gerisher, H. *J. Phys. Chem.* **1984**, *88*, 6096.

258. Herrmann, J. -M. In *Strong-Metal-Support Interactions* (Baker, R. T. K.; Tauster, S. J.; Dumesic, J. A., Eds.). A.C.S. Symposium series, Vol. 298, 1986, American Chemical Society, Washington, D.C., p. 200.

259. Bahl, M. K.; Tsai, S. C.; Chung, Y. W. *Phys. Rev. B.* **1980**, *21*, 1344.

260. Kao, C. C.; Tsai, S. C.; Bahl, M. K.; Chund, Y. W.; Lo, W. J. *Surf. Sci.* **1980**, *95*, 1.

261. Pichat, P.; Disdier, J.; Herrmann, J. -M.; Vaudano, P. *Nouv. J. Chim.* **1986**, *10*, 545.

262. Mu, W.; Herrmann, J. -M.; Pichat, P. *Catal. Lett.* **1989**, *3*, 73.

263. Boreskov, G. K. *Adv. Catal.* **1964**, *15*, 285.

264. Novakova, J. *Catal. Rev.* **1970**, *4*, 77.

265. Dunn, W. W.; Bard, A. J. *Nouv. J. Chim.* **1981**, *5*, 651.

266. Clark, W. C.; Vondjidis, A. G. *J. Catal.* **1965**, *4*, 691.

267. Korsunovskii, G. A. *J. Phys. Chem.* **1965**, *39*, 1139.

268. Krauetler, B.; Bard, A. J. *J. Amer. Chem. Soc.* **1978**, *100*, 4317.

269. Sungbom, C.; Kawai, M.; Tanaka, K. *Bull. Chem. Soc. Jpn.* **1984**, *57*, 871.

270. Sato, S. *J. Catal.* **1985**, *92*, 11.

271. Nakamatsu, H.; Kawai, T.; Koreeda, A.; Kawai, S. *J. Chem. Soc., Faraday Trans. 1* **1985**, *82*, 527.

272. Koudelka, M.; Sanchez, J.; Augustynski, J. *J. Phys. Chem.* **1982**, *86*, 4277.

273. Herrmann, J. -M.; Disdier, J.; Pichat, P. *J. Phys. Chem.* **1986**, *90*, 6028.

274. Borgarello, E.; Serpone, N.; Emo, G.; Harris, R.; Pelizzetti, E.; Minero, C. *Inorg. Chem.* **1986**, *25*, 4499.

275. Herrmann, J. -M.; Disdier, J.; Pichat, P.; Leclercq, C. *Preparation of Catalysts IV* (Delmon, B.; Grange, P.; Jacobs, P. A.; Poncelet, G., Eds.). Elsevier, Amsterdam, 1987, p. 285.

276. Möllers, F.; Tolle, H. J.; Memming, R. *J. Electrochem. Soc.* **1974**, *121*, 1160.

277. Kelly, J. J.; Vondeling, J. K. *J. Electrochem. Soc.* **1975**, *122*, 1103.

278. Stadler, K. H.; Boehm, H. P. *Proc. 8th Intern. Congr. Catal.*, Berlin, Verlag Chemie, Weinheim, Germany, Vol. IV, 1984, p. 803.

279. Jacobs, J. W. M. *J. Phys. Chem.* **1986**, *90*, 6507.

280. Jacobs, J. W. M.; Schryvers, D. *J. Catal.* **1987**, *103*, 436.

281. White, J. R.; O'Sullivan, E. J. M. *J. Electrochem. Soc.* **1987**, *134*, 1133.

282. Yoneyama, H.; Nishimura, N.; Tamura, H. *J. Phys. Chem.* **1981**, *85*, 268.

283. Borgarello, E.; Harris, R.; Serpone, N. *Nouv. J. Chim.* **1985**, *9*, 743.

284. Serpone, N.; Borgarello, E.; Barbeni, M.; Pelizzetti, E.; Pichat, P.; Herrmann, J. M.; Fox, M. A. *J. Photochem.* **1987**, *36*, 373.

285. Fleischauer, P. D.; Alan Kan, H. K.; Shepherd, J. R. *J. Amer. Chem. Soc.* **1972**, *94*, 283.

286. Hada, H.; Tanemura, H.; Yonezawa, Y. *Bull. Chem. Soc. Jpn.* **1978**, *51*, 3154.

287. Hada, H.; Yonezawa, Y.; Saikawa, M. *Bull. Chem. Soc. Jpn.* **1982**, *55*, 2010.

288. (a) Hada, H.; Yonezawa, Y.; Ishino, M.; Tanemura, H. *J. Chem. Soc., Faraday Trans. 1* **1982**, *78*, 2677; (b) Yonezawa, Y.; Kawai, K.; Okai, M.; Nakagawa, K.; Hada, H. *J. Imaging Sci.* **1986**, *30*, 114.

289. Nishimoto, S.; Ohtani, B.; Kajiwara, H.; Kagiya, T. *J. Chem. Soc. Faraday Trans. 1* **1983**, *79*, 2685.

290. Ohtani, B.; Okugawa, Y.; Nishimoto, S.; Kagiya, T. *J. Phys. Chem.* **1987**, *91*, 3550.

291. Herrmann, J. -M.; Disdier, J.; Pichat, P. *J. Catal.* **1988**, *113*, 72.

292. Kobayaski, T.; Taniguchi, Y.; Yoneyama, H.; Tamura, H. *J. Phys. Chem.* **1983**, *87*, 768.

293. Domenech, J.; Curran, J.; Jaffrezic-Renault, N.; Philippe, R. *J. Chem. Research* **1986**, *(5)*, 226.

294. Rufus, B.; Ramakrishnan, V.; Viswanathan, B.; Kuriacose, J. C. *Ind. J. Technol.* **1989**, *27*, 171.

295. Serpone, N.; Borgarello, E.; Pelizzetti, E. In *Photocatalysis and Environment* (Schiavello, M., Ed.). NATO ASI Series, Series C, Vol. 237, Kluwer Academic, Dordrecht, The Netherlands, 1988, p. 527.

296. Van Damme, H.; Hall, W. K. *J. Amer. Chem. Soc.* **1979**, *101*, 437.

297. Herrmann, J. -M.; Disdier, J.; Pichat, P.; Fernandez, A.; Gonzalez-Elipe, A.; Munuera, G. *J. Catal.* **1991**, *132*, 490.
298. Oster, G.; Yamamoto, M. *J. Phys. Chem.* **1966**, *70*, 3033.
299. Cléchet, P.; Martelet, C.; Martin, J.R.; Olier, R. *C.R. Acad. Sci. (Ser. C)* **1978**, *287*, 405.
300. Serpone, N.; Ah-You, Y. K.; Tran, T. P.; Harris, R.; Pelizzetti, E.; Hidaka, H. *Solar Energy* **1987**, *39*, 491.
301. Domenech, J.; Andrés, M. *Gazz. Chim. Ital.* **1987**, *117*, 495.
302. Tanaka, K.; Harada, K.; Murata, S. *Solar Energy* **1986**, *36*, 159.
303. Spanhel, L.; Weller, H.; Henglein, A. *J. Amer. Chem. Soc.* **1987**, *109*, 6632.
304. Spanhel, L.; Weller, H.; Fojtik, A.; Henglein, A. *Ber. Bunsenges Phys. Chem.* **1987**, *91*, 68.
305. Yoshimiva, J.; Kudo, A.; Tanaka, A.; Domen, K.; Maruya, K.; Onishi, T. *Chem. Phys. Lett.* **1988**, *147*, 401.
306. Rabani, J. *J. Phys. Chem.* **1989**, *93*, 7707.
307. Henglein, A.; Gutierrez, M.; Weller, H.; Fojtik, A.; Jirkovsky, J. *Ber. Bunsenges Phys. Chem.* **1989**, *93*, 593.
308. Gruzdhov, Y. A.; Savinov, E. N.; Korolkov, V. V.; Parmon, V. N. *React. Kinet. Katal. Lett.* **1988**, *36*, 395.
309. Ovenston, A.; Walls, J. R. *J. Phys. D* **1985**, *18*, 1859 and references therein.
310. Herrmann, J. M.; Disdier, J.; Pichat, P. *J. Chem. Soc., Faraday Trans 1* **1981**, *77*, 2815.
311. Meissner, D.; Memming, R.; Kastening, B.; Bahnemann, D. *Chem. Phys. Lett.* **1986**, *127*, 419 and references therein.
312. Serpone, N.; Sharma, D. K.; Jannieson, M. A.; Ramsden, J. R.; Graetzel, M. *Chem. Phys. Lett.* **1985**, *115*, 473.
313. Serpone, N.; Borgarello, E.; Barbeni, M.; Pelizzetti, E. *Inorg. Chim. Acta* **1984**, *90*, 191.
314. Pelizzetti, E.; Borgarello, E.; Serpone, N. In *Photoelectrochemistry, Photocatalysis and Photoreactors* (Schiavello, M., Ed.). D. Reidel Publ. Co, Dordrecht, The Netherlands, 1985, p. 293.
315. Borgarello, E.; Serpone, N.; Grätzel, M.; Pelizzetti, E. *Int. J. Hydrogen Energy* **1985**, *10*, 737.
316. Borgarello, E.; Serpone, N.; Liska, P.; Erbs, W.; Graetzel, M.; Pelizzetti, E. *Gazz. Chim. Ital.* **1985**, *115*, 599.
317. Sobczynski, A.; Bard, A. J.; Champion, A.; Fox, M. A.; Mallouk, T.; Webber, S. E.; White, J. M. *J. Phys. Chem.* **1987**, *91*, 3316.
318. Herrmann, J. -M.; Pichat, P. In *Spillover of Adsorbed Species* (Pajonk, G. M.; Teichner, S. J.; Germain, J. E., Eds.). Studies in Surf. Sci. Catal., Elsevier, Amsterdam, vol. 17, 1983, p. 77.
319. Herrmann, J. M. *Proc. 12th Ibero-American Symp. Catal., Inst. Braz. Petrol. ed.* Rio de Janeiro, 1990, vol. 3, p. 414.
320. Pichat, P.; Mozzanega, M. N.; Hoang-Van, C. *J. Phys. Chem.* **1988**, *92*, 467.
321. Pichat, P.; Disdier, J.; Herrmann, J. M.; Mozzanega, M. N.; Hoang-Van, C. *Proc. Symp. Photoelectrochemistry and Electrosynthesis on semiconducting Materials* (Ginley, D. S.; Nozik, A.; Armstrong, N.; Honda, K.; Fujishima, A.; Tanaka, T., Eds.). The Electrochemical Society, Pennington, N.J., vol. 88–14, 1988, p. 50.
322. Fripiat, J. J. In *Surface Properties and Catalysis by Non-metals* (Bonnelle, J. P., et al., Eds.). D. Reidel Publ. Co., 1983, p. 477.

INTERNAL GEOMETRY RELAXATION EFFECTS ON ELECTRON TRANSFER RATES OF AMINO CENTERED SYSTEMS

Stephen F. Nelsen

Advances in Electron Transfer Chemistry
Volume 3, pages 167–189.
Copyright © 1993 by JAI Press Inc.
All rights of reproduction in any form reserved.
ISBN: 1-55938-320-8

1. INTRODUCTION

Any discussion of electron transfer (ET) rates must start with the Marcus theory, for which many excellent reviews are available; we shall only list a few of the more recent ones.[1-4] We simplify this introduction by considering "self"-ET between a neutral molecule $M°$ and its own radical cation M^+ (Eq. 1), so that $\Delta G°$ for the reaction is zero, and work terms do not need to be considered. We shall designate the observed rate constant for this process as k_{ex}. The basic assumption of the Marcus theory is that the activation barrier for self-ET, ΔG^*, is related to the free energy difference λ (also designated E_λ) between the relaxed, individually solvated precursor pair and its "vertical" electron transfer product pair in which neither the solvent surrounding these species nor the internal geometries of $M°$ or M^+ are allowed to relax, by the simple factor of four appearing in Eq. 2. This beautifully simple assumption arises because $\lambda/4$ is the point where identical parabolas representing the initial pair ($\underline{M}°$, M^+) and the relaxed final product pair ($\underline{M}^+$, $M°$) cross when they are displaced along the reaction coordinate in a free energy diagram (see Figure 1). This proportionality factor is rather insensitive to distortions from perfect parabolic wells, and calculated wells arising from m.o. calculations not constrained to being parabolas cross within a few percent of the idealized factor of 0.25.[5] For crossing from the precursor surface to the product surface to occur, there must be an avoided crossing of the precursor and product-pair energy surfaces, representing electronic interaction between the components at the transition state. The separation of the energy surfaces at the transition state is usually designated $2H_{AB}$,[6] and for Eq. 2 to hold it is necessary that $H_{AB} < \lambda$. Equation 2 also only applies to "outer sphere" ET reactions. If bonded intermediates are present between the initial and product pairs, as in ligand transfer ("inner sphere") ET reactions in transition metal systems, k_{ex} can be much larger.[1-4]

$$\underline{M}° + M^+ \rightleftharpoons \underline{M}^+ + M° \qquad (1)$$

$$\Delta G^* = \lambda/4 \qquad (2)$$

Another fundamental assumption of the Marcus theory is that the vertical ET free energy λ may be well approximated by additive terms for *outer* shell (solvent) reorganization and *inner* shell (internal geometry, or vibrational) reorganization, Eq. 3. Exceedingly low self-ET k_{ex} values of under 20 $M^{-1}s^{-1}$ are observed for the transition metal complexes $(H_2O)_6Fe^{2+/3+}$, $(bpy)_3Co^{2+/3+}$, and $(NH_3)_6Co^{2+/3+}$ (see Table 2 in ref. 1). Oxidation for these systems is accompanied by large-metal, ligand-bond length changes (0.14, 0.19, and 0.22 Å respectively), leading to unusually large values for estimated $\Delta G^*_{in} = \lambda_{in}/4$ (8.4, 13.7, and 17.6 kcal/mole respectively). These systems were important in establishing the importance of λ_{in} in determining k_{ex}. In contrast, organic $M°/M^+$ self-ET pairs for which rates had been measured (requiring M^+ to be reasonably stable in the presence of $M°$)

exhibited much larger values of k_{ex}, exceeding $10^8 M^{-1}s^{-1}$ for the cases in Eberson's review.[7] These examples were all characterized by the M^+ charge being delocalized over rather large π systems, leading to rather small geometry changes between neutral and cation radical structures, and small expected λ_{in} values. Rate constants in the range observed, 1 to 2 orders of magnitude below diffusion control, were expected from λ_{out} calculated using the simple approximations of the classical Marcus theory, and it was generally felt that λ_{in} was negligible for such compounds.

$$\lambda = \lambda_{out} + \lambda_{in} \tag{3}$$

2. ESTIMATION OF λ_{in} VALUES

This chapter focuses on ET reactions of amino nitrogen compounds, where clear examples of high λ_{in} ET are frequent.[8] The estimation of λ_{in} has traditionally relied upon summing terms calculated from force constants for vibrations obtained from infrared (IR) spectra, and bond-length displacements between oxidation levels obtained from X-ray crystallography or calculations.[1-4] Large molecules rarely have complete assignments available for their IR spectra, and it is not obvious to us how much of λ_{in} will be obtained by choosing a few bands for analysis. Pyramidalization motions involving many atoms are especially important for amino nitrogen compounds.

An independent way of proceeding for estimation of λ_{in} involves examination of gas phase ionization processes.[5] A traditional Marcus twin parabola diagram, here restricted to internal reorganization components of a self-ET reaction, appears as Figure 1. λ then represents the free energy difference (or perhaps only the enthalpy difference?)[9] between relaxed neutral and radical cation before they have begun to approach the ET transition state (at the bottom of the parabola) and the pair obtained after electron transfer without allowing any relaxation, which is the vertical energy gap to the product parabola. The molecule in its relaxed neutral geometry is represented by $\mathbf{n}$, the relaxed radical cation geometry is represented by $\mathbf{c}$, and the charge present is shown as a superscript.[10] Because neither $\mathbf{n}^o$ nor $\mathbf{c}^+$ has begun the geometry changes which allow ET to occur, the energies of isolated molecules can be used for estimation of λ'_{in}, the enthalpy portion of λ_{in}.[9b] The same species as in Figure 1 are shown in a conventional energy diagram in Figure 2. From Figure 1, λ'_{in} is given by Eq. 4, but as Figure 2 shows, λ'_{in} may also be written as the sum of the relaxation enthalpies of $\mathbf{c}^o$ and $\mathbf{n}^+$ after they are formed by vertical electron transfers from the other oxidation states (Eq. 5). Although we know of no experimental way of determining $\Delta H_{rel}(\mathbf{n}) = \Delta H_f(\mathbf{c}^o) - \Delta H_f(\mathbf{n}^o)$, $\Delta H_{rel}(\mathbf{c}) = \Delta H_f(\mathbf{n}^+) - \Delta H_f(\mathbf{c}^+)$ is the difference between the vertical ionization potential (vIP, available from photoelectron spectroscopy measurements) and the adiabatic ionization potential, aIP. aIP is available directly from PE spectra only if $\Delta H_{rel}(\mathbf{c})$ is rather small, so there is enough overlap of the wavefunctions for $\mathbf{n}^o$ and $\mathbf{c}^+$ to produce observable

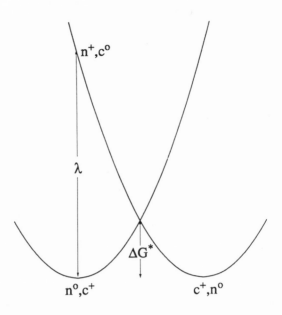

Figure 1. Marcus parabolas for a self-ET reaction.

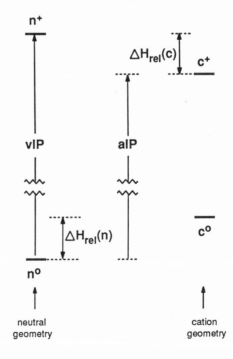

Figure 2. Energy diagram for the isolated components of a self-ET reaction.

intensity. For large $\Delta H_{rel}(\mathbf{c})$ cases, gas-phase equilibration of radical ions may be used. Molecular orbital calculations obviously produce estimated values for ΔH_{rel} as well as geometries and force constants, but one always is using approximations in such calculations, and it is necessary to establish how good a job the calculations used are doing on both oxidation states of a given class of molecules before relying heavily upon the numerical results of such calculations.

$$\lambda'_{in} = \{\Delta H_f(\mathbf{n}^+) + \Delta H_f(\mathbf{c}^o)\} - \{\Delta H_f(\mathbf{n}^o) + \Delta H_f(\mathbf{c}^+)\} \tag{4}$$

$$\lambda'_{in} = \Delta H_{rel}(\mathbf{c}) + \Delta H_{rel}(\mathbf{n}) \tag{5}$$

3. MARCUS AND EYRING ACTIVATION BARRIERS

A source of confusion to us in comparing theoretical and observed ET rate constants has been the distinction between the ΔG^* barrier used in theoretical treatments and the Eyring $\Delta G^{\ddagger}$ barrier usually reported for experimentally determined rate constants. They are quite different in size. Marcus and Sutin have pointed out that switching from the $k = Z\exp(-\Delta G^*/RT)$ rate expression of classical Marcus theory (where Z was usually taken as 1×10^{11} at 25 °C, and k represents the ET rate constant within the precursor complex, in which the components have approached, but not yet started the geometry changes which lead to the transition state) to the Eyring expression of $k_{obs} = [k_bT/h]\exp(-\Delta G^{\ddagger}/RT)$ employed for experimental data leads to the following relationships:[11]

$$\Delta G^{\ddagger} = \Delta G^* - RT \ln(hZ/k_bT) \tag{6}$$

$$\Delta H^{\ddagger} = \Delta H^* - RT/2 \tag{7}$$

$$\Delta S^{\ddagger} = \Delta S^* - R/2 + R \ln(hZ/k_bT) \tag{8}$$

Because of the difference in the temperature dependence of the preexponential terms, and the $(T)^{1/2}$ dependence of Z, the differences in these activation parameters depend slightly on temperature (Table 1):

Table 1. Relationship between Eyring and Marcus Activation Parameters

at T(°C)	−75	−50	−25	0	25	50	75
using $10^{-11}Z=$	0.81	0.87	0.95	0.96	1.00	1.05	1.08
$\Delta G^{\ddagger} - \Delta G^*$, kcal mole^{-1}	1.55	1.77	1.99	2.22	2.45	2.68	2.91
$\Delta H^{\ddagger} - \Delta H^*$, kcal mole^{-1}	−0.20	−0.22	−0.25	−0.27	−0.30	−0.32	−0.35
$\Delta S^{\ddagger} - \Delta S^*$, cal mole^{-1} K^{-1}	−8.79	−8.91	−9.02	−9.11	−9.20	9.28	−9.35

It will be noted that at 25 °C, $\Delta G^{\ddagger}$ is 2.45 kcal/mole greater than ΔG^*, so $\Delta G^{\ddagger}$ is about twice ΔG^* for an ET with k_{ex} of 1.2×10^9 at room temperature. This difference arises from the different apportionment between entropy and the preexponential factor. The enthalpy is usually not known from experiment to the accuracy necessary to distinguish $\Delta H^{\ddagger}$ from ΔH^*. The Eyring equation assumes a "normal" barrier crossing frequency of $k_b T/h = 6.21 \times 10^{12}$ s^{-1} at 25 °C, and places all other factors influencing the observed rate constant into $\Delta S^{\ddagger}$, and hence into $\Delta G^{\ddagger}$. A principal difference between $\Delta G^{\ddagger}$ and ΔG^* is that the entropy cost for M^0 finding M^+ in solution appears in $\Delta G^{\ddagger}$, but has not been incorporated into ΔG^*. All more modern ET theories include for intermolecular ET a precursor complex formation constant K_A as a preexponential multiplier. Sutin points out that K_A is not usually experimentally measurable, and gives several methods of estimating it (ref. 2). Weaver uses K_A in the range 0.25–0.5 for his metallocene self-ET work,[12] corresponding to a ΔS increment of –2.8 to –1.4 cal mole^{-1} K^{-1} in comparing Eyring and Marcus barriers. The entropy cost of bringing two spheres about the size of the molecules under consideration into contact is larger; using Eq. 9 of ref. 13b with the 5.46 Å centers distance that the authors use for the ET transition state of $1^{0/+}$ produces a $\Delta S^{\ddagger}$ increment of –7.3 cal mole^{-1} K^{-1}, close to the observed value. However, this ignores the desolvation which must occur to allow the components to touch, and how to calculate this properly remains unclear. More modern theoretical treatments employ various preexponential transmission coefficients and frequencies, depending upon their sophistication and detail.[1–3] In comparing theoretical expressions with experiment, each change in preexponential term transfers an increment from ΔG^* to the preexponential term, and we find it difficult to focus on how the concept of ΔG^* is supposed to vary as the preexponential term is varied in different theories for ET.[14,15]

4. N,N,N'N'-TETRAMETHYL-p-PHENYLENEDIAMINE SELF-ET

The most thoroughly studied organic self-ET case is that for N,N,N'N'-tetramethyl-p-phenylene diamine, $1^{0/+}$. The radical cation, Wurster's blue, is one of the first cation radicals to be isolated,[16] and X-ray structures are available.[17] Grampp and Jaenicke have recently reported detailed VT-ESR studies over a range of temperatures and solvents for several p-phenylenediamine derivatives,[13] obtaining for $1^{0/+}$ in CH$_3$CN at 25 °C a k_{ex} value of 1.47×10^9 M^{-1}s^{-1} ($\Delta G^{\ddagger} = 4.95$ kcal mole^{-1}), with $\Delta H^{\ddagger} = 2.63 \pm 0.07$ kcal mole^{-1}, and $\Delta S^{\ddagger} = -7.6 \pm 0.1$ cal mole^{-1} K^{-1}.[18] The rate constant is high enough that the observed numbers were corrected in the usual way, employing bulk solvent viscosity.[13] They found that *p*-phenylenediamines follow the "classical" Marcus theory expectation for solvent effects well.[13a] Plots of $\ln(k_{ex})$ versus the solvent parameter γ, which depends only upon bulk properties of pure solvent (Eq. 9) are linear, as expected from the classical equation for λ_{out}, given in Eq. 10. In Eq. 9, n is the refractive index at the sodium D line and ε the static dielectric constant.[19] In Eq. 10 $g(r,d)$ is a distance parameter, to be used with distances in Å to give λ_{out} in kcal mole^{-1} for the constant given.[13] An unusually wide range in γ (0.27–0.53) was available for the 20 °C $1^{o/+}$ data, allowing extrapolation to $\gamma = 0$, where λ_{out} disappears according to Eq. 10. Separation the experimental $\Delta G^*(CH_3CN)$ value of 2.59 kcal mole^{-1} resulted in ΔG^*_{out} and ΔG^*_{in} terms of 2.10 and 0.49 kcal mole^{-1} respectively, so the barrier is 81% determined by the solvent reorganization barrier in acetonitrile, in agreement with the idea that large π system organic molecules have ET barriers determined by solvent reorganization. However, it seems unlikely to us that this ET reaction is actually λ_{out}-dominated. In later papers,[13c,d] Grampp and Jaenicke estimate from force constants and bond length changes from crystal structures that $\Delta G^*_{in}(25 °C)$ is actually far higher, about 1.6 kcal mole^{-1}, or 62% of ΔG^*, when a tunneling correction to the λ_{in}^{∞} value of 11.3 kcal/mole is included. In their most recent analysis Grampp and Jaenicke[13d] conclude that TMPD ΔG^*_{out} and $\Delta G^*_{in}(CH_3CN)$ in are comparable in size and sum up to about 3.3 kcal/mole (depending upon the preexponential factor used), and that self-ET is strongly diabatic, with H_{ab} values under 0.2 kcal/mole.[13d] Fawcett and Foss make slightly different assumptions with the same data, and use ΔG^* at 2.6 kcal/mole.[20] These estimates assume the numerical accuracy of Eq. 10 and of Grampp and Jaenicke's estimate of λ_{in}^{∞}. It is not obvious to us that ΔG^* is actually that low. We have noted that semiempirical AM1 calculations based on Eq. 4 give a significantly higher value for ΔH^*_{in}; the correct AM1 prediction is 4.7 kcal mole^{-1},[21] which would significantly raise ΔG^* and change the conclusion that TMPD undergoes strongly diabatic ET.

$$\gamma = 1/n^2 - 1/\varepsilon \tag{9}$$

$$\lambda_{out} = 332.1 \, g(r,d) \, \gamma \tag{10}$$

5. TRIALKYLAMINE SELF-ET

Trialkylamine radical cations have the smallest π system possible, but studies of their self-ET to gain insight into ET theory have been precluded by two major factors. First, simple trialkylamine radical cations are not stable in the presence of the neutral amines because they rapidly undergo hydrogen transfer to produce protonated amines and very labile α-amino radicals. Although this has usually been described as a proton transfer reaction, it has been pointed out that hydrogen abstraction from the α position of neutral amine by the radical cation might be too fast for the proton transfer to compete kinetically in some cases; both reactions give the same products.[22] Dinnocenzo and Banach have directly studied the kinetics of deprotonation of dianisylmethylamine radical cation by other tertiary amines.[23] The second major difficulty proves more serious. Even if hydrogen transfer is inhibited, a bonded intermediate exists in many cases, so most amine self-ET reactions would still not be examples of outer shell electron transfer. Alder and co-workers have studied several examples of medium ring bridgehead diamines which give colored radical cations which they have shown to be 3e-σ bonded, with an electron in the $\sigma^*(NN)$ antibond.[24] X-ray structures for the tris-$(CH_2)_4$ bridged systems **2** have shown that the N,N distance in the radical cation is 2.30 Å, that the neutral species has inward pyramidalized nitrogens with a 0.31 Å larger N,N distance, and that the dication, with a 2-electron σ bond, has a 0.77 Å smaller N,N distance than the radical cation.[25] With such large geometry changes, λ_{in} must be huge, and Alder and Sessions found that the exothermic electron transfer **2°** to **2²⁺** to produce **2⁺** is remarkably slow, with a rate constant of 0.2 M^{-1} s^{-1} at 25 °C in acetonitrile.[26] The 5-msec lifetime of the less constrained radical cation **3⁺** produced from the bicyclic dication **3²⁺** led to an estimate of about 14.5 kcal/mole for the free energy of breaking its 3e-σ bond. This value falls in the general range of estimates obtained from proton affinites of mono- and bis-bridgehead diamines,[27] although an estimate of the maximum possible bond strength for **2⁺** based on its absorption maximum is 31 kcal/mole.[24] Dinnocenzo and Banach prepared the first 3e-$\sigma(NN)$ bonded radical cation not stabilized by being in a ring, **4⁺**, by $O_2^{•+}$ oxidation of quinuclidine.[28] They failed to observe similar dimer radical cations from several other amines, and suggested that **4⁺** was unique in having an intermolecular 3e-$\sigma(NN)$ bond with a positive bond energy because the nitrogen of quinuclidine cation radical was forced to be pyramidal. Alder suggested that the reason for **4⁺** formation was kinetic rather than thermodynamic,[24] which we believe has been verified experimentally by Gębicki and co-workers' observation of the formation of 3e-σ bonded cations from $R_2N(CH_2)_nNR_2$ (n = 1–3) ionized by pulse radiolysis in methylcyclohexane matrices at 90 K.[29] These radical cations do not have structural constraints forcing the localized radical cation to be nonplanar, but the n = 2 and 3 species rearrange to 3e-$\sigma(NN)$-bonded structures as bond rotations occur to allow approach of the nitrogens. Although bicyclic 3e-$\sigma(NN)$ bonded radical cations from **3²⁺**, **5²⁺**, and other nonbridgehead hexaalkyldiamine dications can be observed by pulse radiolytic reduction at room temperature, oxidation of the neutral compounds does not produce these species, apparently because rapid hydrogen transfers in localized cations intervene.[30] From the low-temperature radiolytic work,[29] conformational changes for certain molecules are apparently enough faster than hydrogen transfers at low temperature to allow formation of the 3e-σ bonded species.

6. TETRAALKYLHYDRAZINE SELF-ET

We have previously reviewed various aspects of the hydrazine, hydrazine radical cation electron transfer equilibrium,[31] and will not discuss much of this material here. Neutral tetraalkylhydrazines are characterized by strongly pyramidal nitrogens and a rather weak conformational preference for a 90° lone pair, lone-pair dihedral angle θ. Conformational changes of hydrazines show a significant dependence of the barrier for inversion at one nitrogen on the configuration at the noninverting nitrogen.[32] For example, the barrier for inversion at the equatorially methylated nitrogen in the axial, equatorial conformation of 1,2-dimethylhexahydropyridazine, **6ae**, which produces the diaxial conformation **6aa** through a "non-passing" transition state with θ ≅ 90° is 7.6 kcal/mole (−100 °C). In contrast the inversion of the axially methylated nitrogen to produce the diequatorial conformation **6ee** through a "passing" transition state with θ ≅ 0° has an estimated barrier of 12.7 kcal/mole (+2 °C). The principal cause of the 5 kcal/mole difference in inversion barrier is electronic, arising from the difference in transition state θ value.[32b] Hydrazine radical cations have substantially different geometries about the nitrogens than their neutral forms. The nitrogens are considerably flattened, and there is a strong electronic preference for coplanar lone pairs (θ = 0 or 180°), tending to eclipse the alkyl substituents.[33] This large geometry change leads to unusual thermodynamics for the hydrazine, hydrazine radical cation ET couple, characterized by an unusually large range in formal potential (E°′ value) and extremely poor correlations of solution E°′ with gas-phase vIP. The latter arises because E°′ is a relative measure of the free energy for adiabatic ionization in solution, and there is an unusually large effect of changing alkyl substituents on vIP - aIP.[31] All hydrazines show significantly lower E°′ in plots of E°′ versus vIP than do fused-ring aromatic hydrocarbons and alkylated benzenes, which exhibit nearly linear plots. The deviations observed correlate at least roughly with vIP - aIP. The large geometry change between the two oxidation states leads to very slow electron transfer even in intramolecular cases. Bis-hydrazine **7**, in which the nitrogens of the two hydrazine units are held in proximity by two bridging

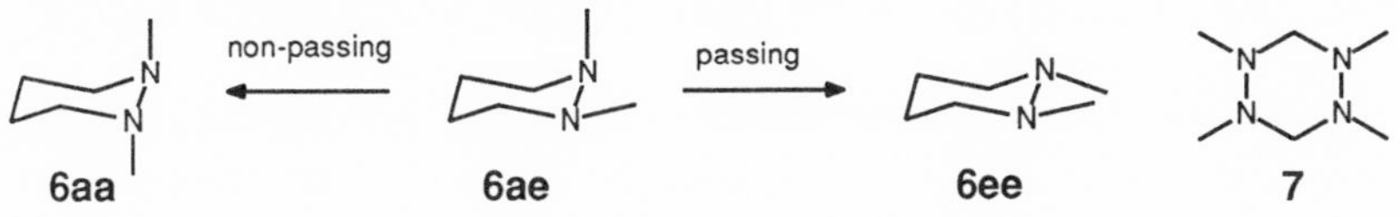

| 6aa | 6ae | 6ee | 7 |

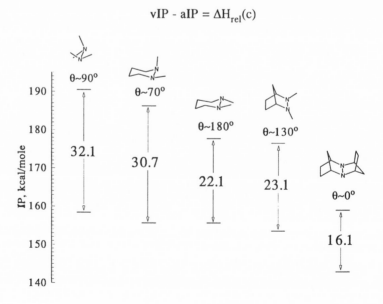

methylene groups clearly has the spin localized in one-hydrazine unit, and electron transfer is slow on the ESR time scale.[34] Dialkylhexahydropyridazine radical cations have been shown by ESR and optical absorption studies to be more planar at nitrogen than most tetraalkylhydrazine radical cations,[35] and a large geometry change must be required to reach the similar geometry at the oxidized and neutral hydrazine units required for ET.

The pattern observed for vertical ionization potentials of hydrazines is very complex because vIP is significantly influenced by the size of the substituent, pyramidality at nitrogen, and θ for rotation about the NN bond.[36] More recently, Meot-Ner(Mautner), Nelsen, and Rumack have measured aIP values for over 30 tetraalkylhydrazines by equilibrating the hydrazine radical cations with each other and with aniline standards, using high-pressure mass spectrometry over a wide range of temperature so that ΔH values could be extracted from the ΔG values measured.[37] Figure 3 gives a graphical representation of some of these data, and Table 2 compares experimental and calculated values for **6, 8–10**, and two acyclic

$$vIP - aIP = \Delta H_{rel}(c)$$

Figure 3. Plot of experimental values of vIP and aIP for tetramethylhydrazine, **6ae**, **6ee**, **9**, and **10**. The numbers shown are values of vIP-aIP.

Table 2. Comparison of Experimental and AM1 Calculations of $\Delta H_{rel}(\mathbf{c})$ (kcal mole^{-1})

Compound	vIP – aIP exper.[42a]	$\Delta H_f(n^+) - \Delta H_f(c^+)$ AM1 calc.	calc. – obs.
8	34.6	36.0	1.4
Me$_2$NNMe$_2$	32.1	33.2	1.1
tBuMeNNMe$_2$	19.6	24.2	4.6
6ae	30.7	30.2	–0.5
6ee	22.1	17.6	–4.5
9	23.1	19.4	–3.7
10	16.1	15.8	–0.3

hydrazines. This work showed that as expected from the pattern of E$^{o\prime}$ versus vIP plots, vIP - aIP = $\Delta H_{rel}(\mathbf{c})$ is sensitive to the θ value of the hydrazine. Note that the *gauche* lone-pair conformation **6ae** has a 39% larger $\Delta H_{rel}(\mathbf{c})$ value than does the θ~180° **6ee** conformation of the same compound. The θ~0° sesquibicyclic system **10** has the lowest $\Delta H_{rel}(\mathbf{c})$ value of any compound studied.[38] It was unfortunately the only such compound which could be examined because of molecular weight and low vIP problems (increasing bicyclic ring sizes in sesquibicyclic hydrazines lowers vIP significantly, which would cause the gaps in the equilibration ladder to be too large for measurement of equilibrium constants).

Although the general characteristics of hydrazine radical cation geometries were clear from CV and spectroscopic work, actually determining their structure required isolable compounds. Tetramethylhydrazine radical cation and most others persist for hours at millimolar concentration, but they do not prove isolable. The problem appears to be that electron transfer disproportionation in concentrated samples produces tiny amounts of dication, neutral pairs. These pairs must proton transfer so rapidly that decomposition competes with back electron transfer, which is on the order of one-volt exothermic, and thus quite rapid. The C$_\alpha$–H bonds of alkyl groups attached to oxidized nitrogen are clearly weakened by hyperconjugation, in which electrons from these bonds mix with the electron deficient p-rich nitrogen orbital. Although removal of the α hydrogens by changing two of the substituents to *tert*-butyl groups is synthetically easy, the strain introduced by tertiary substituents and the resulting weaker N-*tert*-C$_\alpha$ bonds proved to shorten both radical cation and dication lifetimes.[39] We solved the α-deprotonation problem by building all four alkyl groups into N,N-bicyclic ring systems, which hold the C$_\alpha$–H bonds nearly perpendicular to the p-rich nitrogen orbital. Bis-*N,N*-dialkylhydrazines indeed have long-lived dications in the presence of their neutral compound (as shown by reversible second oxidation CV waves), and the radical cations are isolable.[40] Synthesis of these compounds proved to be easy, because bicyclic N-chloroamines couple efficiently when treated with *tert*-butyllithium.

11

The geometrical feature which solves the oxidized form lifetime problem, "Bredt's rule kinetic protection",[40] also prevents the dehydrochlorination which most N-chloramines undergo when treated with base. We will only discuss here the compound which gave us the most information, 8,8'-bi(8-azabi-cyclo[3.2.1]oc-tane), **11**.[41] Neutral **11** exists only in $\theta = 180°$ conformations because of its large alkyl substituents. Both the neutral compound and the radical cation crystallize in the bis-axial-NR_2 substituted piperidine ring *anti* conformation illustrated. The NN bond shortens 0.146 Å (9.9%) and the average of the bond angles at N, α(av), changes 7.7° of the 10.5° range between a tetrahedral and a planar atom, or 73%. More favorable solvation/ion pairing for the *syn* form of the dication cause the *syn* and *anti* **11**$^+$,**11**$^{2+}$ CV waves to be resolved, and flatter nitrogens in the *syn* than the *anti* radical cation cause a 19-nm shift in their absorption spectra, allowing analysis of mixtures by two methods. The ability to analyze mixtures of **11**$^+$ along with that to enrich the *anti*-rich equilibrium mixture (*syn*-**11**$^+$ is 1.45 ± 0.15 kcal/mole higher in free energy at 22 °C) in the *syn* isomer by photolysis allowed measurement of the barriers to rotation about the 3e-π bond. *syn*-**11**$^+\rightarrow$*anti*-**11**$^+$ has $\Delta G^{\ddagger}$ 22.0 ± 0.06 kcal mole^{-1} ($\Delta H^{\ddagger}$ = 20.2 ± 1.6 kcal mole^{-1}, $\Delta S^{\ddagger}$ –5.9 ± 5.5 cal mole^{-1} K^{-1}) and *anti*-**11**$^+\rightarrow$*syn*-**11**$^+$ has $\Delta G^{\ddagger}$ = 23.4 kcal mole^{-1}. Because NN rotation involves loss of the π bonding between the nitrogens, these barriers constitute a measurement of the resonance stabilization in the 3e-π bond of **11**$^+$, which is also consistent with the 19 kcal/mole estimated for 9-NMe$_2$-9-azabicyclo[3.3.1]-nonane radical cation by comparing its E°′ value with those of other 9-substituted compounds.[40a] The significance of an approximate 20-kcal/mole 3e-π bond resonance stabilization for ET considerations is that 3e-π bonding is more stabilizing than 3e-σ bonding in trialkylamine radical cation dimers, so that in contrast to amines, hydrazines should undergo the outer shell electron transfer which ET theories directly address, and bonded intermediates should not be present on the ET reaction coordinate. An estimate of k_{ex} for **11**$^{0/+}$ was also available, because neutral **11** catalyzes the conversion of *syn*-**11**$^+$ to *anti*-**11**$^+$. Both the scan rate dependence for CV curves of **11** oxidation and measuring the increase in isomerization rate for solutions enhanced in *syn*-**11**$^+$ after addition of neutral **11** gave reasonable fit for the reaction *syn*-**11**° + *anti*-**11**$^+ \rightleftharpoons$ *syn*-**11**$^+$ + *anti*-**11** at 22 °C using $\Delta G°$ = 1.67 kcal mole^{-1}, k_f = 260 M^{-1}s^{-1}, k_b = 4500 M^{-1}s^{-1}, for a roughly estimated k_{ex}($\Delta G°$ = 0) value of about 10^3 M^{-1}s^{-1}.[41a]

Self-ET is too slow for bis-*N,N*-bicyclic tetraalkylhydrazines to allow measurement of k_{ex} by magnetic resonance line broadening techniques. No broadening was detected in the ESR spectrum (as expected from the small k_{ex} value estimated above), and distinct line broadening effects could not be established for the rather

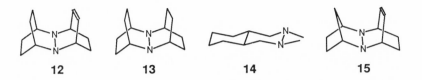

12   **13**   **14**   **15**

broad ¹H-NMR signals of these bicyclic hydrazines. The discovery that cyclic dienes add to protonated bicyclic azo compounds to produce Diels-Alder adducts allows the preparation of bis-*N,N'*-bicyclic ("sesquibicyclic") hydrazines including **12** and its hydrogenation product, **13**.[42] These compounds are restricted to θ values near 0° by the near 0° CNNC dihedral angles imposed by the bicyclic rings. The bulky alkyl groups prevent the nitrogens in neutral sesquibicyclic compounds from being as pyramidal as those of bis-*N,N*-bicyclic hydrazines, and their Bredt's rule protection allows isolation of the radical cations. In some cases, X-ray structures were obtained, and Table 3 summarizes some hydrazine NN bond length and α(av) values, also including data for the diequatorial diazadecalin derivative **14**,[43] and sesquibicyclic hydrazine **15**.[44] These X-ray data show that electron loss from **12** decreases the NN bond length by 0.148 Å (9.9%) and increases α(av) 5.6° (53% of 10.5°). The structure for **13⁺OTs⁻** (determined after failing to obtain solvable structures for the BF₄⁻ and PF₆⁻ salts) is not of high quality (R_w is 11%), making the 11.2% decrease in NN bond length and 6.3° increase in α(av) (60% of 10.5°) obtained for electron removal from **13** less certain. Although the change in d(NN) is very similar for **11⁰/⁺** and **12⁰/⁺**, the sesquibicyclic compound exhibits a smaller change in pyramidality at nitrogen upon electron removal. Self-ET is significantly faster for **12⁰/⁺**,[45] as shown by NMR line broadening studies which yield k_{ex} (25 °C CD₃CN) of $1.2_1 \times 10^4$ M⁻¹s⁻¹, roughly 11× larger that of **11⁰/⁺**, but a factor of

Table 3. Comparison of Crystallographic and Calculated Geometries for Hydrazines and their Oxidized Forms

Compound	d(NN),Å X-ray	d(NN),Å AM1	α(av),° X-ray	α(av),° AM1
	Neutral Hydrazines			
14°	1.486	1.421	107.6	111.3
11°	1.469	1.398	107.9	109.9
12°	1.497	1.409	112.1	112.3
13°	1.492	1.406	112.8	113.6
15°	1.500	1.424	110.0	110.4
	Oxidized Hydrazines			
11⁺	1.323	1.299	115.6	116.2
12⁺	1.349	1.323	117.7	117.9
13⁺	1.325	1.322	118.3	118.1
13²⁺	1.270	1.273	120.0	120.0

8.3×10^{-5} smaller than that of $\mathbf{1}^{o/+}$ under the same conditions. ET is even slower for $\mathbf{13}^{o/+}$, although k_{ex} cannot be measured as accurately because of radical cation decomposition; k_{ex} (23 °C CD$_3$CN) was estimated at 700 M^{-1}s^{-1}.[5] The large geometry change between neutral and cation radical being significantly involved in causing the small k_{ex} value is indicated by the fact that $\mathbf{13}^{+/2+}$ shows faster self-ET, k_{ex} (25 °C CD$_3$CN) 2.1$_3(\pm 0.1_3) \times 10^4$ M^{-1}s^{-1}, despite the work term which must be present for approach of a radical cation to a dication. The geometry change between radical cation and dication is clearly much smaller, d(NN) decreasing about 0.055 Å (4.2%), and α(av) increasing about 1.7° (16% of 10.5°), so λ_{in} must be considerably smaller for removal of the second electron from $\mathbf{13}$ than the first.

The effect of solvent on $\mathbf{12}^{o/+}$ k_{ex} values has been studied.[45a] In contrast to the behavior of $\mathbf{1}^{o/+}$, plots of ln(k_{ex}) versus the solvent parameter γ are significantly nonlinear. Interestingly, however, plots of ln(k_{ex}) for $\mathbf{12}^{o/+}$ versus Kosower Z values[46] or Dimroth–Reichardt $E_T(30)$ values,[47] are nearly linear, including the alcohol solvents.[48] These solvent parameters are known to correlate well with rate constants for a wide variety of reactions studied by organic chemists,[47a] which are characterized by considerably higher barriers than most self-ET reactions. We suggested that it is significant that eight dimeric Ru$^{II/III}$ mixed-valence complexes in Creutz's review article,[49] which have λ_{max} in the near-IR range (transition energies 22–31 kcal mole^{-1} in acetonitrile) exhibit linear transition energy (E_{op}) versus γ plots, and that p-phenylenediamines such as $\mathbf{1}^{o/+}$,[13] which have 4$\Delta G^{\ddagger}$ in the same general range (19.6 kcal mole^{-1} for $\mathbf{1}^{o/+}$), show linear ln(k_{ex}) versus γ plots. Z and $E_T(30)$ are defined by the solvent effects on transition energies for charge transfer bands which occur at 71.3 and 46.0 kcal/mole respectively, and the solvent dependence of these transition energy is clearly different from γ. 4$\Delta G^{\ddagger}$ for $\mathbf{12}^{o/+}$ is 47.6 kcal mole^{-1}, and shows a similar solvent effect on its activation barrier. Because k_{ex} for $\mathbf{1}^{o/+}$ is within a factor of 10 of diffusion control, intermolecular self-ET reactions which are very much faster than this cannot be found. Several examples of intermolecular self-ET reactions which also do not give linear plots of ln(k_{ex}) versus γ, but exhibit solvent dynamical ("solvent friction") effects, where the tumbling rate of the solvent clearly influences the rate of electron transfer, are now known.[50] These reactions are characterized by low intrinsic barriers λ, and solvent dynamical effects not seen for self-ET occur for heterogeneous ET for several cases, including $\mathbf{1}^{o/+}$.[51] The heterogeneous ET barrier should be half that for homogeneous ET.[50] Because linear γ dependence is now known not to occur for both high and low λ processes, it seems unwise to place much reliance upon the numerical accuracy of Eq. 10. Even if Eq. 10 is not numerically accurate, the idea that λ_{out} will disappear in nonpolar solvents might still work reasonably well. Because $\mathbf{12}^{o/+}$ ln(k_{ex}) values are linear with $E_T(30)$ or Z, extrapolation of the observed ΔG^* values to hexane ($E_T(30)$ = 31.0, Z ~52.6$_6$) might be argued to provide an experimental estimate for ΔG^*_{in}. Excluding the clearly anomalous chloroform point,[48] extrapolations to hexane give estimated ΔG^*_{in} values of about 6.9 $\pm$ 0.1$_5$.[52]

7. SEMIEMPIRICAL MO CALCULATIONS ON HYDRAZINES

The polycyclic structures of sesquibicyclic hydrazines severely limit the geometries possible for these compounds, and probably as a direct result, molecular orbital (MO) calculations work surprisingly well for describing changes in the thermodynamics of electron loss. Table 2 shows that AM1 calculations do not do a good job of predicting either NN bond length or pyramidality in the less constrained hydrazines **14** or **11**, but that they are much more successful at obtaining α(av) for the sesquibicyclics. Furthermore, AM1 calculations suggest that pyramidality at nitrogen measured by α(av) is the only geometric parameter which is important for determining the thermodynamics of electron loss from these compounds. For 13 sesquibicyclic hydrazines which contain all 5- to 7-membered rings, AM1 gives the unanticipated result that $\Delta H_f(\mathbf{c}^+)$-$\Delta H_f(\mathbf{n}^\circ)$ versus α(av) of $\mathbf{n}^\circ$ is linear (average deviation 5%).[37] This behavior does not occur for other classes of hydrazines. More surprisingly, a plot of observed $E^{\circ\prime}$ values (which cover a 0.58 V = 13.4 kcal mole^{-1} range) vs calculated α(av) of $\mathbf{n}^\circ$ gives parallel lines for the saturated and unsaturated compounds with the unsaturated compound line lying 0.11 V higher in $E^{\circ\prime}$. Within each series the calculations reproduce the effect of pyramidalization on $E^{\circ\prime}$. This result requires that approach of solvent/counterion to the nitrogen atoms is not a significant factor affecting the thermodynamics of electron removal for sesquibicyclic hydrazines because approach to the N atoms is far easier for **10** than for **13**. Measurements of solvation energies for smaller hydrazine $\mathbf{n}^\circ,\mathbf{c}^+$ couples achieved by plotting aIP versus $E^{\circ\prime}$ show correlation of the solvation energy of the alkyl groups with their effectiveness at lowering gas-phase vIP when pyramidality or θ differences are not a factor.[37] Both of these phenomena are consistent with the positive charge being delocalized over the alkyl groups, and no significant amount of positive charge being present on the $\mathbf{c}^+$ nitrogens, which is also the result calculated by AM1. For the second electron transfer, ease of solvent/counterion approach to the nitrogens definitely is a factor determining $E^{\circ\prime}$, and the dications are calculated to have significant positive charge on nitrogen.[38]

The best experimental test of whether AM1 calculations are able to calculate useful λ'_{in} values is to see how well they reproduce experimental values of $\Delta H_{rel}(\mathbf{c})$, which as pointed out above, ought to be roughly half of λ'_{in}. Table 2 compares observed and calculated numbers. The AM1 calculations are certainly not perfect, but they do predict the large $\Delta H_{rel}(\mathbf{c})$ values for θ~90° hydrazines rather well, and a significantly lower value for **6ee** than **6ae**, although they get the difference too large.[53] Although the AM1 calculations do successfully predict a lower value for t-BuMeNNMe$_2$, with its flattened nitrogen, than for other acyclic hydrazines, experiment gets $\Delta H_{rel}(\mathbf{c})$ even lower. AM1 predicts too low a $\Delta H_{rel}(\mathbf{c})$ value for **8** by 3.7 kcal mole^{-1}, but does successfully predict a significantly lower value than for acyclic hydrazines. Finally, AM1 does an excellent job of predicting the value for **10**, the only sesquibicyclic compound for which an experimental number is

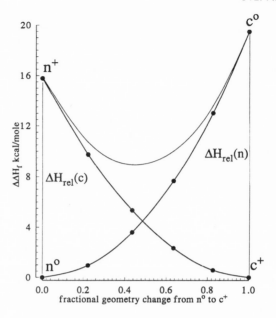

Figure 4. AM1-calculated energy surface for the inner shell enthalpy changes for $10^{o/+}$ self-ET.

available. We find the agreement in Table 2 very encouraging for useful prediction ΔH^*_{et} for sesquibicyclic hydrazines.

The Marcus parabola plot of Figure 1 introduces an apparent symmetry for self-ET which is not quite present. ΔH^*_{in} should represent the point on the energy surface in which the geometries of $\mathbf{n^o}$ and $\mathbf{c^+}$ have both been changed until they are the same, and the sum of the energies of deformation of each species are minimized. Figure 4 shows an AM1 calculation of enthalpy changes for $\mathbf{10^{o/+}}$ self-ET (the diagram is similar to that shown as Figure 3 of ref. 5, where $\mathbf{13^{o/+}}$ self-ET is illustrated). To construct Figure 4, each of the 42 internal coordinates which determine the geometry of $\mathbf{n^o}$ in C_s symmetry was stepped in 20% increments towards that of $\mathbf{c^+}$, producing the $\mathbf{n^o \rightarrow c^o}$ increase in enthalpy. The same process was repeated for $\mathbf{c^+ \rightarrow n^+}$, producing a somewhat smaller increase in enthalpy, because the radical cation structure is more easily deformed than the neutral structure. The geometry coordinate used in constructing the x axis of Figure 4 is the fractional change in $\alpha(av)$. Smooth curves which are clearly very close to half-parabolas (although the calculations are not constrained to produce parabolas) have been added between the calculated points, as well as one showing the sum of the $\Delta\Delta H_f$ values calculated at each geometry. Because $\Delta H_{rel}(\mathbf{n})$ and $\Delta H_{rel}(\mathbf{c})$ are calculated not to be the same (nor should they be in general), the crossover of the half-parabolas does not come at exactly the 0.5 fractional geometry change, nor

Table 4. Eyring and Classical Marcus Energy Barriers for Sesquibicyclic Hydrazine Self-ET (25°C, in CD_3CN) Compared with ΔH_{in}^* in Values Calculated by AM1 (kcal mole^{-1})

	$\Delta G^{\ddagger}$	$\Delta H^{\ddagger}$	ΔG^*	ΔH_{in}^* (AM1)
	Unsaturated Compounds			
$12^{0/+}$	11.9	~7.4	9.4	8.4
$15^{0/+}$	12.8	~6.2	10.4	9.8
	Saturated Compounds			
$11^{0/+}$	~13.4	—	~10.9	8.7
$13^{0/+}$	13.6	—	11.1	8.2
$10^{0/+}$	11.6	~5.9	9.2	8.8
$13^{+/2+}$	11.5	~8.5	9.1	3.7

does the minimum of the sum curve come at exactly the crossover point. However, the calculated minimum is very close to $\lambda'_{in}/4 = [\Delta H_{rel}(\mathbf{n}) + \Delta H_{rel}(\mathbf{c})]/4$; Eq. 2 is indeed an exceptionally good approximation according to these m.o. calculations.

Table 4 compares free energy barriers from the experimental self-ET k_{ex} values with the enthalpy barriers calculated by AM1, including yet unpublished data for $10^{0/+}$[54] and $15^{0/+}$.[55] The $\Delta H^{\ddagger}$ obtained in variable temperature experiments are also included, but both the scatter in the values observed for $12^{0/+}$ in various solvents[45a] and the lower enthalpy but higher free energy barrier for $15^{0/+}$ compared to $12^{0/+}$ do not give much confidence that systematic errors have been properly controlled.[54] As noted above, ΔG^*_{in} estimated from the solvent effect on the $12^{0/+}$ barrier is about 6.9 kcal mole^{-1}, which is 73% of the measured ΔG^*, and probably within experimental error of the observed $\Delta H^{\ddagger}$. AM1 clearly is successful at predicting the high ΔH^*_{in} value for $12^{0/+}$, which appears to be a principal cause of its low k_{ex} value. However, the experimentally observed increase in barrier upon saturation of **12** to give **13**, and decrease upon ring contraction from **13** to **10** are not yet understood. The AM1 calculations suggest that changes in ΔH^*_{in} are not likely to be the cause of these observations.

One thing we have little evidence upon is how important diabaticity (presence of a small electron transfer transmission coefficient, κ_{el}) may be in hydrazine self-ET reactions. The nitrogens appear to be rather completely surrounded by the alkyl substituents, especially in **12** and **13**, and it seems unlikely that direct interaction of the two-atom π systems of these compounds can occur. The surface of sesquibicyclic hydrazines is unusually convex for organic ET systems and consists mostly of saturated methylene groups. One might well wonder whether effective orbital overlap could occur, and could even ask whether tetraalkylhydrazines are simply so different structurally from compounds usually investigated in ET work that results involving them have little bearing on the general problem. A major structural difference between hydrazines and the metal-centered systems which have formed the core of ET studies involves charge and spin distribution. The metal atom is clearly the center of both the charge and the spin introduced upon electron removal from a metallocene, for example. Weaver and co-workers[12] argue for negligible λ_{in} values for all metallocenes, and attribute the significantly smaller k_{ex} observed for ferrocenes than cobaltacenes entirely to smaller electron transmission coefficients, κ_{el}, for the ferrocenes. Delocalization of spin and charge from the cyclopentadiene rings to the iron atoms are significantly worse than to the cobalt atoms of cobaltacenes, resulting in small H_{AB} values and therefore low κ_{el} values for ferrocenes. Methylation at the cyclopentadiene rings increases k_{ex} for metallocenes due to increased charge and spin delocalization onto the cyclopentadienyl groups, resulting in better π interactions at the transition state and an increase in H_{ab}.[12] Electron removal from hydrazines is characterized by a large λ_{in} value, and although there is obviously some delocalization of spin onto the alkyl groups, most of the spin density remains in the NN 3e-π bond. As discussed above, the charge introduced by electron removal is principally if not entirely on the alkyl groups in the radical cation. Despite these large structural differences, hydrazines and ferrocenes undergo "mixed" electron transfer close to the rates predicted by classical Marcus cross-rate theory,[11] as demonstrated in work by Pladziewicz and co-workers for **12** and **15** undergoing ET with three ferrocene derivatives.[55] It seems likely that κ_{el} is relatively small and varies with structure for sesquibicyclic hydrazines. No data which allows determination of how adiabatic sesquibicyclic hydrazine self-ET might or might not be are currently available, although the rather negative $\Delta S^{\ddagger}$ values observed (twice as negative for **12**$^{0/+}$ as for **1**$^{0/+}$) are consistent with the low preexponential factor low κ_{et} values would cause. If all of the near 2 kcal mole^{-1} difference between ΔG^{*} and ΔH^{*} were attributed to a small κ_{el} value, it could be as low as 0.03, although presumably part of the difference is caused by K_A, and ΔH^{*} is probably not known very accurately.

8. ACYLATED AMINO NITROGEN COMPOUNDS

Amides are thermodynamically difficult to oxidize, and their radical cations are unstable in solution, even when the alkyl groups are Bredt's rule protected by being incorporated into a 9-azabicyclo[3.3.1]nonane system.[40a] Acylated alkylhydrazines are easier to oxidize, and several give radical cations which are long-lived on the CV time scale.[56] N-acylated hydrazines have restricted N–CO rotation and much more planar nitrogen atoms than do entirely alkylated hydrazines, but incorporation of the N–CO unit into a five-membered ring results in detectable deviations from planarity at N. Although the nitrogens are nearly planar for the nearly perpendicular NN bond compound **16** (two conformations in the crystal with θ = 95.2 and 91.9° have α(av) 119.1 and 119.3° respectively), compounds con-

16 **17**

strained to conformations with large lone pair, lone-pair overlap such as **17** (α(av) = 115.8°) are significantly pyramidalized at N.[56-58] Even *N,N'*- diacylated hydrazines show significantly lower $E^{o'}$ than aromatic compounds in $E^{o'}$ versus vIP plots, suggesting significant relaxation energies upon electron removal. For example, both **16** and **17** show 18.1 kcal/mole deviations, which are 46 and 62% of the values for the related tetraalkylhydrazines with both C=O groups replaced by CH_2. Even when the nitrogens of the neutral diacylhydrazine are essentially planar, as in **16**, significant relaxation energies upon electron removal must be present.[56] A large change in bonding accompanies electron loss because the amide-like N–CO bond of the neutral compound is replaced by a 3e-π(NN) bond in the radical cation, and relatively little delocalization of spin onto the CO groups occurs, even when the geometry enforces π,π overlap. Unfortunately, semiempirical m.o. calculations such as AM1 treat N–CO bonds rather poorly. Although AM1 calculations predict that *p*-urazine derivative **18** would give a radical cation with the charge located in one hydrazine unit (analogous to **7** in this respect) this does not appear to be the case. Structure **18⁺** exhibits ET between its hydrazine units which is certainly rapid on the ESR time scale at low temperature, its X-ray structure is consistent with a delocalized structure, and despite significant dependence of its optical absorption maximum on solvent (λ_m 862 nm in CH_3CN, 890 nm in $CHCl_3$) it was concluded that **18⁺** probably has its charge delocalized over both hydrazine units.[59]

Hydroxylamine radical cations require Bredt's rule protection of their O-alkyl groups (but not the N alkyl groups) for observation of a reversible CV oxidation wave, despite significantly more spin density at N than O.[60] The 2-oxa-3-azabicyclo[2.2.2]octane group provides protection of the O and one N alkyl group, and long radical cation lifetimes are observed even when electron-withdrawing acyl groups are substituted at oxygen. The radical cation from the carbonyl-bridged bis-hydroxylamine **19** is the only N-centered radical cation yet reported for which an intramolecular ET rate constant has been measured. The ESR spectrum of **19⁺** shows splittings for two equivalent N atoms at room temperature, but ET between the hydroxylamine units clearly becomes slower at lower temperatures, and VT-

18 **19**

ESR measurements gave $\Delta G^{\ddagger}_{et}(CH_3CN)$ near 3.5 kcal/mole between –84 and –30 °C.[61] This radical cation exhibits anomalous near-IR absorption, because three bands are present, each narrower than Hush theory predicts for an intervalence transfer band, at 1250, 980, and 850 nm. The solvent sensitivity of the longest wavelength band is less than for organometallic intervalence complexes, but linear dependence of the transition energy upon γ was observed for five solvents, and extrapolation to $\gamma = 0$ gave an intercept of 20.5 kcal/mole. This is consistent with expectation for λ_{in} for this system based on $E^{o\prime}$ versus vIP plots, and the 22-kcal/mole transition energy in CH_3CN compared with the observed $\Delta G^{\ddagger}$ is consistent with an H_{AB} value of 2.3 kcal mole^{-1} for $\mathbf{19^+}$.[61] Unfortunately, analogous compounds with larger bridging groups causing greater distance between the N atoms proved to give radical cations which were too labile for similar measurements to be made.

9. CONCLUSION

It would obviously be desirable to study intramolecular ET in N-centered radical cations for which the electronics of electron loss are better understood, and where the geometry of the charge-bearing units and the bridging units are controlled and varied. Such experiments are underway, and it is hoped that they will result in greater insight into the factors which control electron transfer kinetics.

ACKNOWLEDGMENTS

We thank the National Institutes of Health for financial support under grant GM-29549, and both NIH and NSF for support of the experimental work on which the hydrazine work is based. We are extremely grateful to the large number of UW graduate students and collaborators from other laboratories who carried out the experimental work on which this paper is based; their names appear in the references.

REFERENCES AND NOTES

1. Marcus, R. A.; Sutin, N. *Biochim. Biophys. Acta* **1985**, *811*, 265–322.
2. Sutin, N. *Prog. Inorg. Chem.* **1983**, *30*, 441.
3. Sutin, N. *Acct. Chem. Res.* **1982**, *15*, 275.
4. Eberson, L. *Adv. Phys. Org. Chem.* **1982**, *18*, 79–185.
5. Nelsen, S. F.; Blackstock, S. C.; Kim, Y. *J. Am. Chem. Soc.* **1987**, *109*, 677.
6. Also designated H_{12}. The factor of two arises because H_{AB} is the energy separation between ΔG^* and the crossover point, $\lambda/4$. H_{AB} is almost invariably described as the "electronic coupling matrix element," which seems unfortunate, because there is no 2×2 matrix in an electronic description of a real system sophisticated enough to actually calculate the energy surface gap at the transition state.

7. Ref. 4, Table 7. The tetramethyl-*p*-phenylene diamine in D_2O value of 2.5×10^4 (Bruce, C. R.; Noerbert, R. E.; Weissman, S. I. *J. Chem. Phys.* **1956**, *24*, 473) is obviously incorrect. The problem was that the NMR line broadening data were analyzed assuming the usual pseudo-first order conditions ($[\mathbf{M^o}]>[\mathbf{M^+}]$) and assuming the concentration of $\mathbf{M^o}$ was that added to the vessel; the experiment was conducted at pH = 4, where most of the $\mathbf{M^o}$ is protonated, and does not exchange.

8. For discussion of nuclear reorganization barriers in transition metal compounds, see Sutin, N; Brunschwig, B. S.; Creutz, C.; Winker, J. R. *Pure & Appl. Chem*, **1988**, *60*, 1817.

9. (a) Despite the patient explanation by Marcus and Sutin addressing the question of E_{op} (the transition energy in solution, which is equated to λ) being a free energy energy instead of an enthalpy (ref. 1, footnote ¶, p.281), we still have trouble understanding how the λ_{in} portion of it is a free energy. E_{op} is an experimentally measured number, which may well be a free energy, but λ appears to be a vertical gap on an energy diagram to us. (b) We believe that the λ_{in} obtained by summing $\Delta H_{rel}(\mathbf{c})$ and $\Delta H_{rel}(\mathbf{n})$ is an enthalpy, because vertical ionization is accepted to represent an enthalpy. There is no change in entropy during a vertical transition, and the energy for vertical removal of an electron from $\mathbf{n}^o$ contains will not contain any entropy difference between $\mathbf{n}^+$ and $\mathbf{n}^o$, and putting an electron vertically into $\mathbf{c}^+$ will not contain any entropy difference between $\mathbf{c}^o$ and $\mathbf{c}^+$. The sum of these energies will be an enthalpy, λ'_{in}, and one quarter of it will be ΔH^*_{in}. We do not see how ΔH^*_{in} could be equal to ΔG^*_{in} unless ΔS^*_{in} were negligibly small. We doubt that it is for amino nitrogen compounds, where bonding interactions are considerably changed by removal of an electron, and direct equilibration of hydrazine radical cations demonstrates that there are large differences in entropy between hydrazine radical cations of different types.[36]

10. Previously we designated $\mathbf{n}^+$ as **cng** and $\mathbf{c}^o$ as **ncg**.[5]

11. Marcus, R. A.; Sutin, N. *Inorg. Chem.* **1975**, *14*, 213.

12. (a) Nielson, R. M.; McManis, G. E.; Safford, L. K.; Weaver, M. J. *J. Phys. Chem.* **1989**, *93*, 2152; (b) McManis, G. E.; Nealson, R. M.; Gochev, A.; Weaver, M. J. *J. Am. Chem. Soc.* **1989**, *111*, 5533.

13. (a) Grampp, G.; Jaenicke, W. *Ber. Bunsenges. Phys. Chem.* **1984**, *88*, 325; (b) Grampp, G.; Jaenicke, W. *Ibid.* **1984**, *88*, 335; (c) Grampp, G.; Jaenicke, W. *J. Chem. Soc., Faraday Trans. 2* **1985**, 1035. We thank Günter Grampp for informing us that a factor of 2 in λ_{in} is missing throughout this paper (there are two molecules in the ET reaction); the estimated numbers here include both molecules. d. Grampp, G.; Jaenicke, W. *Ber. Bunsenges. Phys. Chem.* **1991**, *95*, 904.

14. For example, consider the beautiful work in which Miller and Closs convincingly demonstrated the existence of the long-sought "inverted region" by measuring rate constants k_{et} across a 3-α-substituted androstane skeleton, from 16-equatorially substituted biphenyl radical anions generated by pulse radiolysis to a series of eight acceptors substituted at the 3-equatorial position.[13] They use the "Golden Rule" equations shown below, observing excellent fit to k_{et} using the physically reasonable parameters:

$$k_{et} = PRE \, \Sigma(e^{-S}S^w/w!)exp[-(\lambda s + \omega h v)^2/\lambda_s 4RT]$$

$$S = \lambda_v/hv, \quad PRE = (2\pi/\hbar)\,|V|^2(4\pi\lambda_s RT)^{-1/2}, \text{ where}$$

$\lambda_s = 0.75$ eV, $\lambda_v = 0.45$ eV, $hv = 1500$ cm^{-1}, $V = 6.2$ cm^{-1}. Although w is to be summed from 0 to ∞, at $\Delta G^o = 0$, the w = 0 term produces 96% of k_{et} using the above parameters. We may therefore write the rate expression at $\Delta G^o = 0$ as:

$$k_{et}(\Delta G^o = 0) = [2\pi/\hbar]\,|V|^2[4\pi\lambda_s RT]^{-1/2}exp[-(\chi\lambda+\lambda_s)/4RT], \quad \chi = 4RT/hv.$$

Making the nomenclatural changes of $\lambda_v = \lambda_{in}$, $\lambda_s = \lambda_o$, and $|V| = H_{AB}$, this is nearly the same as a Marcus theory equation, stated to apply to non-adiabatic cases (Eq. 17, ref. 1) except that λ in the preexponential factor has been replaced by λ_s, and in the exponential factor by $\chi\lambda_v+\lambda_s$. Because χ is numerically 0.549, only 55% of λ_v is employed in the exponential term. The Golden Rule expression only appears to give the Marcus theory expression as λ_v approaches zero. Tunnelling is included in the Golden Rule expression, and has been incorporated into more modern versions of Marcus theory.[1-3]

15. (a) Calcaterra, L. T.; Closs, G. L.; Miller, J. R. *J. Am. Chem. Soc.* **1983**, *105*, 670; (b) Miller, J. R.; Calcaterra, L. T.; Closs, G. L. *Ibid.* **1984**, *106*, 3047; (c) Liang, N.; Miller, J. R.; Closs, G. L. *Ibid.* **1989**, *111*, 8740; (d) Liang, N.; Miller, J. R.; Closs, G. L. *Ibid.*, **1990**, *112*, 5353; (e) Closs, G. L.; Miller, J. R. *Science* **1988**, *240*, 440.

16. For a description of the beginnings of organic radical ion chemistry, see Roth, H. D. *Tetrahedron*, **1986**, *42*, 6097.

17. (a) Tanaka, J.; Mizumo, M. *Bull. Soc. Chem. Japan* **1969**, *42*, 1841; (b) Tanaka, J.; Sakabe, N. *Acta Crystallogr.* **1968**, *B24*, 1345; (c) De Boer, J. L.; Vos, A.; Huml, K. *Acta Crystallogr.* **1968**, *B24*, 542; (d) De Boer, J. L.; Vos, A. *Acta Crystallogr.* **1972**, *B24*, 835.

18. This rate constant at 25 °C was calculated from the 20 °C value of $1.1_7 \times 10^9$ (corresponding to $\Delta G^{\ddagger}(20 \,°C) = 4.99$),[13a] extrapolated using the -7.6 cal mol^{-1} K^{-1} $\Delta S^{\ddagger}$ value;[11b,c] the value obtained from the activation parameters reported is $1.6_0 \times 10^9$(25 °C), 9% larger.

19. Alternatively, n^2 is designated the optical dielectric constant, D_{op} when the static dielectric constant is designated D_s in stating Eq. 9.

20. Fawcett, W. R.; Foss, C. A. Jr. *J. Electroanal. Chem.* **1989**, *270*, 103.

21. Our published number[5] of 2.06 kcal/mole for this quantity is in error. The energy surface for TMPD ionization is far more complex than has been discussed, when the possibility of aromatic C–N bond rotations is properly considered. The present number is that from *syn* C_{2v} (untwisted) neutral TMPD, which is not the AM1 minimum energy species (a 36° twisted C_i species is 0.6 kcal/mole more stable), but is that which leads to smallest ΔH^*_{in}.

22. Nelsen, S. F.; Ippoliti, J. T. *J. Am. Chem. Soc.* **1986**, *108*, 4879.

23. Dinnocenzo, J. B.; Banach, T. E. *J. Am. Chem. Soc.* **1989**, *111*, 8646.

24. (a) Alder, R. W. *Tetrahedron* **1990**, *46*, 687; (b) Alder, R. W. *Acct. Chem. Res.* **1983**, *16*, 321; (c) Alder, R. W.; Sessions, R. B. *The Chemistry of Amino, Nitroso, and Nitro Compounds and their Derivatives* (Patai, S., Ed.). Wiley, New York, 1982, Chapter 18, p. 763.

25. Alder, R. W.; Orpen, A. G.; White, J. M. *J. Chem Soc. Chem. Commun.* **1985**, 949.

26. Alder, R. W.; Sessions, R. B. *J. Am. Chem. Soc.* **1979**, *101*, 3651.

27. (a) Nelsen, S. F.; Alder, R. W.; Sessions, R. B.; Asmus, K.-D.; Hiller, K.-O.; Göbl, M. *J. Am. Chem. Soc.* **1980**, *102*, 1429; (b) Alder, R. W.; Arrowsmith, R. J.; Casson, A.; Sessions, R. B.; Heilbronner, E.; Kovač, T.; Huber, H.; Taagepera, M. *Ibid.* **1981**, *103*, 6137.

28. Dinnocenzo, J. B.; Banach, T. E. *J. Am. Chem. Soc.* **1988**, *110*, 971.

29. Gębicki, J.; Marcinek, A.; Stradowski, C. *J. Phys. Org. Chem.* **1990**, *3*, 606.

30. Nelsen, S. F.; Ippoliti, J. T.; Petillo, P. A. *J. Org. Chem.* **1990**, *55*, 3825.

31. (a) Nelsen, S. F. *Molecular Structures and Energetics* (Liebman, J. F.; Greenberg, A., Eds.). VCH, Deerfield Beach, FL, 1986, Vol. 3, Chapter 1; (b) Nelsen, S. F. *Acct. Chem. Res.* **1981**, *14*, 131.

32. For reviews of neutral hydrazine conformational changes see (a) Nelsen, S. F. *Acyclic Organonitrogen Stereodynamics* (Lambert, J. B.; Takeuchi, Y., Eds.). VCH, New York, 1992, Chapter 3, p. 89; (b) Nelsen, S. F. *Acct. Chem. Res.* **1978**, *11*, 14; (c) Shvo, Y. *Chemistry of Hydrazo, Azo, and Azoxy Groups* (Patai, S., Ed). Wiley, New York, 1975, Part 2, Chapter 21, p. 1017.

33. For a review of N-centered radical cation conformational changes see Nelsen, S. F. *Acyclic Organonitrogen Stereodynamics* (Lambert, J. B.; Takeuchi, Y., Eds.). VCH, New York, 1992, Chapter 7, p. 245.

34. Nelsen, S. F.; Hintz, P. J.; Buschek, J. M.; Weisman, G. R. *J. Am. Chem. Soc.* **1975**, *97*, 4933.

35. Nelsen, S. F.; Yumibe, N. P. *J. Org. Chem.* **1985**, *50*, 4749.

36. Nelsen, S. F. *J. Org. Chem.* **1984**, *49*, 1891.

37. (a) Nelsen, S. F.; Rumack, D. T.; Meot-Ner(Mautner), M. *J. Am. Chem. Soc.* **1988**, *110*, 7945; (b) Nelsen, S. F.; Rumack, D. T.; Meot-Ner(Mautner), M. *Ibid.* **1987**, *109*, 1373.

38. (a) Nelsen, S. F.; Frigo, T. B.; Kim, Y. *J. Am. Chem. Soc.* **1989**, *111*, 5387; (b) Redetermination of the photoelectron spectrum for **10** in ref. 37a showed that the vIP used in ref. 36a was .12 eV too

large, lowering $\Delta H_{rel}(c)$ from the 18.6 kcal mole^{-1} quoted earlier to the 16.1 kcal mole^{-1} quoted here.

39. Nelsen, S. F.; Parmelee, W. P. *J. Org. Chem.* **1981**, *46*, 3453.

40. (a) Nelsen, S. F.; Kessel, C. R.; Brien, D. J. *J. Am. Chem. Soc.* **1980**, *102*, 702; (b) Nelsen, S. F.; Kessel, C. R. *Ibid.* **1977**, *99*, 4461.

41. (a) Nelsen, S. F.; Cunkle, G. T.; Evans, D. H.; Haller, K. J.; Kaftory, M.; Kirste, B.; Clark, T. *J. Am. Chem. Soc.* **1985**, *107*, 3829; (b) Nelsen, S. F.; Cunkle, G. T.; Evans, D. H.; *J. Am. Chem. Soc.* **1983**, *105*, 5928.

42. Nelsen, S. F.; Blackstock, S. C.; Frigo, T. B. *J. Am. Chem. Soc.* **1984**, *106*, 3366.

43. Nelsen, S. F.; Hollinsed, W. C.; Calabrese, J. C. *J. Am. Chem. Soc.* **1977**, *99*, 4461.

44. Nelson, S. F.; Wang, Y.; Powell, D. R.; Hiyashi, R. K. *J. Am. Chem. Soc.* **1993**, *115*, 5246.

45. (a) Nelsen, S. F; Kim, Y.; Blackstock, S. C. *J. Am. Chem. Soc.* **1989**, *111*, 2045; (b) Nelsen, S. F.; Blackstock, S. C. *Ibid.* **1985**, *107*, 5548.

46. (a) Kosower, E. M. *Molecular Biochemistry.* McGraw-Hill, London, 1962, pp. 180–195; (b) Kosower, E. M. *J. Am. Chem. Soc.* **1958**, *80*, 3253.

47. (a) Reichardt, C. *Solvents and Solvent Effects in Organic Chemistry*, 2nd ed. VCH, Weinheim, 1988, p. 365; (b) Dimroth, K.; Reichardt, C.; Siepmann, T.; Bohlman, F. *Justus Liebigs Ann. Chem.* **1963**, *661*, 1.

48. k_{ex} is anomalously low in CDCl$_3$, the least polar solvent studied. Something special must be involved, and we have ignored this solvent in the discussion. Deviations in linearity of ln(k_{ex}) vs solvent parameter plots would presumably always occur at low enough solvent polarity because as ion pairing becomes important, the counterion must be transferred at the transition state, which ought to slow k_{ex}. Solubility problems make study of this effect difficult.

49. Creutz, C. *Prog. Inorg. Chem.* **1983**, *30*, 1.

50. For a recent review of dynamical solvent effects on ET, see Weaver, M. J.; McManis, G. E., III *Acct. Chem. Res.* **1990**, *23*, 294.

51. Opallo, M. *J. Chem. Soc., Faraday Trans. 1* **1986**, *82*, 339.

52. A plot of ΔG^* in 9 solvents[44a] vs. $E_T(30)$ has a correlation coefficient of 0.970, and extrapolates to ΔG^*(hexane) 6.75 kcal mole^{-1}, while the plot vs. Z gives r = 0.979 and 6.88 kcal mole^{-1}. Removal of the poorest fitting point, CD$_3$NO$_2$, raises the values to 6.82 and 6.95 respectively.

53. AM1 calculations clearly do a poor job on the relative energies of **6** conformations, getting the order of stability exactly backwards by predicting **6ee** to be the least stable chair form and **6aa** to be the most, so it is perhaps not surprising that they would get the **6ee** $\Delta H_{rel}(c)$ value rather far off.

54. Unpublished data of Yichun Wang.

55. Nelsen, S. F.; Wang, Y.; Ramm, M. T.; Accola, M. A.; Pladziewicz, J. R. *J. Phys. Chem.* **1992**, *96*, 10654.

56. (a) Nelsen, S. F.; Blackstock, S. C.; Petillo, P. A.; Agmon, I.; Kaftory, M. *J. Am. Chem. Soc.* **1987**, *109*, 5724; (b) Nelsen, S. F.; Blackstock, S. C.; Frigo, T. B. *Ibid.* **1983**, *105*, 3115.

57. (a) Agmon, I.; Kaftory, M.; Nelsen, S. F.; Blackstock, S. C. *J. Am. Chem. Soc.* **1986**, *108*, 4477. (b) Kaftory, M.; Agmon, I. *Ibid.* **1984**, *106*, 7785.

58. Nelsen, S. F.; Petillo, P. A.; Chang, H.; Frigo, T. B.; Dougherty, D. A.; Kaftory, M. *J. Org. Chem.* **1991**, *56*, 613.

59. Nelsen, S. F.; Kim, Y.; Neugebauer, F. A.; Krieger, C.; Siegel, R.; Kaftory, M. *J. Org. Chem.* **1991**, *56*, 1045.

60. (a) Nelsen, S. F.; Thompson-Colón, J. A.; Kirste, B.; Rosenhouse, A.; Kaftory, M. *Ibid.* **1987**, *109*, 7128; (b) Nelsen, S. F.; Thompson-Colón, J. A. *J. Org. Chem.* **1983**, *48*, 3364.

61. Nelsen, S. F.; Thompson-Colón, J. A.; Kaftory, M. *J. Am. Chem. Soc.* **1989**, *111*, 2809.

SEQUENTIAL ELECTRON TRANSFER REACTIONS CATALYZED BY CYTOCHROME P-450 ENZYMES

F. Peter Guengerich and Timothy L. Macdonald

Advances in Electron Transfer Chemistry
Volume 3, pages 191–241.
Copyright © 1993 by JAI Press Inc.
All rights of reproduction in any form reserved.
ISBN: 1-55938-320-8

1. INTRODUCTION

1.1. Background

Cytochrome P-450 (P-450) enzymes are hemoproteins that were discovered in the 1950s because of their unique spectral properties and their roles in the oxidation of a broad variety of drugs, pesticides, carcinogens, steroids, and fat-soluble vitamins. This diversity of substrates has made the study of the P-450 enzymes popular among many biological scientists. For some perspective on the significance of these reactions see recent reviews by Guengerich[1] and by Porter and Coon[2] and references therein. The P-450s comprise a large family of enzymes, probably at least 40 in each animal species; a guide to the classification of the genes and their primary sequences is presented elsewhere.[3] These enzymes are not strictly defined as cytochromes, in that they do not transfer their electrons on to biological acceptors. The classification of a hemoprotein as a P-450 is operational and is based on a $Fe^{2+} \bullet CO$ complex showing an absorption maximum for the Soret band near 450 nm. This property is the result of axial ligation of the iron by a cysteinyl thiolate. Most of the P-450s are monooxygenases and catalyze reactions with the stoichiometry shown in Scheme 1 where NAD(P)H denotes reduced nicotinamide adenine dinucleotide (3′ phosphate) and R is an organic substrate (Scheme 1). As we will see later, P-450s enzymes occasionally catalyze one-electron reductions of substrates; in these instances, a substrate can serve as a "surrogate" oxygen species. In addition, some P-450s catalyze the rearrangement of substrates that are already in higher oxidation states, and the mechanism involves formal hypervalent iron states.[4,5]

Before proceeding to our discussion of electron transfer in P-450 oxidations, it is appropriate to review the steps by which oxygen is activated in the P-450s. A scheme is presented below (Scheme 2). The process is initiated by binding of the substrate to the enzyme, which appears to be a rapid process. The site of binding is considered to always be somewhere near the distal ligand site so that oxidation can ultimately take place. The next step is electron transfer to heme. Electrons flow from the cofactor NADH or NADPH through another flavoprotein, NADPH P-450

$$NAD(P)H + H^+ + O_2 + R \longrightarrow RO + H_2O + NAD(P)^+$$

Scheme 1.

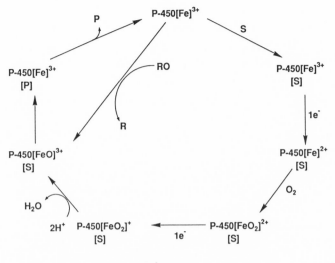

Scheme 2.

reductase, moving to one flavin (FAD) and then on to another (FMN) before going to the hemoprotein. In some P-450s found in bacteria and mitochondria, the pathway of electron transfer is from the pyridine nucleotide to a flavoprotein to an iron-sulfur protein and then to the P-450 heme. The process of transfer of electrons into the heme is relatively long range, and any details regarding tunnels through aromatic residues have not yet been delineated. Reduction of the ferric protein is often facilitated by binding of the substrate; this may be a thermodynamic[6] or kinetic[7] effect depending upon the situation. In some cases binding of the substrate also leads to exclusion of water at the distal ligand site and a shift in the iron spin state from low to high. After the iron is reduced to the ferrous state, oxygen is bound and generates the $Fe^{2+} \bullet O_2$ complex. This is formally equivalent to the oxygen-bound entity in oxyhemoglobin or oxymyoglobin, but is not stable in the case of P-450. However, this complex can be observed with the use of rapid reaction or cryoenzymology techniques.[8–10] A second electron now enters the system, either from the same source as the first (flavoprotein or iron-sulfur protein) or from cytochrome b_5, another hemoprotein. The $E_{1/2}$ for this second electron transfer into the heme is about 0.3 V higher than the first, and is thermodynamically more favorable.[7]

After this point the putative intermediates have not been detected in the normal enzyme cycle and our knowledge is based on various models. Two protons must be added and H_2O is released, giving the formal $(FeO)^{3+}$ entity, which is the basis for most of the reactions to be discussed here. There is an alternate "shunt" for generating the $(FeO)^{3+}$ complex, involving so-called "oxygen surrogates," such as iodosylbenzene or alkyl hydroperoxides. Heterolytic cleavage of these compounds

$$RIO \ + \ Fe^{3+} \ \rightleftharpoons \ (FeO)^{3+} \ + \ RI$$

$$ROOH \ + \ Fe^{3+} \ \longrightarrow \ (FeOH)^{3+} \ + \ RO^{-}$$

$$ROOH \ + \ Fe^{3+} \ \longrightarrow \ (FeOH)^{2+} \ + \ RO\cdot$$

Scheme 3.

yields an active entity thought to be identical to the putative endogeneous $(FeO)^{3+}$ species, which can support oxidative catalysis. However, homolytic cleavage of hydroperoxides can predominate, resulting in chemistry that is dominated by the alkoxy radical formed by this process (Scheme 3).[11]

There have been reports of transient spectral species thought to be the $(FeO)^{3+}$ species generated by the addition of iodosylbenzene or a hydroperoxide to P-450s.[12-14] When such oxidants are added to model metalloporphyrins, such as mangano(III) tetraphenylporphin, relatively stable oxo forms result.[15-17] Some of these have been extensively characterized and information from these biomimetic models has been incorporated into the fabric of our current views on P-450 mechanisms. It should be pointed out here that most, but not all, of the various P-450 reactions have been demonstrated with such models,[16,17] but that the mechanisms of these models may not be the same as those of the enzymes.

1.2. Early Suggestions of Electron Transfer Reactions in the P-450s

The early work on the chemistry of P-450 catalysis was influenced by views on the importance of "mobile" oxidizing species, such as singlet oxygen and super-oxide anion.[18-20] Another concept was that of an "oxene-like" oxidant.[21] With the successful demonstration that oxygen surrogates, such as cumene hydroperoxide, IO_4^-, and iodosylbenzene, could support typical P-450 reactions[22-24] came the view that P-450 iron-oxygen complexes might serve as intermediates in catalysis.[25] Peroxidase intermediates, such as Compound I (formally $[FeO]^{3+}$) and Compound II (formally $[FeO]^{2+}$), had been recognized for some time.[26] These species were known to catalyze the oxidation of amines and phenols,[27] but these transformations were not considered to be typical P-450 reactions for several reasons; these reactions, unlike those with P-450s, generated stable radicals, and the oxygen in aldehydes generated from amines by peroxidases is derived from the medium, in contrast to molecular oxygen (O_2) as the source in the case of the P-450s.[28] Nevertheless, the possibility was considered that intermediates akin to peroxidase Compounds I and II might be relevant to P-450 catalysis. Indeed, the addition of cumene hydroperoxide to P-450 in the presence of aminopyrine did generate aminopyrine cation radicals,[29] but this reaction may be the result of cumenyl alkoxy radicals.[11]

Groves et al.[30] demonstrated that a mixture of iodosylbenzene and iron(III) tetraphenylporphin could produce the oxygenation of methylene and olefin moieties, with the metalloporphyrin serving as a catalyst. Such biomimetic models were developed further. With mangano and chromium porphyrins, the hypervalent metal species are relatively stable and could be shown to be reduced when the "substrate" or another reductant was added.[15,17,31] The study of these models has proliferated, and it is now clear that such models can utilize a variety of different metals (e.g., Fe, Mn, Cr, Ru, Ni), oxidants (e.g., iodosylbenzenes, IO_4^-, ClO_2^-, alkyl hydroperoxides), anionic ligands (e.g., OH^-, Cl^-, imidazole), and ligand structures (e.g., porphyrins, bleomycin).[16,17,32–39] Most of the P-450 oxidation reactions have been demonstrated by these (apparently) biomimetic models, with the exception of some in which the products are highly unstable (e.g., secondary amine oxygenation) (Hammons, Kadlubar, and Guengerich, unpublished results). The $(FeO)^{3+}$ species can be generated electrochemically, and has been described by electrochemical parameters, with $E_{1/2}$ in the range of $+1.0–1.8$ V for both the $(FeO)^{3+}$ to $(FeO)^{2+}$ and the $(FeO)^{2+}$ to Fe^{3+} steps, depending upon the porphyrin, ligands, and solvent.[40,41] These $E_{1/2}$ values may be compared to those estimated (with dyes) for the analogous reductions of horseradish peroxidase Compounds I and II, which are $+0.75$ and $+0.73$ V.[42] Thus, in all cases the potentials for the two steps are rather similar. In the case of the model metalloporphyrins and peroxidases, there is considerable evidence that the formal $(FeO)^{3+}$ entities usually involve a $Fe^{IV} = O$ structure and a radical cation dispersed in the porphyrin,[43] or sometimes localized in an aromatic amino acid residue of the protein (e.g., cytochrome *c* peroxidase[44]). The view is now commonly held that the $Fe^{IV} = O$/porphyrin radical cation structure exists in the P-450s,[45] but there is little in the way of concrete evidence for or against this in the enzymes.

Our own investigations into the potential for electron transfer events in P-450 oxidations began in 1979, and most of our results are mentioned in this review. Our initial interest involved the mechanism of oxidation of halogenated hydrocarbons, and our current views of this matter are discussed under the heading of "Iodine Oxidation." The known ease of oxidation of amines and sulfur compounds suggested that they might be readily oxidized by one-electron transfer reactions, and we have investigated the matter using substrates that undergo diagnostic rearrangements when radical intermediates are involved. This approach has proven to be of great utility in elucidating the mechanisms by our research groups and others,[45,46] and has been extensively employed in the investigation of other enzymes that undergo single electron transfer processes. Another approach we proposed involved the analysis of rates of oxidation of series of similar compounds of varying $E_{1/2}$.[47,48] We have also utilized approaches with kinetic hydrogen isotope effects in our studies.[49–51]

2. AMINE N-DEALKYLATION AND OXYGENATION

2.1. Early Studies Involving Isotopic Labeling

Amine N-demethylation reactions were some of the first oxidations attributed to what are now recognized as P-450 enzymes.[52,53] The mechanism has taken some time to understand. In principle there are several chemical precedents for the reaction (Scheme 4). One possibility is N-oxygenation followed by rearrangement and loss of the aldehyde; that is, a biological Polonowski reaction.[54-56] However, this reaction involves ferrous, not ferric iron, and is acid-catalyzed; there does not appear to be a good biological example of this, as discussed later under N-hydroxylation reactions (*vide infra*). The oxidation of alkylamines in electrochemical and some chemical systems involves the sequential abstraction of an electron, proton, and electron, and the products are derived from the hydrolysis of the imine. Peroxidases oxidize alkylamines via such a mechanism, and this type of pathway has been proposed for P-450 as well.[29] However, in such a mechanism the oxygen atom is derived from water; P- 450-catalyzed N-dealkylation reactions are characterized by the incorporation of label from $^{18}O_2$ into the aldehyde (if it is rapidly reduced *in situ* to prevent exchange).[28,57-59] The oxygen-labeling pattern would be consistent with a mechanism involving a carbinolamine intermediate, which was proposed much earlier by Brodie et al.[60] (Scheme 5). Most carbinolamines are unstable but some, such as that of *N*-methylcarbazole, can be isolated; labeling studies showed that label was incorporated into these from $^{18}O_2$ and not from $H_2^{18}O$.[61,62]

The most widely accepted mechanism for N-dealkylation involves a sequence of electron transfer from the nitrogen atom to give an aminium radical, followed by deprotonation of the α-hydrogen atom to produce an amine-substituted carbon-centered radical, and subsequent oxygen rebound (radical recombination) to the radical, a process which has precedent in the work of Groves[25] (*vide infra*). The carbinolamine decomposes to yield the amine plus carbonyl compound (Scheme

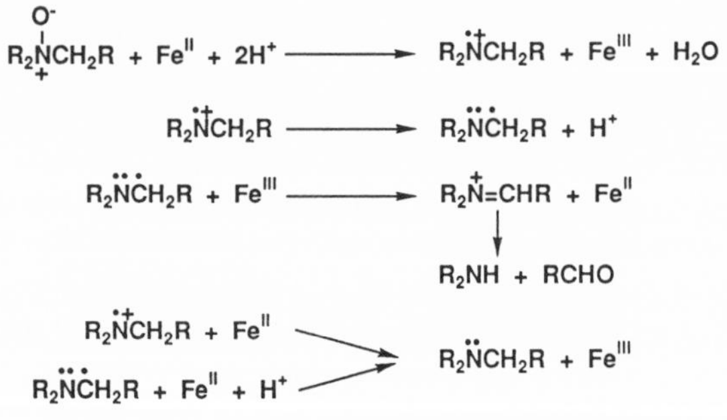

Scheme 4.

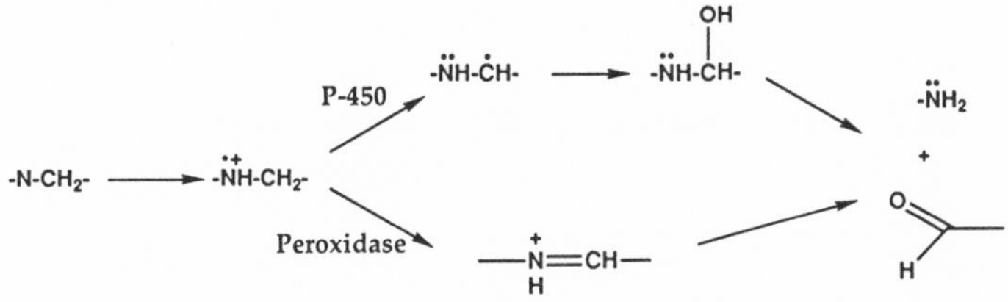

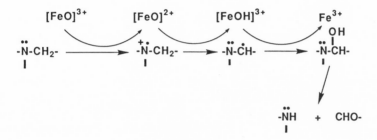

Scheme 5.

$$[FeO]^{3+} \quad [FeO]^{2+} \quad [FeOH]^{3+} \quad Fe^{3+}$$

Scheme 6.

6). Several lines of evidence for this general view follow, along with discussion of limitations of our knowledge.

2.2. Dihydropyridines

1,4-Dihydropyridine structures are of interest because they appear in the pyridine nucleotide cofactors as well as numerous drugs and model compounds. Over the years considerable evidence has accrued that these are oxidized by overall two-electron processes.[63] These species undergo oxidation and reduction processes via both direct and sequential overall two-electron processes, and the mechanism appears to be dependent upon the nature of the partner in the oxidation/reduction reaction. At present, it appears that most 1,4-dihydropyridine drugs are oxidized via sequential electron transfer pathways (Scheme 7).

Dihydropyridines with 4-alkyl substituents were known to be mechanism-based inactivators of P-450 enzymes, and Ortiz de Montellano and his associates had demonstrated that the process involved N-alkylation of the porphyrin.[64] A seminal contribution was made by Augusto et al.,[65] who demonstrated that the oxidation of 1,4-dihydro-2,6-dimethyl-4-ethyl-3,5-pyridinedicarboxylic acid diethyl ester involved the release of ethyl radicals. These radicals could be trapped with nitrones, and the adducts were characterized by their ESR and mass spectra. These findings were interpreted in support of a mechanism in which P-450 abstracts one electron

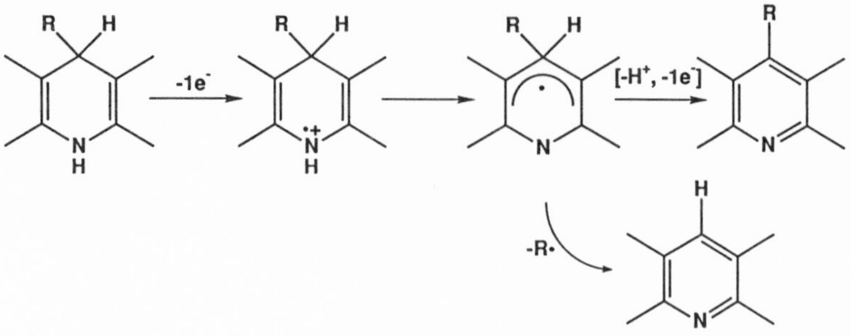

Scheme 7.

from the nitrogen atom and the resultant aminium radical cation subsequently ejects the alkyl radical to produce a stable pyridinium structure (Scheme 8). The events in this sequence of dihydropyridine aromatization may be viewed as having mechanistic parallels to the N-dealkylation of amines, with the exception that the radical cation intermediate does not undergo "rebound" of the iron-bound oxygen, but instead undergoes alkyl radical elimination (Scheme 8). The reason for this behavior may be a function of the driving force provided by the alkyl radical-scission pathway by aromatization (Scheme 8).

These conclusions have been strengthened by mechanistic studies involving 4-aryl substituted 1,4-dihydropyridines. In the oxidation of these compounds the

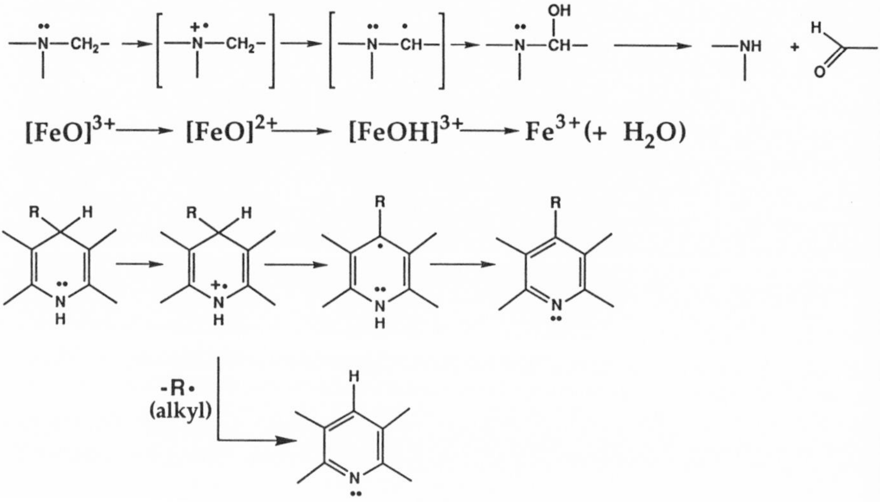

Scheme 8.

4-aryl substituent is not lost.[66,67] A single P-450 enzyme found in human liver, P-450 3A4, has been shown to be the catalyst most active in oxidizing a great variety of such structures.[68] The observations that a great variety of compounds containing the 1,4-dihydropyridine nucleus are oxidized by the enzyme, and that both enantiomers of chiral 1,4-dihydropyridines are oxidized at similar rates, can be used as arguments that deprotonation at the 4-position of the aminium radical is a relatively rapid and facile process. Further, the kinetic deuterium isotope effects for the removal of that proton are very low and consistent with its acidity.[49] Other arguments against mechanisms that involve hydroxylation at the 4-position and subsequent dehydration have been advanced.[69] The oxidation of 1,4-dihydro-1,2,6-trimethyl-3,5-pyridine dicarboxylic acid dimethyl ester has been examined in detail. The dehydrogenation and N-demethylation reactions were both found to occur at rather similar rates; both reactions are hypothesized to proceed from an aminium radical intermediate.[49]

Although, as noted above, the kinetic isotope effects for dihydropyridine aromatization are generally reported to be small; a high kinetic deuterium isotope effect has been reported for the dehydrogenation of an N-phenyl substituted dihydropyridine and the basis remains unclear.[69] There has also been a report of a high kinetic deuterium isotope effect for the dehydrogenation of the drug nifedipine {1,4-dihydro-2,6-dimethyl-[4-(2-nitrophenyl)]-3,5-pyridinedicarboxylic acid dimethyl ester},[70] although this could not be confirmed in our own work.[50] Recently a report has appeared[71] which questions the conclusions of the study of Augusto et al.[65] and suggests that the radicals trapped in the microsomal studies might have had their origin in peroxides generated from lipids due to the presence of adventitious metals. In our work we have found that the treatments used in that study had no effect on rates of oxidation of 1,4-dihydropyridines substituted at the 4-position with alkyl or aryl substituents.[68]

The oxidation of 1,4-dihydropyridines by P-450s is a peroxidase (or sometimes called "peroxygenase") reaction in the sense that an amine is oxidized without oxygen rebound (although peroxides are not used). Further, peroxidases apparently devoid of capability for oxygen rebound (e.g., horseradish peroxidase[72]) are capable of catalyzing the dehydrogenation of 1,4-dihydropyridines.[50,73]

2.3. Cycloalkylamine Rearrangements

Studies with strained cycloalkylamines have also provided mechanistic information. These compounds have long been utilized as mechanistic probes in physical organic chemistry because of their propensity to rearrange when radicals are located α to the ring; the dependence of the rates of these rearrangements on the ring substituents has enabled the "clocking" of the lifetimes of the radical intermediates.[74] Further, these compounds have proven to be useful mechanism-based inhibitors of redox-active enzymes, such as monoamine oxidase.[75]

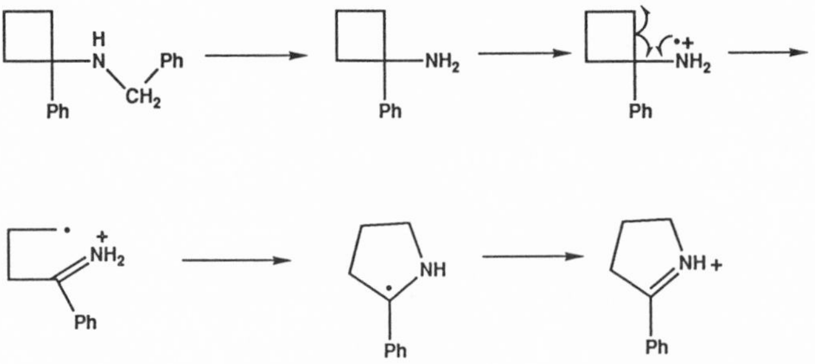

Scheme 9.

The first studies with P-450 and cycloalkylamines were reported by Hanzlik et al. who found that *N*-benzylcyclopropylamine was a mechanism-based inactivator of microsomal P-450.[76] They interpreted the inhibition in terms of sequential hydroxylation, dehydration, and reaction of the Schiff base with the protein. However, imines are not particularly reactive species, and a more likely explanation appears to be that the amine undergoes a one-electron oxidation, and the aminium radical rapidly rearranges to a methylene radical which can alkylate the protein or prosthetic group. Indeed, we found that *N*-(1-methyl)cyclopropyl-*N*-benzylamine was as effective an inhibitor of P-450 as was the desmethyl compound.[77] Hanzlik's group has reinterpreted their results in favor of an aminium radical mechanism.[78]

Another piece of evidence supporting an aminium radical mechanism in the case of monoamine oxidase is the ring expansion of cyclobutylamine to a pyrrolidine.[79] Similar results were found with P-450 and the results are interpreted in terms of a mechanism involving the generation of an aminium radical, ring opening, formation of an expanded ring, and finally removal of a proton and electron to yield the product (Scheme 9).[80] Cyclobutylamines are also mechanism-based inactivators of P-450s, but are not as efficient as their cyclopropyl analogs because of slower rates of ring opening.[74]

In principle, it should be possible to detect products of oxygen rebound to radical sites arising from rearrangement of the putative aminium radicals, as has been done in the case of cyclopropanone hydrate monomethyl ether (*vide infra*). However, there is no record of any attempts to find such products, that is, glycolaldehyde and the amine. Another deficiency in this area is the lack of characterization of the putative conjugates formed by reaction of the radicals with the protein or prosthetic group. In the case of horseradish peroxidase, cyclopropane hydrate is oxidized; the resulting radical rearranges, and a heme adduct is formed in which the *meso* methine group has been modified.[81] Attempts to find such a product in P-450 2B1 which has been inactivated by cyclopropane hydrate or by substituted cyclopropylamines have yielded negative results.[80] The main type of adducts resulting from these cycloalkylamine inactivations appear to be protein-porphyrin cross-linked products[80] whose chemical nature and mechanism of formation remain elusive.[82,83]

2.4. N-Hydroxylation Reactions

N-hydroxylation reactions are of considerable interest in the metabolism of drugs and carcinogens. Such oxidations usually render tertiary amines more hydrophilic and readily excreted from the body. In the case of primary and some secondary arylamines, hydroxylation and subsequent esterification introduce a leaving group that facilitates formation of a (formal) nitrenium ion, which may react with DNA to form an adduct, and thereby ultimately initiate carcinogenesis.[84] A flavoprotein, termed flavin-containing monooxygenase [EC 1.14.13.8, dimethylaniline monooxygenase (N-oxide-forming)], is involved in many of these amine oxygenations; its catalytic mechanism involves reaction of the nitrogen (or sulfur) atom with a C-4a hydroperoxide; apparently a strictly heterolytic mechanism.[85] Distinguishing between roles for this flavin-containing monooxygenase and P-450 enzymes in crude enzyme preparations is not always a simple task.[85] It should be emphasized that tertiary N-oxides are generally rather stable molecules, and the early suggestion that they might be intermediates in the overall mechanism of N-dealkylation[86,87] is generally not tenable. For instance, the N-oxides of sparteine are stable indefinitely and cannot be invoked as intermediates in the P-450-catalyzed conversion of the compound to imines.[88] However, there are some cases in which N-oxides, or their equivalents, do break down to yield products characteristic of N-dealkylation reactions. In conjugated structures, such as the drug verapamil, a Cope elimination of the N-oxide is facilitated,[89,90] and some nitrones decompose to aldehydes.[91]

In the past, the view was held that P-450 did not catalyze many N-oxygenation reactions.[92] If aminium radicals are intermediates in both N-dealkylation and N-oxygenation reactions, as would seem reasonable, then when an α-proton is present it should be lost rapidly, because of its acidity (pKa = ~ 3–8), to generate a carbon-centered radical which is the site for oxygen rebound.[92] Thus, we previously viewed P-450-catalyzed N-oxygenation reactions as falling into three categories (Scheme 10):

1. Arylamines have no α-protons to lose and many are recognized to be N-oxygenated or ring hydroxylated by P-450 enzyme.[84]
2. When the nitrogen atom is at a true bridgehead position, the energetic cost of α-hydrogen abstraction is prohibitive because of steric restrictions imposed by Bredt's rule. Thus, quinidine is converted to an N-oxide by P-450.[93]
3. With azo dyes, the redox potential is low due to the capability for electron donation by the neighboring atom to stabilize the putative aminium radical, and the α-proton acidity is presumably higher than in aminium radicals. Thus, azoprocarbazine is oxidized to both possible azoxy products by P-450s.[94]

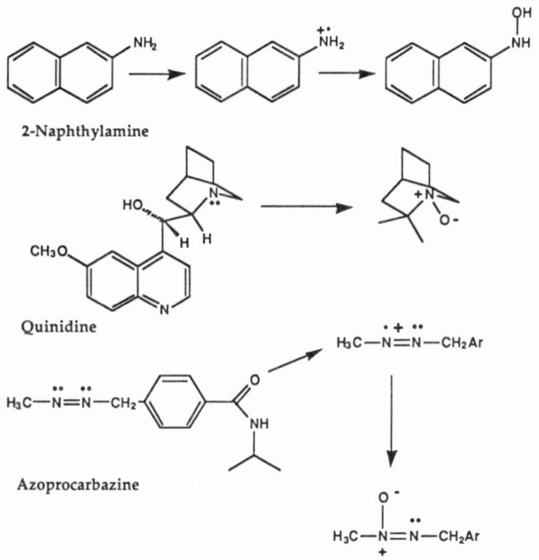

Scheme 10.

However, in recent years there have been some new developments that influence how we think about these reactions. First, studies with cycloalkyl radical clocks suggest that rates of radical recombination (oxygen rebound) in P-450 reactions are even faster than previously thought, being $\sim 2 \times 10^{10}$ s^{-1} [95] (*vide infra*). Second, there has been considerable reevaluation of the acidity of protons adjacent to aminium radicals.[96–99] Indeed, these are not as acidic as previously believed. These new findings may have considerable implications for some of the current dogma regarding the interpretation of kinetic hydrogen isotope studies observed in P-450-catalyzed N-dealkylation (*vide infra*).

Several examples of P-450-catalyzed N-oxygenation reactions outside of the categories discussed under Scheme 10 have now been clearly demonstrated. The prototype pyrrolizidine alkaloid is oxidized by P-450s to both an N-oxide and a formal N-dealkylation product, the pyrrole, which can be viewed as resulting from rearrangement (Scheme 11).[100] Further, evidence has been obtained that a single P-450 enzyme generates both products, and it is reasonable to consider elements of a common mechanism for both reactions.[100,101] These oxidations are of particular interest in that the N-oxide is considered a detoxicated product, but the pyrrole is conjugated with proteins and DNA via nucleophilic attack at the two benzylic methylene carbons, which usually bear good leaving groups.[102] Another example comes from our own work with *N*-(1-phenyl)cyclobutyl-*N*-benzylamine, which is oxidized by P-450 to yield products of both N-dealkylation and N-oxygenation.[80]

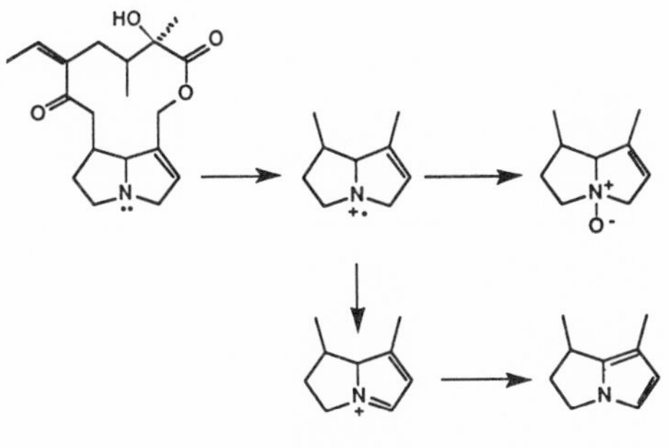

Scheme 11.

The N-dealkylation products, benzaldehyde and 1-phenylcyclobutylamine, predominate. However, both *N*-hydroxy,*N*-(1-phenyl)cyclobutyl-*N*-benzyl and *N*-(1-phenyl)cyclobutyl phenyl nitrone are also formed (Scheme 12). The nitrone appears to emanate from the hydroxylamine, and not from a benzylidene intermediate. Amphetamines are also oxidized to hydroxylamines by P-450 enzymes.[103] These appear to be further oxidized to C-nitroso compounds, which complex tightly to ferrous P-450 heme and give unusual spectra.[104]

It is intellectually satisfying to view N-oxidation reactions in terms of a common aminium radical intermediate, with the partitioning of products governed by the various rates of deprotonation and recombination (Scheme 13). These rate constants should be functions of the details of each active site, particularly spatial details related to the geometry of the transition states in each P-450 enzyme and each substrate. However, there are some other proposals and results regarding N-oxygenation reactions that must be considered. When ring-substituted aniline

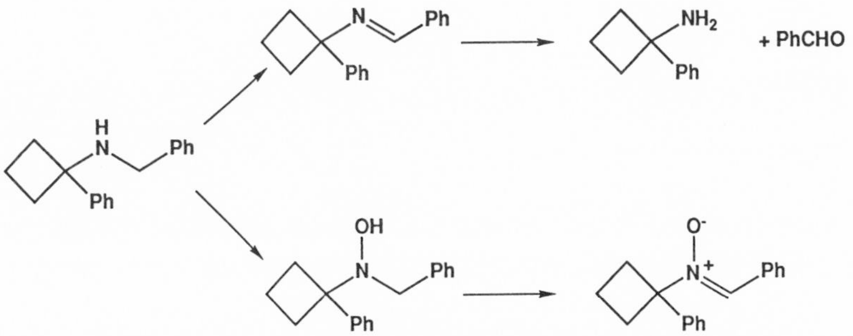

Scheme 12.

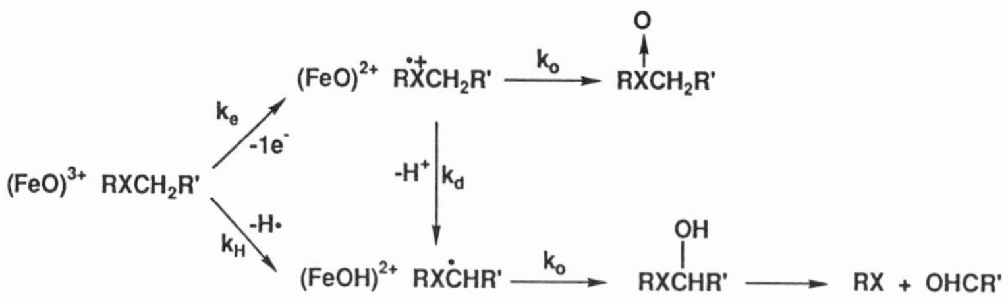

Scheme 13.

compounds are considered, the same Hammett plots are not seen for N-dealkylation and N-hydroxylation reactions,[105] which may be indicative of alternate mechanisms (*vide infra*). The possibility that an arylaminium radical is further oxidized to a nitrenium (or the nitrenium is formed directly) has been considered; the process would have to proceed by concomitant electron transfer to yield a formal $(FeOH)^{3+}$ complex that collapses via a heterolytic scission to the hydroxylamine product.[106] Hückel molecular orbital calculations suggest that the C-hydroxylation products resulting from the oxidation of 1- and 2-naphthylamine and 2-aminofluorene are better explained by such a heterolytic mechanism than one in which the charge of the aminium radical is dispersed into the ring structure.[106] However, it is not clear that these C-hydroxylation products emanate from the same oxidized nitrogen intermediate as opposed to more direct attacks at the carbons themselves. Further studies are necessary to unambiguously resolve this issue.

2.5. Kinetic Hydrogen Isotope Effects

Studies involving kinetic hydrogen isotope effects have been undertaken in efforts to elucidate mechanisms of amine N-dealkylation by P-450 enzymes.[49,50,57,69,70,107–112] These studies have utility in that an α-hydrogen must be lost in the reaction, and the magnitude of the isotope effect may provide insight into the mechanism, particularly with regard to whether hydrogen atom or electron transfer reactions are involved. The currently accepted dogma is that high isotope effects (> 5) are associated with a hydrogen atom abstraction mechanism, and low isotope effects (1–3) are associated with a mechanism involving rate-limiting one-electron oxidation of the nitrogen atom, rapid α-deprotonation of the aminium radical species, and oxygen rebound to yield a carbinolamine (Scheme 14).[113] In both mechanisms the C–H bond must be cleaved but the relative kinetic rates of individual steps determine the apparent isotope effect for the N-dealkylation reaction. Excluded from consideration is a mechanism in which P-450 abstracts the elements of a hydrogen atom from the aminium radical, for this mechanism of imine formation is ruled out by ^{18}O labeling studies in all cases examined to date.[61,62] However, aminium radical hydrogen *atom* abstraction may occur in

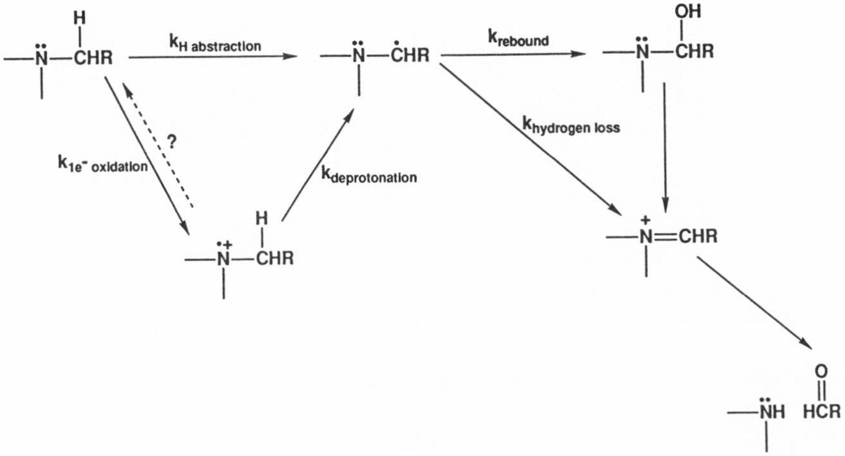

Scheme 14.

reactions catalyzed by some peroxidases, although the question of whether the hydrogen is abstracted as a hydrogen atom is not fully resolved and is actively under investigation.

The paradigm discussed above has been useful in understanding aspects of P-450-mediated N-dealkylation reactions. The basis for this view is that many chemical reactions for which there is good evidence that hydrogen atom abstraction is important show high kinetic *intermolecular* isotope effects; for example, permanganate-catalyzed N-dealkylation of a primary amine ($^Dk = 7.0$).[114] High *intramolecular* kinetic deuterium isotope effects are often observed for C-hydroxylation reactions catalyzed by P-450 enzymes.[51,115–118] Further, several chemical reactions for which one-electron oxidation is either obligatory, or for which strong evidence exists, show low isotope effects; for example, anodic N-dealkylation (1.8),[119] chlorite-mediated N-dealkylation (1.3–3.0),[120] permangate-mediated N-dealkylation of a tertiary amine (1.8),[121] ferricyanide-mediated N-dealkylation (3.6),[122–124] and photochemical-initiated N-dealkylation (2.2).[125] When P-450-catalyzed N-dealkylations have been examined, the intramolecular and intermolecular kinetic deuterium isotope effects are almost invariably low.[49,80,111,117] These low values are consistent with some theoretical calculations.[109] Further, ethers and esters, which have considerably higher $E_{1/2}$ values than amines and are not readily oxidized by one-electron processes, often show high isotope effects (> 10).[117,118,126,127] Amides are not as difficult to oxidize as ethers, but they are still more difficult than amines because of their higher $E_{1/2}$ values; amide N-dealkylation reactions are catalyzed by P-450 enzymes, but these are generally associated with higher isotope effects.[110] Indeed, the P-450 mechanism appears to shift from one of initial one-electron transfer to one of hydrogen abstraction when the $E_{1/2}$ of the substrate becomes > ~ 1.5–2 V in a series;[113] this trend is seen not only in groups of amines/amides, but also in alcohols/ethers,[117,118,128] phenyl halides,[129] and even hydrocarbons.[130]

While the work with kinetic isotope effects discussed above seems to have been useful in understanding mechanisms, there are a number of caveats which need to be considered, and some revision of concepts is required. Foremost, it should be emphasized that kinetic deuterium isotope effects estimated for enzymatic reactions are ratios of complex equations,[131] and P-450 is no exception in this regard. One way of simplifying the analysis is to reduce the complexity of the isotope effect by estimating the *intrinsic* kinetic deuterium isotope effect, which is the value applicable to the C–H bond breaking event itself. However, gaining access to the isotopically sensitive step in a complex catalytic sequence in order to obtain the intrinsic isotope effect (^Dk) can be extremely challenging. A strategy employing unsymmetrically labeled substrates, such as $RN(CH_3)(CD_3)$, can provide the intramolecular isotope effect; a value for the H/D discrimination at the isotopically sensitive step. The intramolecular isotope effect, while not the intrinsic isotope effect, represents both mechanistically and numerically the intrinsic effect better than the kinetic (intermolecular) isotope effect. In only a few cases have such parameters been reported for P-450 reactions.[49–51,117,132]

It is important to recognize that severe limitations exist on the interpretation of both kinetic and intramolecular isotope effects in P-450-catalyzed reactions as a consequence of precise mechanistic details of these transformations. For example, for full expression of intramolecular isotope effects on the α-hydroxylation of $RN(CH_3)(CD_3)$, it is required that rapid exchange occurs for the two methyl groups, which are presumably in different enzyme-active site pockets with different intrinsic rates for C–H(D) bond breakage. Such "equilibration" could occur either by rapid and reversible substrate entry onto the "activated" enzyme, or by simple rotation of the R–N bond *in the enzyme* at the active site (e.g., irreversible enzyme-substrate complex and subsequent rapid equilibration). Although it is generally assumed that equilibration occurs, it is not yet known which of these mechanistic options transpires. The sequence of events does, however, have extremely important consequences in interpreting isotope effect data; for example, if the second mechanism is operative, amines, with low energy barriers for rotation, might be expected to provide fundamentally different values than amides, which have high energy barriers for rotation. Another important question that remains unresolved is whether the initial electron transfer step (k_e– in Scheme 14) is reversible. Such reversibility in the electron transfer step, expected to occur as the redox potential of the substrate approaches and surpasses the enzyme redox potential, would have the effect of fully expressing the intramolecular isotope effect on a kinetically sensitive step. Thus, differentiating between a direct H abstraction mechanism and a reversible electron transfer/deprotonation mechanism cannot be unambiguously ascertained through isotope effect studies.

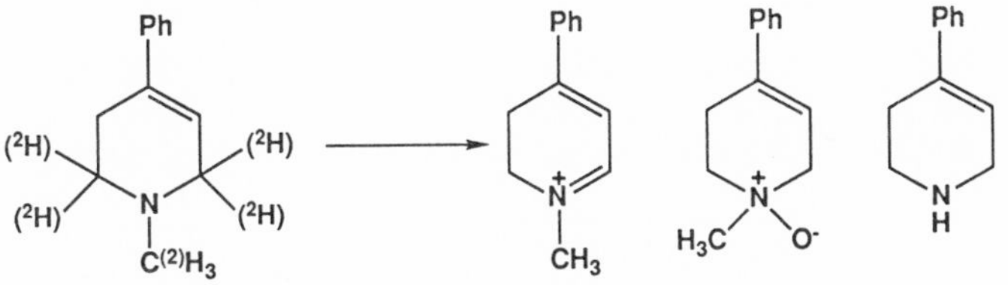

Scheme 15.

Another problem arises with the meaning of the high kinetic deuterium isotope effects observed for the N-demethylation of *N,N*-dimethylanilines by horseradish peroxidase and some other peroxidases.[111] These high isotope effects have been interpreted as evidence for a hydrogen atom abstraction mechanism.[111] However, with horseradish peroxidase there is some evidence that the iron and oxygen atoms of the Compound I and II forms are not accessible for hydrogen abstraction and oxygen rebound reactions.[72,133] Oxidations have been postulated to proceed instead through electron transfer reactions at the "heme edge" of the porphyrin; that is, a *meso* methine position.[72] Thus, the interpretation of the high kinetic deuterium isotope effects seen for N-dealkylation catalyzed by horseradish peroxidase may require reconsideration.

Related work on the mechanism of oxidation of 1-methyl-4-phenyl-1,2,3,6-tetra-hydropyridine by the flavoprotein monoamine oxidase may provide some insight[112] (Scheme 15). Oxidation to the dihydropyridine by this flavoprotein shows intermolecular (non-competitive) isotope effects of $^DV = 4.4$ and $^D(V/K) = 9.4$,[134] although there is considerable evidence from other experiments suggesting that the enzyme operates via sequential electron transfer mechanisms with other substrates.[135–137] The same reaction is also catalyzed by P-450 1A1 with $^DV = 3.0$ and $^D(V/K) = 1.0$.[112] It is difficult to evaluate the significance of the values obtained for the P-450 reaction in the absence of more data (because of the potential for attenuation); estimation of the intrinsic isotope effect would be useful in this regard. Nevertheless, the isotope effect for the monoamine oxidase-catalyzed reaction is definitely quite high. The point has been made that the electron transfer pathway is compatible with the high isotope effect if the initial one-electron oxidation is readily reversible, thus compensating an aspect of forward commitment to catalysis and providing a "release" for the aminium radical (to block attenuation of the isotope effect).[112] This hypothesis does not seem unreasonable and may apply to the high isotope effects seen in N-dealkylation reactions catalyzed by horseradish peroxidase.[111] However, in the case of the oxidation of the tetrahydropyridine oxidation catalyzed by P-450, the N-oxide is also a prominent product.[112] If both the dihydropyridine and N-oxide products proceed from a common aminium radical intermediate, then a higher kinetic isotope effect (for dihydropyridine formation) might have been expected, with "metabolic switching" (or "isotopically sensitive branching"[132]) to yield more of the N-oxide (Scheme 14). Kurebayashi[57] found a small increase in the level of the N-oxygenation product associated with a small kinetic isotope effect for deamination of cyclohexylamine; in the case of 1-methyl-4-phenyl-1,2,3,6-tetrahydropyridine the effect of deuterium substitution on N-oxide formation was not reported. In general, it does not appear that the existence of a branched pathway is sufficient by itself to generate high kinetic deuterium isotope effects for P-450-catalyzed reactions (see also Jones et al.[132]).

Another point requiring further consideration is the acidity of protons associated with radicals. The proton α to an aminium radical has long been considered to be highly acidic and lost rapidly after formation of the amine radical cation. In 1,4-dihydropyridines this site (C4) is separated by a double bond and is chemically analogous; the kinetics of deprotonation at this position have been measured and appear to be quite rapid in both aromatic-flanked systems[138] and unsubstituted dihydropyridines.[123,124] However, little direct information regarding the (thermodynamic) acidity of aminium radical α-protons has been available. Nelsen and Ippoliti[99] have suggested that α-aminium protons are not as acidic as previously thought on the basis of calculations using the Nicholas and Arnold equations.[139] One of the reasons that rapid deprotonation of aminium protons was seen in earlier studies (e.g., electrochemistry) was that a high concentration of base was usually present in the bulk solution.[99] More recently, Dinnocenzo and Banach[98] have prepared the tertiary amine cation radical N,N-di(4-anisidinyl)-N-methylamine as the arsenic hexafluoride salt and examined its reactions with quinuclidine. The kinetic data were interpreted to rule out all but a rate-limiting proton-transfer mechanism in which the kinetic deuterium isotope effect was 7.7. From their data, Dinnocenzo and Banach calculated a pKa of 10 for the methyl protons of the aminium radical (in acetonitrile). Further, they suggest that the pKa for loss of a methyl proton from the N,N-dimethylanilinium radical should be ~ 9, and that for 1-benzyl-1,4-nicotinamide cation radical (4-H) to be ~ 3.5. These considerations, along with an estimate of the pKa of 7 for the proton bonded to the *nitrogen* of an aniline radical cation,[97] may have considerable importance for the way in which we view P-450-catalyzed oxidations of amines.

A model is presented for unifying mechanisms of amine oxidation by P-450 enzymes which rationalizes the kinetic hydrogen isotope effects seen with these hemoproteins (Scheme 16). A key tenant of this proposal is specific base catalysis; that is, rapid removal of the α-proton by a group in the enzyme or by the $(FeO)^{2+}$ species itself. Several different P-450 enzymes of differing structure have all been found to show low kinetic deuterium isotope effects for amine N-dealkylation. It is certainly possible that the proximity of an electron pair near the aminium radical (through-space) would stabilize it. It seems highly unlikely that a single residue would be correctly positioned to accept the base in every case, given the variation in amino acid sequence.[3] However, since the mechanism of P-450 catalysis almost universally involves oxygen rebound, the $(FeO)^{2+}$ entity should be accessible in every case and positioned close to the proton. In addition, since the $(FeO)^{2+}$ species

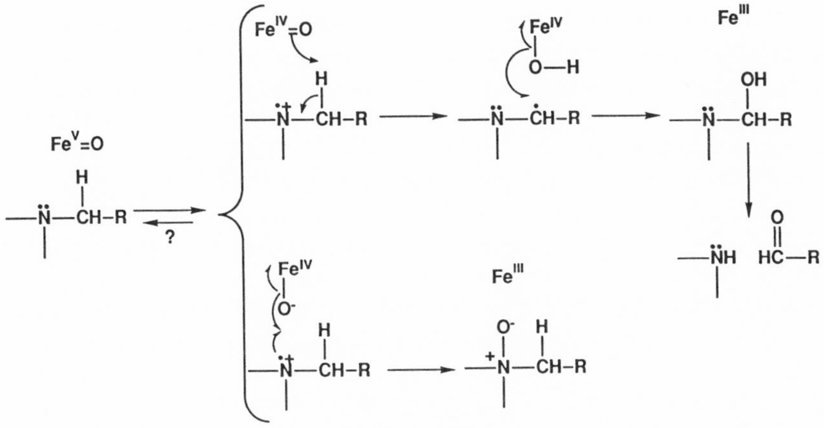

Scheme 16.

possesses a *net* negative (–1) charge, this species would be expected to be a basic moiety with high excess electron density around the oxygen atom. In some instances, the geometry within the active site is probably such that oxygen rebound to the nitrogen atom itself is feasible and competitive with α-hydrogen abstraction. This view would be strengthened if more evidence for partitioning from N-dealkylation to N-oxygenation products could be obtained, either through isotopic or group substitution. It would also be of interest to examine how other non-P-450 oxidases fit into this scheme, since several are known to catalyze heteroatom oxygenation and dealkylation reactions.[27,140–142]

The case of oxidation of 1,4-dihydropyridines may also be considered. In a sense this transformation is a sort of peroxidase reaction catalyzed by P-450 enzymes, in that a dehydrogenation equivalent to N-dealkylation occurs in the absence of oxygen rebound (*vide supra*[142]). Sinha and Bruice[123] have examined the oxidation of model dihydropyridines by obligate one-electron oxidants and found that the kinetic deuterium isotope effect seen for the removal of the hydrogen at the 4-position may be modulated by the strength of the oxidant (affecting the rate of initial one-electron abstraction from the nitrogen) and the base in the medium (affecting removal of the proton from the 4-position). Thus, a high isotope effect is observed when the initial one-electron oxidation is fast and removal of the proton is slow, and a low isotope effect is observed when electron transfer is slow and proton transfer is fast. To date, only low kinetic isotope effects have been seen in the dehydrogenation of 1,4-dihydropyridines by P-450 enzymes, except in one case with an N-phenyl-substituted compound[49,50,69] (one report of a high isotope effect[70] has not been reproduced[50]). It is conceivable that the $(FeO)^{2+}$ might be positioned above the plane of the dihydropyridine ring in such a way as to be in proximity to both the nitrogen atom and the 4-proton, but it is unlikely that such a feature would be seen in all of the different proteins.[142,143] The reason for the low kinetic deuterium isotope effects may well lie in an inherently lower pKa for the aminium radical,[98] thus facilitating rapid release even in the absence of specific base catalysis. In this regard, a limited study has shown that low kinetic tritium isotope effects are seen even with horseradish peroxidase.[111] The paradigm in Scheme 14 can explain many phenomena; it remains to be tested further.

2.6. Oxidations of Thioanisoles and *N,N*-Dimethylanilines

Substituted aromatic substrates are excellent diagnostic tools for investigating the influence of electronic effects in organic reactions. In this capacity, these classes of compounds have been utilized by a number of investigators to examine the role of electron transfer in oxidation reactions of P-450s and other enzymes.

Galliani et al.[144,145] examined rates of N-demethylation of a series of 12 *para*-substituted *N,N*-dimethylanilines in rat liver microsomes. When log k_{cat} was used as the parameter for comparison to σ, a ρ value of -0.94 was obtained. This value is in good agreement with the values obtained for the oxidation of some cyclic tertiary amines in aqueous solution by ferricyanide and ClO_2^-,[146,147] reactions considered to proceed via one-electron abstraction. The units regarding rates of N-demethylation in this work are unclear and difficult to reconcile with other studies, and no consideration was given to interference with the N-oxides which might be produced in such a crude system. However, subsequent work (*vide infra*) has verified their initial conclusions, and it is also of interest to note that much less dependence of the reaction rate upon lipophilicity and steric bulk factors was seen than on electronic influences.[144] In contrast, Burstyn et al.[105] also examined rates of N-demethylation of a series of *N,N*-dimethylanilines in rabbit liver microsomes and found little dependence of rates upon either σ_r or π, a lipophilicity parameter. However, the rate constants were rather low and it is difficult to evaluate the implications of these results.

Lindsay Smith and Mortimer[148] determined the rates for oxidation of a series of *meta*- and *para*-substituted *N,N*-dimethylbenzylamines in a chemical system utilizing iodosylbenzene and Fe^{III} or Mn^{III} tetraphenylporphins (solvent unspecified). They reported small negative ρ values of -0.41 for the Fe^{III} system, and -0.22 for the Mn^{III} system (absolute rates were not given).[148]

Watanabe, Iyanagi, and Oae have determined the rates of oxygenation of a series of *para*-substituted thioanisoles in various systems containing P-450s.[149–152] Using an enzyme preparation that is probably the rabbit P-450 2B4 enzyme, they found a ρ^+ value of -0.16 for the sulfoxidation of this series of thioanisoles.[152] The logarithms of the rates of sulfoxide formation were also correlated with the one-electron oxidation potentials (across the range of 1.26–1.85 V, vs. SCE). In addition, a ρ^+ value of -0.2 was obtained when the conversion of a similar series of sulfoxides to sulfones was examined, and a linear plot of log k_{cat} versus E_p was seen in the range of 1.75–2.05 V (vs. SCE).[149] However, in these studies rates of product formation were inferred from the substrate-dependent rates of NADPH consumption. This approach is not reliable in that many compounds are known to exacerbate the "upcoupling" of the enzyme system; that is, reduction of O_2 to H_2O_2 and H_2O without product formation.[153]

In our own studies we measured the rates of N-demethylation of a series of *para*-substituted *N,N*-dimethylaniline derivatives by purified rat P-450 2B1.[47,48] Two systems were used to support oxidation in this enzyme system. In the first, the enzyme was reconstituted in the usual manner with the flavoprotein NADPH-P-450 reductase supported by NADPH and O_2. In the other system, P-450 was mixed with iodosylbenzene in order to circumvent limitations in rates of substrate oxidation imposed by the overall process of enzyme reduction and oxygen activation. This latter system yielded rates an order of magnitude greater than seen with the normal system. When the *para* substituent in the series was varied across the range of -CH_3 to -NO_2 the ρ values were –0.61 and –0.74.[47] In this study it was demonstrated that any reactions leading to N-oxide could be neglected in the analysis because of low rates of N-oxygenation.

At this point mention should be made of studies on the oxidation of such series of compounds by other oxidases. Burstyn et al.[105] reported that varying the *para* substituent did not influence rates of N-hydroxylation in a series of anilines (the primary amines) in rabbit liver microsomes (probably due to P-450 2B4, *vide supra*), but the reported rates were rather low. Also, in the same study, no influence of substitution on rates of N-demethylation of *N,N*-dimethylanilines was seen (*vide supra*). Sakurada et al.[154] examined rates of reduction of horseradish peroxidase compound II and reported ρ values of –5.75 for anilines and –3.78 for phenols. In a paper by Kobayashi et al.,[155] the rates of S-oxygenation of *para*-substituted thioanisoles were not found to correlate with σ^+ for horseradish peroxidase, while a value of ρ = –1.40 was found for the same reaction catalyzed by chloroperoxidase. This ρ value is considered to be indicative of a radical cation/oxygen rebound mechanism. However, there are some caveats to interpretation of this study. Product formation was monitored only by changes in the UV absorbance at selected wavelengths. Another dilemma is that there is only limited evidence that the $(FeO)^{2+}$ entity is accessible for oxygen rebound reactions in horseradish peroxidase, and the enzyme is considered by some investigators to catalyze oxidations through electron transfer at the heme edge.[72]

Oae et al.[156,157] have examined rates of N- and S-oxygenation of substituted *N,N*-dimethylanilines and thioanisoles (respectively) by the biomimetic model 5-ethyl-4a-hydroperoxy-3-methyllumiflavin and flavin-containing monooxygenase. The model flavin hydroperoxide reactions yielded ρ values of –1.23 and –1.47 for the *N,N*-dimethylanilines and thioanisoles, respectively. It should be emphasized again that the reaction is thought to involve a rate-determining electrophilic attack of the hydroperoxide on the electron-rich heteroatoms,[85] a mechanism quite distinct from that now generally accepted for the dealkylation reactions catalyzed by P-450. Nevertheless, similar ρ values are observed and illustrate the complexity of interpreting data from rate-equilibria studies, such as the Hammett analysis, particularly when complex enzymatic sequences are involved. For example, no dependence of the rate of thioanisole oxidation on the substituent (or σ) was observed with the flavoprotein, presumably because of the rate-limiting nature of other steps in the enzyme reaction mechanism.[157]

It is critically important to recognize that significant limitations exist in the application of linear-free energy relationships (LFER), such as the Hammett analysis, or nonlinear free energy relationships, such as the Marcus approach (see below), to the study of complex enzymatic transformations such as those catalyzed by P-450s. The nature of these limitations is very similar to those encountered in the analysis of enzymatic kinetic isotope effects, although the theoretical bases for the isotope effect and LFER derive from distinct fundamental molecular properties. For example, an isotope effect or free energy relationship may be expressed only on V_{max} or on V_{max}/K_m, and the kinetically sensitive step (to isotope effect or LFER) may be "masked" by other rate-determining contributions to the overall catalytic sequence. "Dissecting out" the free energy relationship from a study utilizing the Hammett approach on the V_{max} of a P-450-catalyzed transformation is necessarily more complex and correspondingly more tenuous than the related analysis of isotope effects. This is a consequence of the requirement for analysis of a series of often structurally and electronically different compounds, which potentially possess different K_m values. Often the analysis of V_{max}/K_m for the transformation—undertaken in part to accommodate differences in substrate binding to the enzyme—will not yield interpretable data, because of the constraints on expression of V_{max}/K_m effects imposed by the catalytic sequence (analogous to the isotope effect analysis of V_{max} and V_{max}/K_m).[131]

A more recent approach to gaining insight into the correlation between kinetic and thermodynamic aspects of electron transfer reactions involves the nonlinear free energy relationship pioneered by Marcus and adapted by others. The use of these analyses in biological systems has been limited, usually restricted to aspects of electron transfer between well-characterized redox partners.[158-160] Because a large body of evidence had been obtained supporting the view that electron transfer is involved in the oxidation of amines by P-450, we sought to analyze the data in terms of classical outer-sphere electron transfer using the Marcus equation.[48] The analysis assumes that the putative initial one-electron transfer reaction must be at least partially rate-limiting. Of course it is also assumed that no appreciable factors other than electronic ones will influence differences in reaction rates seen among the N,N-dimethylanilines in the series. We found that the best fit of the k_{cat} values, where $k_{cat} = Ae^{-(\Delta G^+/RT)}$ was obtained with the Rehm–Weller expression:

$$\Delta G^+ = \Delta G°/2 = [(\Delta G°/2)^2 + (\lambda/4)^2]^{1/2}$$

and Agmon–Levine expression:

Table 1. Summary of the Analysis of the NADPH-P-450 Reductase-Supported and Iodosyldenzene-Supported Demethylation Data by Rehm–Weller and Agmon–Levine Approaches[48]

	$A^a(min^{-1})$	λ^a $(kcal\ mol^{-1})$	$\{E^{1}\!/_{2(app)}{}^a\ [V\ (SCE)]$
	NADPH-P-450 Reductase-Supported Systems		
Rehm–Weller	130 ($\pm$30)	22.6 ($\pm$1.6)	1.70 ($\pm$0.06)
Agmon–Levine	24 ($\pm$3)	21.8 ($\pm$1.0)	1.78 ($\pm$0.04)
	Iodosylbenzene-Supported Systems		
Rehm–Weller	2400 ($\pm$1200)	25.8 ($\pm$3.1)	1.96 ($\pm$0.09)
Agmon–Levine	370 ($\pm$40)	23.9 ($\pm$0.6)	1.97 ($\pm$0.05)

Note: Numbers in parentheses represent one standard deviation.

$$\Delta G^{+} = \Delta G^{\circ} = \frac{\lambda}{4\ln2}\ln[1 + e^{-4\ln2(\Delta G^{\circ}/\lambda)}]$$

which have been shown to be special cases of a more general equation for predicting the barriers to electron transfer.[161] (T is the absolute temperature, R is the gas constant, A is a pre-exponential factor, ΔG^{+} is the free energy of activation for electron transfer within the association complex, ΔG° is the standard free energy for electron transfer, and λ is the reorganization energy.) Estimates of values of the parameter λ were 23 kcal mol^{-1} and in the range of 1.7 to 2.0 for $E_{1/2}$ (Table 1).[48]

The curves fitted for the Agmon–Levine and Rehm–Weller equations were nearly superimposable and both passed near all points.[48] The two expressions yielded virtually identical λ and $E_{1/2(app)}$ values with significant differences appearing only in the pre-exponential term A. The difference in $E_{1/2(app)}$ for the reductase- and iodosylbenzene-supported systems is not significant, and an average of 1.85 V (SCE) is considered to represent both systems. The basic Marcus equation itself:

$$\Delta G^{+} = \frac{\lambda}{4}\left(1 + \frac{\Delta G^{\circ}}{\lambda}\right)^{2}$$

did not yield suitable fittings, demanding values of λ ranging from 70 to 100 kcal mol^{-1} and $E_{1/2(app)}$ from 4 to 5 V, primarily because of the inverted region of the parabola. The point has been made that the accuracy of these estimates of λ and $E_{1/2(app)}$ is limited because the data points lie close to the plateau region and that the analysis would be enhanced by consideration of more data points at lower values of k_{cat} and higher $E_{1/2(sub)}$.[162] However, what now appears to be clear is that with increasing $E_{1/2(sub)}$ the P-450 mechanism tends away from electron transfer and towards hydrogen atom abstraction, as we have previously emphasized.[92,113] Indeed we originally tried to include data points for anisole O-demethylation in the analysis but found the scatter considerable, as would be expected when electron transfer is not involved.

The differences in the values for the pre-exponential term A, derived from the Rehm–Weller and Agmon–Levine equations, are a function of the equations and are not relevant. However, it is significant that the difference in the three parameters between the reductase- and iosodylbenzene-supported activation protocols lies primarily in the pre-exponential factor, since this finding is consistent with the assumption in our analysis that oxygen activation is a pre-exponential contribution to substrate oxidation, and that substrate oxidation is rate-determining subsequent to this process. The Marcus equation is derived from statistical mechanics and requires that as the electron transfer becomes extremely exothermic ($G^+ \ll G°$), the rate of reaction becomes slower.[158,163] This requirement for the Marcus "inverted region," with which experimental observations seldom agree (see refs. 160,164,165), has led to the development of empirical approaches to correlating the rate of electron transfer with the difference in potentials of the donor-acceptor pair and an additional extrakinetic parameter. The Agmon–Levine and Rehm–Weller equations are representative of semiempirical relationships in which the predicted rate of electron transfer approaches an asymptote as the calculated exergonicity of the reaction becomes greater; the asymptote of these equations is thought to represent the diffusion-controlled rate for electron transfer in intermolecular reactions.[160,166] The asymptote that is approached in the case modeled here is a function of the equilibrium constants of the equilibria preceding electron transfer, and depends critically on the method of oxygen activation.

The "fit" of the k_{cat}-redox potential data to the semiempirical Agmon–Levine and Rehm–Weller equations is extremely good for studies of this complexity. However, it must be emphasized that suitable interpretation of the parameters derived from these equations depends upon several assumptions. Perhaps the most critical assumption is the requirement that the electron transfer step be the rate-limiting step subsequent to oxygen activation (see above); no chemical process or physical step, such as enzyme "off-rate", may contribute to rate determination. For example, should the off-rate of the enzyme-product complex partially contribute to the rate-determining step, the asymptote of the rate-$E_{1/2}$ relationship would represent the enzyme off-rate. If such a case were to obtain, the parameters derived from the Rehm–Weller and Agmon–Levine analysis would be slightly altered. Although we cannot accurately calculate the exact magnitude of the alterations in the reorganization energy and the apparent enzyme redox potential (see below), it is clear that an increasing contribution of the enzyme off-rate to the rate-determining step would serve to *lower* the enzyme $E_{1/2 (app)}$. Despite considerable efforts to assess the contribution of the enzyme product off-rate to the "postoxygen activation" enzyme catalytic rate (which collectively suggest the contribution is minimal),[48] we cannot unambiguously rule out this possibility. Nonetheless, this approach has provided, for the first time, data regarding the redox potential of the enzyme-activated oxygen species; further research will be required to confirm this estimate.

The analysis provides average values for λ of 23 kcal mol^{-1} and for $E_{1/2(app)}$ of +1.85 V (SCE) (Table 1). Support for this reorganization energy value (λ) for the electron transfer between *para*-substituted *N,N*-dimethylaniline substrates and a heme-oxo species in an enzyme active site can be derived from model studies. From a variety of chemical studies it is known that the λ values for *N,N*-dimethylaniline/ *N,N*-dimethylaniline radical cation electron self-exchange are low (< 15 kcal mol^{-1}).[167,168] The reorganization energies for electron transfer in proteins with porphyrin prosthetic groups span a considerable range. Although theoretical treatments of electron transfer in porphyrin redox proteins suggested that the reorganization energies for these processes might be low (< ~ 8 kcal mol^{-1})[169], recent experimental data suggest that the λ value for electron transfer in natural and model bis(porphyrin) protein couples is larger (~ 15–20 kcal mol^{-1};[170,171] ~ 50 kcal mol^{-1};[172] e.g., see McLendon et al.[173] and Peterson-Kennedy et al.[174]). However, all of these electron-transfer processes are "long-range," in contrast to that proceeding in the P-450 active site, and the reorganization energy is distance-dependent.[158] Assuming an intermediate range (~ 30 kcal mol^{-1}) for the self-exchange of the P-450 $(FeO)^{3+}/(FeO)^{2+}$ couple, a reorganization energy of ~ 23 kcal mol^{-1} for the P-450-catalyzed oxidation of *N,N*-dimethylaniline substrates would seem to be in excellent agreement.

The estimated $E_{1/2(app)}$ of 1.85 V (SCE) for the P-450 $(FeO)^{3+}/(FeO)^{2+}$ couple is higher than the values measured for model mangano[175] and iron porphyrins[40,41] and the hypervalent forms of horseradish peroxidase.[42] These values range from 1.0 to 1.8 in the model metalloporphyrins, depending upon the solvent and ligands. The $E_{1/2(app)}$ values for oxidation-reduction of horseradish peroxidase Compounds I and Compound II are accessible by dye equilibration methods with the reported (pH-dependent) values being 0.75 and 0.73 V (vs. SCE) at pH 7.0. Unlike some other peoxidases, horseradish peroxidase does not appear to involve protein radicals and the measured $E_{1/2(app)}$ is believed to reflect redox changes of the heme-oxo/porphyrin species.

We propose that subsequent to electron transfer, coulombic interactions between the charged heme-oxo radical anion/*N,N*-dimethylaniline radical cation pair can account for the enhanced $E_{1/2(app)}$ of P-450. In our treatment, the $E_{1/2(app)}$ is the sum of an intrinsic oxidation potential, $E_{1/2(int)}$, for the $(FeO)^{3+}$ porphyrin core of the enzyme and a coulombic factor [$E_{(cf)}$], which arises from the gain or loss of electrostatic free energy upon transfer of the electron in the transition state.[176] The free energy for the electrostatic contribution, $G_{(cf)}$, is expressed below as:

$$G_{(cf)} = (Z_1 - Z_2 - 1)e^2f/r_{1/2}D$$

$$E_{1/2(app)} = E_{1/2(int)} + E_{(cf)}$$

$$E_{(cf)} \text{ (expressed in V)} = +14.4/r_{1/2}D,$$

where Z_1 and Z_2 are the initial charges of the enzyme active oxidant and the substrate, respectively, e is the electronic charge, f is a factor expressing the influence of ionic strength (μ) on the coulombic interaction that becomes unity near zero ionic strength, r is the center-center internuclear distance in the transition state between the nuclei of Z_1 and Z_2 (in Å), and D is the static dielectric constant of the enzyme active site. Assuming that the active site of P-450 has negligible ionic strength ($\mu \cong 0$), and because the substrate and iron-oxo core of the enzyme are uncharged [due to the heme-axial thiolate ligation of the $(FeO)^{3+}$ species, *vide infra*], $E_{(cf)}$ collapses to 14.4 Å/$r_{1,2}$D by substitution in the Nernst equation. Thus, the internuclear distance for electron transfer and the dielectric constant of the P-450 active site assume crucial importance in determining the magnitude of $E_{(cf)}$.

Neither the internuclear distance ($r_{1,2}$) nor the static dielectric constant (D) of the active site are known, although reasonable estimates of these critical parameters can be advanced. For illustration, a distance of 5 Å between the charged centers (the nuclei of the negatively charged oxygen atom and the positively charged nitrogen atom) and a dielectric constant of 3 are considered. Support for a 5 Å distance approximation derives from substrate-free and substrate-bound bacterial P-450 101 (P-450$_{cam}$) in which the distance between the $(FeO)^{3+}$ oxygen and the hydrogen abstracted in camphor is postulated to be less than half of this proposed distance.[177] Presumably, distances substantially greater than 5 Å (e.g., 20 Å) are not realistic in the active site. Controversy currently surrounds the estimation of dielectric constants in proteins (for conflicting viewpoints, see Gilson and Honig,[178,179] Warshel and Russell,[180] and Warshel[181]), since the range of possible values (2.0–80 or higher) has a dramatic impact on the magnitude of electrostatic interactions in proteins, which are thought to affect both protein structure and function. To illustrate the size of the postulated coulombic interaction in P-450 electron transfer, we utilize a dielectric constant of 3. This value has been theoretically estimated to be the effective dielectric constant for a folded protein at internal distances greater than 5 Å from the protein-water interface,[178] and is intermediate between the 2.0[179] and 3.5 values[182] typically employed for proteins. At these postulated values for the distance of electron transfer and the active-site dielectric constant, $E_{(cf)}$ corresponds to 1 V, which would serve to enhance the intrinsic oxidation potential of the enzyme. A decrease in the dielectric constant to 2 increases $E_{(cf)}$ to ~ 1.5 V; increases in either the distance or dielectric constant serve to correspondingly reduce the magnitude of $E_{(cf)}$.

We therefore conclude that in P-450 catalytic processes initiated by electron transfer, coulombic attraction of the ion pair produced by this reaction contributes significantly to the $E_{1/2(app)}$ of the enzyme. At present, the contributions to the effective oxidation potential of P-450 by the intrinsic oxidation potential of the $(FeO)^{3+}$ species and the coulombic factor cannot be rigorously dissected. Nonetheless, the coulombic contribution must be significant if model studies are our guide. For example, if the 1.0- to 1.3-V estimates of the oxidation potential of the $(FeO)^{3+}$ core of the enzyme derived from model studies are reasonable, the magnitude of the electrostatic interaction is postulated to be approximately 0.5 to 0.8 V. The proposal of a coulombic interaction in P-450-mediated electron transfer reactions is dependent upon the overall charge neutrality imparted to the $(FeO)^{3+}$ species by the cysteinyl thiolate ligand and the heme dianionic ligand. Thus, the unique chemistry of P-450, long presumed to be a function of the unusual axial thiolate ligand of this hemoprotein,[45,183,184] may be in part due to the overall charge neutrality of the high-valent heme-oxo species afforded by this ligand. The axial thiolate ligation would also serve to localize anionic charge on the ferryl oxygen subsequent to electron transfer,[183,184] which would additionally serve to enhance the coulombic interaction. These properties of the P-450 $(FeO)^{3+}/(FeO)^{2+}$ couple conferred by the axial thiolate ligand have not been previously recognized and imply an additional role for this ligand in the chemistry of P-450. While it has long been thought that a critical role of the thiolate anion ligand is to facilitate heterolytic cleavage of the O–O bond of the bound oxygen,[183,184] recent studies have argued that specific base catalysis at the distal face of the porphyrin may be more important in this regard.[185]

The proposal of an effective oxidation potential for P-450 of ~ 1.85 V (SCE), and of a significant contribution to this potential from a coulombic interaction, has important implications for the mechanisms of P-450-catalyzed oxidations of heteroatom-containing and unsaturated substrates and for model systems that mimic the enzyme. For example, such electrostatic interactions may not be significant in the reactions of model systems or in enzymatic processes that proceed at the protein-water interface, since the oxidant may not be neutral or the dielectric constant of the media may be sufficiently large (> 20). Thus, the effective oxidation potential of these oxidants would not be enhanced by $E_{(cf)}$ and consequently not represent ideal models of the enzymatic transformations of P-450. In addition, the proposal of an $E_{1/2(app)}$ of ~ 1.85 V for P-450 dramatically broadens the scope of substrates that are potential substrates for enzymatic oxidation by initial electron transfer. To quote the concluding remarks of Eberson:[162] "It is tempting to assume that with decreasing reactivity of the substrate, we have a range of mechanisms beginning with fully developed ET (electron transfer) and radical cation formation and ending with a synchronous electron/proton or electron/oxgyen transfer step (or classically, a hydrogen or oxygen atom transfer step). Then it is only the timing between the successive steps that determines whether the typical reactions of radical cations will be exhibited or not." This view is in concert with our own paradigm for the P-450 mechanism,[92,113] which is generally accepted by others[45]: when P-450s encounter substrates of low $E_{1/2}$ they tend to function in an electron transfer mode; when the $E_{1/2}$ is higher, P-450s shift to a hydrogen atom abstraction mode. In this regard the effective $E_{1/2}$ is a function of the distance between the reacting centers of interest, of course. Experimental data are consistent with a role of P-450 in the single electron transfer oxidation of substrates with a range of oxidation-reduction potentials for amines ($E_{1/2} = ~ 1.0$ V) to alcohols and iodides ($E_{1/2} = < ~ 2$ V).[128,129] Interestingly, there appears to be a shift in the magnitude of the kinetic deuterium isotope effect near this $E_{1/2}$.[113]

3. OXIDATIONS OF OTHER ATOMS

3.1. Sulfur and Phosphorus Oxidation

The general reaction with sulfur compounds is S-oxygenation, and S-dealkylation reactions are also observed but are not predominant.[157] This situation contrasts with the analogous nitrogen compounds, where N-dealkylation is usually predominant over N-oxygenation when α-hydrogens are present (*vide supra*). This difference can be attributed to the greater stability of sulfur radical cations compared to nitrogen. In many cases unstable S–O compounds are formed and undergo unusual rearrangements.[186]

Phosphines have even lower $E_{1/2}$ values than sulfides and are oxygenated by P-450s.[187] These can also be formed from $(FeO)^{2+}$ compounds; for example, reaction of triphenylphosphine with model oxo-metalloporphyrin systems.[188]

3.2. Alkane Oxidation

Alkane oxidation is generally understood in terms of C-hydroxylations, which are thought to involve hydrogen atom abstraction and oxygen rebound (Scheme 17). However, with some diagnostic and even natural substrates, rearrangements of the sort already mentioned in regard to cycloalkylamines are encountered.

Some of the evidence for the view that hydrogen atom abstraction is a common theme of alkane oxidation comes from the loss of stereochemistry which is observed in hydroxylations at methyl and methylene carbons. A high degree of scrambling is seen when a (chiral) $-C^1H^2H^3H$ derivatized methyl group is hydroxylated,[189] with only a small degree of retention of configuration. Even with some methylene hydroxylations extensive racemization occurs in forming the hydroxylated product.[190–192]

Studies with substituted cyclohexenes indicate the rehybridization of a putative allylic methylene radical as an explanation for the products observed in oxidations catalyzed by P-450 or biomimetic models (Scheme 18).[192] The oxidation of the terpene pulegone to menthofuran is also rationalized by the oligate rearrangement of an allylic radical.[193] The transformations of alkene allylic hydroxylation is further discussed under the heading of "Oxidation of π-Bonds."

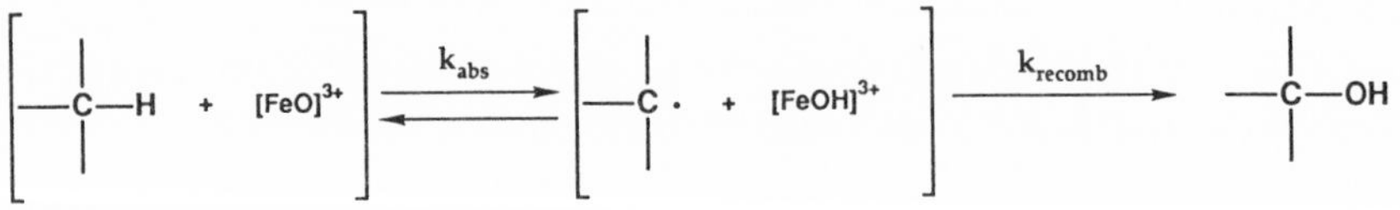

Scheme 17.

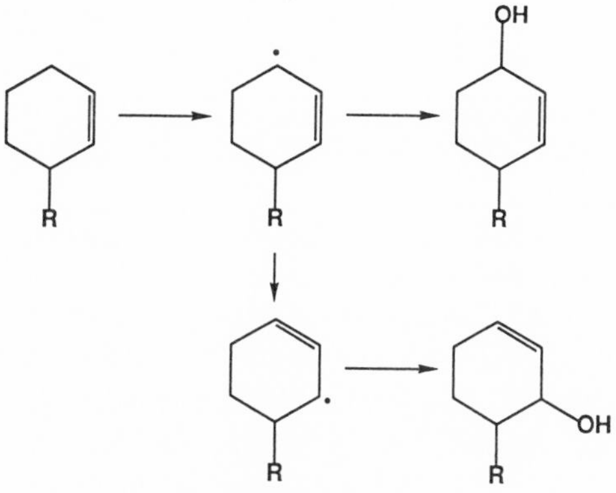

Scheme 18.

A number of cyclopropyl "radical clock" substrates have been utilized with P-450s to provide further evidence for the intermediacy of radicals, and to estimate the rate of radical recombination. Nortricyclane and methylcyclopropane were oxidized to their hydroxylated derivatives without rearrangement.[130] However, oxidation of bicyclo[2.1.0]pentane yielded a 7:1 mixture of *endo*-2-hydroxybicyclo[2.1.0] pentane and 3-cyclopenten-1-ol (Scheme 19). Further, considerable (mechanism-based) inactivation of the P-450 was seen during the course of the enzymatic incubations, suggesting that the putative radical was reacting with the protein. On the basis of these observations, Ortiz de Montellano and Stearns

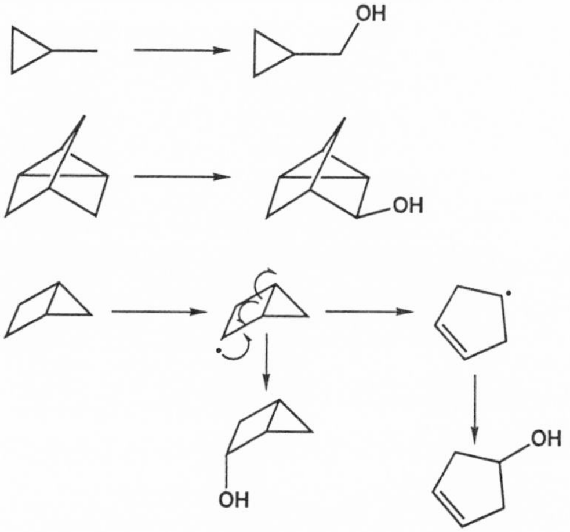

Scheme 19.

estimated that the radical pair formed in P-450-catalyzed hydroxylations collapses at a rate $> 10^9$ s^{-1}.[130]

Ingold has recently extended some earlier studies on rates of oxygen rebound in P-450s using a series of cyclopropanes of which the rates of ring opening and closing (of the radicals) have been estimated by nitroxide radical trapping.[95] Three of these compounds [cis- and trans-1,2-dimethylcyclopropane and bicyclo (2.1.0.) pentane] gave mixtures of products derived from direct rebound and from a mechanism involving ring opening, which were used to derive an estimate of the rate of radical recombination of 2×10^{10} s^{-1}.[95] The studies were done with liver microsomal preparations derived from male phenobarbital-treated rats, as was the work of Ortiz de Montellano.[130] The work is suggestive that the enzyme being investigated is P-450 2B1, although more definitive proof is lacking. The extent to which the rate of oxygen rebound will vary among the different P-450s (and other monooxygenases where similar phenomena have been observed[194,195]) is unknown. As pointed out by Bowry and Ingold,[95] cyclopropane radical clocks should not be used indiscriminantly with enzymes. In their work, evidence was provided that rates of ring opening in the enzyme cavity (of P-450 2B1 ?) are similar to those in solution, but this is not necessarily the case in all situations, particularly those compounds with bulkier groups which may either resist rearrangement or which after rearrangement may have the radical placed in a position which cannot recombine with the $(FeOH)^{3+}$ complex. Moreover, all of these reactions are reversible, and the oxygen rebound step may serve to "trap" a less stable (putatively) unrearranged species due to physical constraints on the rebound step. Thus, rapid (ring opening/closing) reversibility may falsely imply a lack of rearrangement.

As pointed out earlier, oxidations of alkanes are generally considered to involve initial abstraction of hydrogen atoms (Schemes 13, 14, 17–19). However, some strained alkanes have low $E_{1/2}$ values and are easily oxidized by one-electron transfer reactions. Quadricyclane ($E_{1/2}$ 0.9 V vs. SCE) is oxidized by P-450 to the bicyclo [3.1.0.] hexene aldehyde derivative by P-450, a reaction characteristic of a radical cation intermediate, as well as to nortricyclanol (the formation of which is not P-450 dependent).[196] The products are rationalized in Scheme 20. Thus, the continuum between abstraction of electrons and hydrogen atoms as a function of $E_{1/2}$ is seen in unsubstituted alkanes[46,196,197] as well as heteroatoms (vide supra).

Desaturation reactions are observed in P-450 reactions and have been clearly shown not to involve alcohol intermediates in several cases now. The list of substrates which undergo desaturation includes valproic acid,[198] testosterone,[199] lovastatin,[200] ethyl carbamate,[201] lindane,[202] dihydroergosterol,[203] warfarin,[204] and probably ethyl bromide.[201] (Not included here are nitrogen-containing systems such as the 1,4-dihydropyridines or acetaminophen (vide supra).) Desaturation does not appear to be favored over hydroxylation to the extent seen in some other monooxygenases more devoted to dehydrogenation of natural substrates such as fatty acids,[205] but the catalytic mechanism is probably similar and follows the

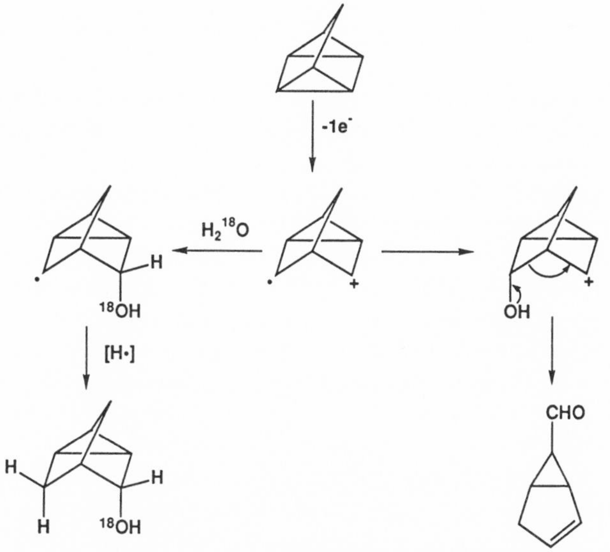

Scheme 20.

course shown in Scheme 21. Both hydroxylation and desaturation show high kinetic deuterium isotope effects.[198,201] In the case of valproic acid, only deuterium substitution at the terminal (methyl) carbon yields an isotope effect, suggesting initial oxidation at that site and rapid loss of the second formal hydrogen atom.[198] In most of the cases seen to date some alcohol formation is observed.

At this time, the partitioning between oxygen rebound and abstraction of a (formal) second hydrogen atom cannot be understood in discrete physical terms and must be attributed to factors in the different transition states of the various enzyme/substrate complexes that are not understood.[206]

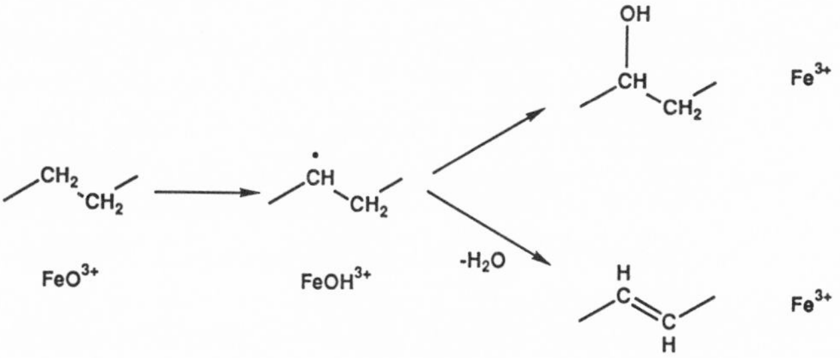

Scheme 21.

$$(FeO)^{3+} \ ROCH_2R' \longrightarrow (FeO)^{2+} \ RO\overset{\cdot}{C}HR' \longrightarrow$$

$$Fe^{3+} \ ROCH(OH)R' \longrightarrow Fe^{3+} \ ROH \ OHCR'$$

Scheme 22.

The O-dealkylation of ethers and carboxylic acid esters is probably best understood in terms of hydrogen atom abstraction from the carbon atom, oxygen rebound, and rearrangement of the hemiacetal-like product (Scheme 22). The $E_{1/2}$ values for one-electron abstraction from ethers and especially carboxylic acid esters are relatively high for P-450 oxidation through an electron abstraction mechanism.[113] Also, very high intrinsic kinetic deuterium isotope effects have been measured for ethers[117] and esters.[51,126] Conceivably these high-isotope effects may reflect reversible electron transfer and slow deprotonation. This possibility is discussed in the section under "alcohol oxidation" (*vide infra*).

3.3. Iodine Oxidation

In the late 1970s we considered the possibility that, if the oxygen species activated by P-450 were electrophilic as often thought at the time, then it might react with the outer shell electrons of organic halides and give rise to haloso entities (Scheme 23).[207] However, the only stable haloso compounds known were iodosylbenzene derivatives, and these are readily oxidized by protein sulfhydryl groups.[208]

An approach was used in which the reaction was supported by iodosylbenzene and [^{125}I]iodobenzene was the substrate. Thus the excess of iodosylbenzene present served to trap any radiolabeled product which would have been formed (Scheme 24): P-450 enzymes were clearly demonstrate to catalyze the reaction.[207] However, evidence for the oxidation of iodine atoms in the normal catalytic system (sup-

$$RX \longrightarrow RX\text{-}O$$

$$RCH_2XR' \longrightarrow XR' + RCHO$$

Scheme 23.

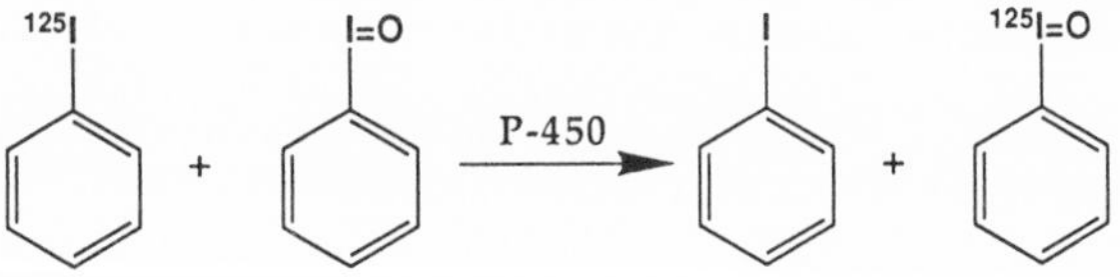

Scheme 24.

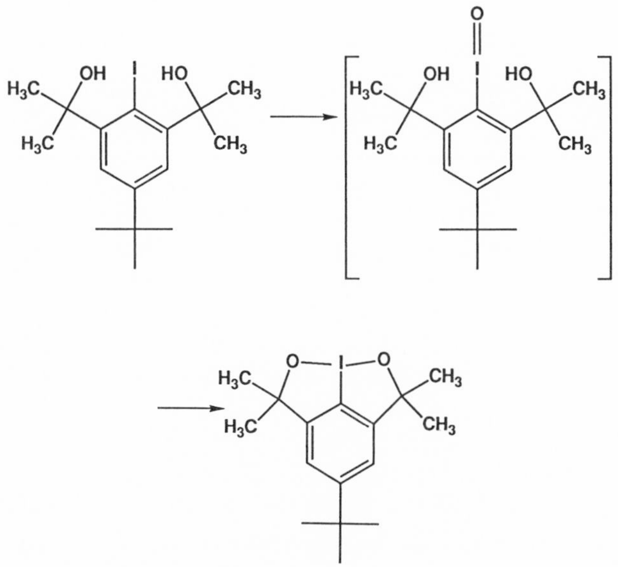

Scheme 25.

ported by NADPH and O_2) remained elusive. The availability of 4-*tert*-butyl-2,5-bis [1-hydroxy-1-(trifluoromethyl)-2,2,2-trifluoroethyl] iodobenzene provided a possible solution; the oxidized product (the iodinane) is relatively stable and more than one-half could be recovered after addition to the protein system under consideration, and the substrate and product were readily separated by HPLC (Scheme 25).[129] The oxidation of 4-*tert*-butyl-2,5-bis [1-hydroxyl-1-(trifluoromethyl)-2,2,2-trifluoroethyl] iodobenzene to 10-*tert*-butyl-3,3,6,6-tetrakis(trifluoromethyl)-4,5,6-benzo-1-ioda-7,8-dioxabicyclo[3.3.1]octane by P-450 was clearly demonstrated in the presence of NADPH and oxygen. Oxidation of the corresponding bromide was not detected, consistent with the measured difference in $E_{1/2}$ (0.8 V). Oxidation of the iodide could also be demonstrated with horseradish peroxidase or model metalloporphyrins, but was considerably slower.[129] It is not actually possible to distinguish between a reaction involving oxygen transfer and dehydrogenation (cf. upper path of Scheme 5), or net two-electron transfer (with abstraction of protons, lower path of Scheme 5).

It is our view that this oxidation reaction proceeds by one-electron transfer. The $E_{1/2}$, +2.0 V vs. SCE, is within the range of the P-450 $(FeO)^{3+}$ species and even the model metalloporphyrins.

3.4. Oxidation of π-Bonds

The π-bonds of alkenes, alkynes, and aromatic compounds undergo several distinct metabolic pathways. Although the P-450-catalyzed transformations of these systems are often considered individually, these processes can be collectively considered to be variations on a common "mechanistic theme". The variations in mechanism are caused by the wide diversity in electronic characteristics, such as the redox potential, the steric features of each system, the particular structure in which the π-bond exists, and the thermodynamic "driving force" of aromaticity, when applicable. It is this structural and electronic diversity in π-systems that serves to produce a variety of possible metabolites. An additional complicating matter is that the pathway of P-450-catalyzed oxidation of many π-bonded structures is often influenced by the particular P-450 enzyme.

Despite the caveats stated above, it has long been recognized that the "normal" pathway of P-450-catalyzed oxidation of olefins results in epoxides (Scheme 26) and the "normal" pathway for unsubstituted aromatic compounds results in phenols (Scheme 27). The "normal" oxidative pathway for alkynyl compounds is more complex and results in both "suicide inhibition" and carboxylic acids, if the substrate is a terminal alkyne (Scheme 28).

Although the precise mechanistic details of these transformations have not been completely resolved, it has become clear in recent years that the mechanisms of monooxygenation of all of these π-bonded systems involves not a concerted oxygen transfer, but instead a stepwise sequence of oxygen addition to the π-system and subsequent collapse to the product. Such a stepwise reaction scheme has enabled rationalization of the array of alternate metabolites often derived from P-450-catalyzed transformation of π-bonded systems. The differences in product profiles for a particular compound that are often obtained for various P-450 enzymes are thought to be a consequence of subtle differences in active site

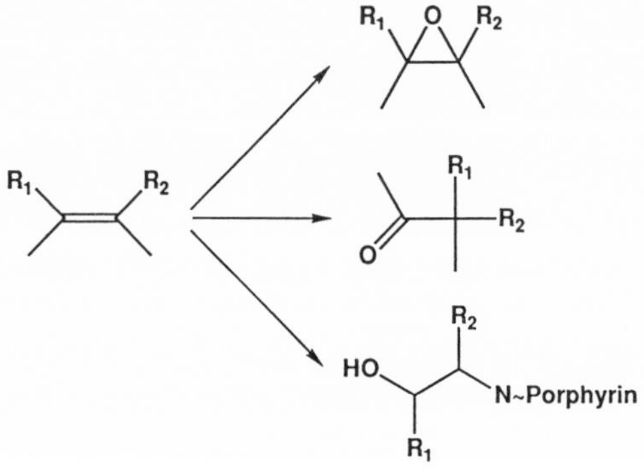

Scheme 26.

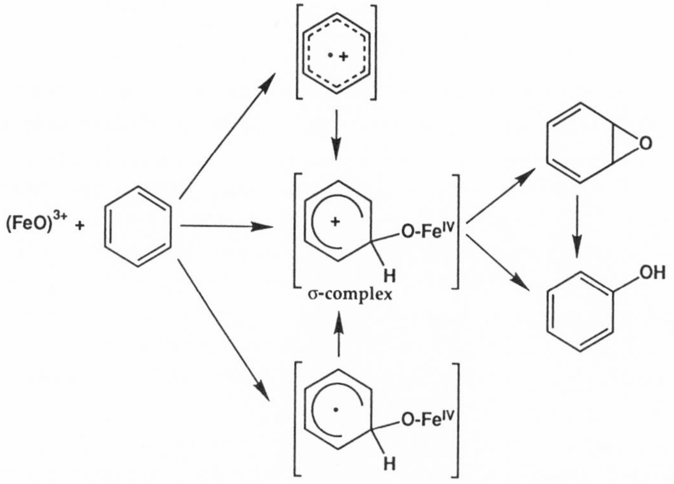

Scheme 27.

geometries that influence the various competing rate constants for ring closure, rearrangement, and reaction with active site nucleophiles of the initial iron oxy-π-system adduct or σ-complex. Importantly, current mechanistic hypotheses propose that all of the observed products are derived from collapse of the initial iron oxy-π-system σ-adduct. The discussion that follows will address current mechanistic thinking regarding formation of the enzyme-π-bond σ-complex.

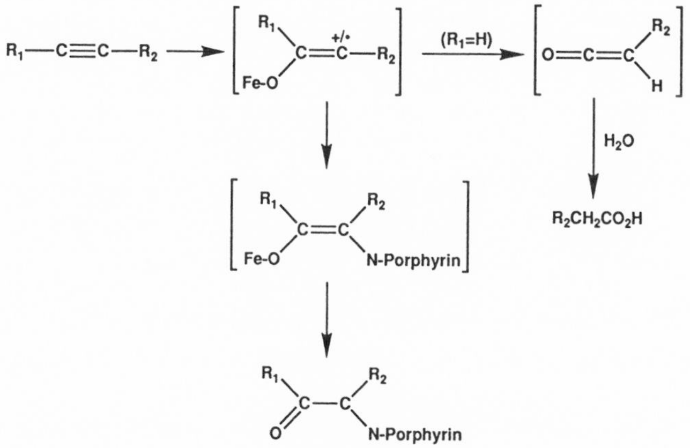

Scheme 28.

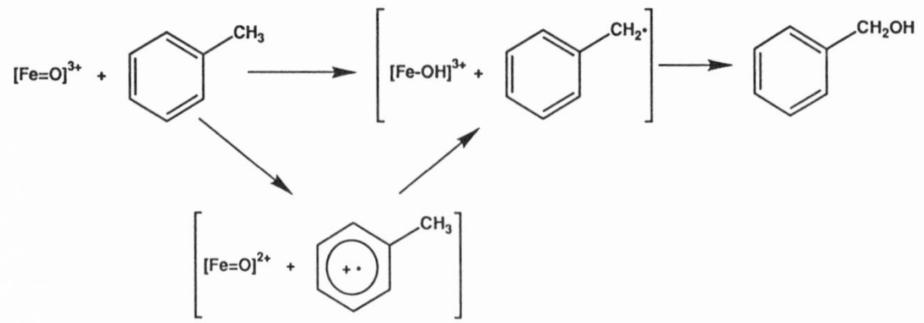

Scheme 29.

If the π-bonded system possesses an alkyl substituent, hydroxylation adjacent (or α) to the π-bond can occur. This pathway is often the exclusive pathway for alkylated aromatics, such as toluene (Scheme 29), and tetrasubstituted alkenes. Whether this α-hydroxylation pathway is best viewed as an alkane hydroxylation process that is independent of (but facilitated by) the π system, or as a process that is initiated by and integral to oxidation of the π-system, has not been resolved. Based on the limited data available regarding the oxidation of π-systems by P-450, we have advanced a unified and general scheme for the oxidation of all π-bonded systems (Scheme 30).

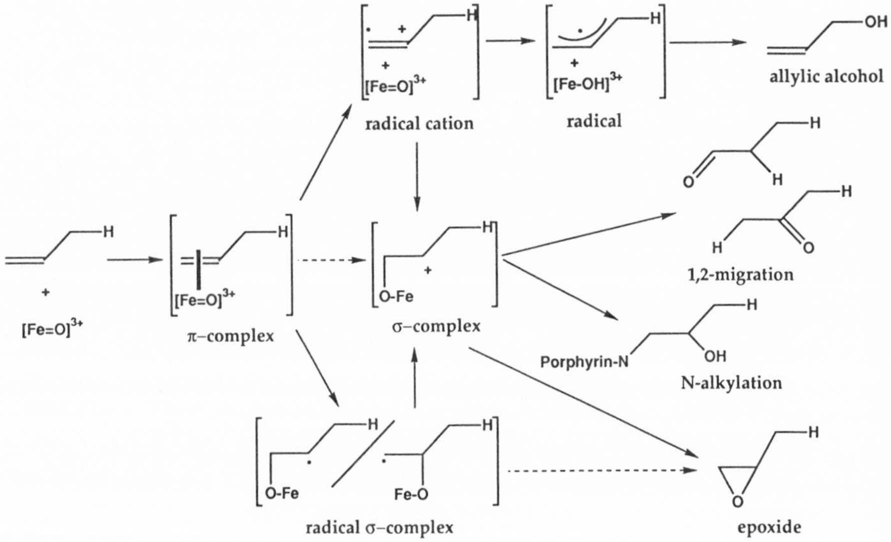

Scheme 30.

Scheme 30 proposes that P-450 initiates both π-oxidation and α-hydroxylation through initial π-complexation followed by either electron transfer or radicaloid addition to the unsaturated moiety. This scheme also proposes that if the π system exhibits a sufficiently low redox potential (< 2.0 V vs. SCE), electron transfer will transpire leading to a substrate radical cation and enzymic iron-oxy radical anion. The radical ion pair derived from electron transfer is proposed to collapse to either the classical σ-complex via C–O bond formation or to undergo deprotonation to generate the intermediate derived from H-atom abstraction. Such deprotonation reactions are well characterized for aromatic radical cations, such as toluene for which the pKa is estimated to be –6, and may well be facilitated by the active site oxy anion base generated upon electron transfer. The alternative pathway of radicaloid addition of the iron-oxo species directly to the π-system (presumably the pathway of choice for substrates with inaccessible redox potentials) would produce an initial "radical σ-complex" would be anticipated to undergo internal electron transfer to produce the "classical" cationic σ-complex. The "classical" σ-complex would proceed to products via well established reaction mechanisms.

At present, this "unified mechanistic scheme" for the oxidation of π-systems by P-450 must remain a hypothesis. Although a body of mechanistic studies have served to confirm the stepwise nature of π-system oxidation by P-450, these studies have not provided convincing insight into the precise nature of the steps leading to the presumed σ-complex intermediate. In addition, the considerable volume of studies using metalloporphyrin models of the enzyme have not illuminated the mechanistic details of the enzymic oxidation sequence; in fact, these models have often served to illustrate the differences between the models and the enzyme and to demonstrate a lack of mimicry of the enzyme (*vide infra*). Nonetheless, limited evidence suggests that P-450 can oxidize some π-systems by electron transfer.

The most direct evidence for the involvement of electron transfer oxidation of π-bonded systems by P-450 derives from the work of Cavalieri, Rogan, and their colleagues.[209–211] In a series of studies lasting over a decade, these researchers have demonstrated that P-450 can oxidize polynuclear aromatic hydrocarbons, such as benzo[*a*]pyrene, to radical cations, which are characterized through the identification of radical cation-specific products.[209,210] With benzo[*a*]pyrene, a potent environmental contaminant and carcinogen, an important alternative pathway of arene epoxidation has also been demonstrated to occur,[212] although the precise sequence of chemical steps has not been elucidated. The nature of the P-450-mediated electron transfer pathway in the metabolism of polynuclear aromatics, such as benzo[*a*]pyrene, can be disputed, however. Since P-450 is an overall two-electron oxidant, it is *a priori* implausible that the enzyme would release a highly reactive, single-electron oxidized radical cation from its active site. Cavalieri et al. suggest that the radical cation pathway may represent a nonactive site, "heme edge" oxidative process akin to that observed for peroxidases with polynuclear aromatic compounds that have low oxidation potentials.[209,210] Nonetheless, these data demonstrate that P-450 can oxidize aromatic compounds via single-electron transfer processes, although the demonstration of an active site-mediated redox process remains clouded by the chemistry that would be anticipated for an enzyme-mediated electron transfer process.

In a study of the rates of hydroxylation of monosubstituted (H, F, Cl, Br, and I) benzenes by phenobarbital-induced liver microsomes (which consisted primarily of the 2B1 isotype), a linear correlation of the k_{cat} for total oxidation with the σ^+ of the halogen substituent was observed.[213] A similar correlation was not observed for the individual rates of hydroxylation at the *ortho* and *para* positions, which underwent a monotonic change in *ortho/para* ratio as a function of the halide oxidizability. These data were interpreted in terms of a rate-dependent oxidation step involving either single-electron transfer (to the radical cation) followed by collapse at the *meta* position, or direct iron-oxy addition to the *meta* position. Subsequent collapse of the *meta*-σ-adduct, which must be dictated by the halide substituent, was proposed to lead to the observed *ortho/para* phenol product ratio. Thus, the product-determining step is postulated to be independent of the rate-determining step, and since σ^+ correlates with $E_{1/2}$, the rate-determining step was proposed to involve arene oxidation via single-electron transfer.

In another structure-activity relationship study utilizing liver microsomes from phenobarbital-induced rats (in which the major enzyme involved was P-450 2B1), the overall rate (k_{cat}) for oxidation of a series of methylated benzenes correlated with increasing methyl substitution (Macdonald and Plucinski, unpublished data). Both the substrate oxidation potential ($E_{1/2}$) and lipophilicity (Hansch π value) correlate with increasing arene methylation; although the correlation of rate with oxidation potential was better than with lipophilicity, separation of these parameters and unambiguous identification of the critical correlate(s) proved challenging. For all substrates, the predominant product(s) (> 95%) were the corresponding benzyl alcohols, often as isomeric mixtures if unsymmetrical polymethylated benzenes were utilized. In this study (as in the investigation of the rates of halobenzene metabolism), the redox parameter correlated better with the overall rate of substrate oxidation rather than with individual isomeric metabolites. Moreover, although phenol was the exclusive metabolite of benzene, the rate of benzene oxidation was slower than the rate of toluene oxidation, and consistent with the rate predicted based on its $E_{1/2}$ value. Although these data are subject to multiple possible interpretations, an analysis consistent with the emerging data is that for the P-450 2B1 enzyme, an enzyme with low substrate specificity, oxidation of arenes involves a rate-determining step sensitive to the $E_{1/2}$ of the π-system. That the correlation with $E_{1/2}$ remains despite the formation of different metabolites (isomeric phenols vs. isomeric benzyl alcohols) suggests that the redox-sensitive step may be electron transfer.

Insight into the detailed electronic mechanism of alkene and alkyne oxidation by P-450 has also been challenging. Current mechanistic rationale suggests that these monooxygenation processes involve either radicaloid or possibly electrophilic addition to the π-bond to produce a σ-complex which subsequently collapses to product (*vide infra*). Alkenyl and alkynyl α-hydroxylation is thought to be a facilitated alkane hydroxylation mechanism (due to resonance stabilization of the radical intermediate). Many of the arguments against the intermediacy of π-complexes or radical cations in the oxidation of π-bonded systems are based on the lack of correlation of the rates of oxidation with free energy parameters, such as substrate $E_{1/2}$ or Hammett σ values. The oxidation potential of P-450 has also been considered to be insufficient to oxidize alkenes and alkynes by single-electron transfer. However, since most of these π-systems possess redox potentials in the 1.6–2.5 V (vs. SCE) range, such systems could prove to be a "test" of the effective oxidation potential obtained from the studies described above. A confounding variable in obtaining free energy relationships for P-450-catalyzed oxidations of alkenes is the apparent sensitivity of the process to steric features of the alkene. Examples of P-450-catalyzed epoxidations of tetrasubstituted alkenes are limited (or none). Since the facility of oxidation, as assessed by free energy parameters (e.g., σ or σ$^+$) or redox potential ($E_{1/2}$), generally parallel increased alkene substitution (e.g., enhanced steric hindrance), the balance between steric and electronic features may well obscure an underlying free energy relationship.

In order to probe the possible intermediacy of an alkene radical cation in epoxidation and/or α-hydroxylation, the metabolism of hexamethyl Dewar benzene by P-450 2B1 was investigated. Hexamethyl Dewar benzene has an $E_{1/2}$ of 1.8 V (vs. SCE) and the derived radical cation undergoes rapid rearrangement to the hexamethylbenzene radical cation ($k_{rearr} > 10^8$ s^{-1}) (Scheme 31). In addition, our studies of the oxidation of hexamethylbenzene suggested that the observed metabolite (regardless of the intermediacy of the radical cation) would be the corresponding benzyl alcohol. If a radical cation is an intermediate in the oxidation of hexamethyl Dewar benzene by P-450, the rapid rate of this rearrangement might be able to compete with the "rebound" step of the $[FeO]^{2+}$ species to the oxidized alkene. In addition, in studies that appeared during the course of our investigations, Ortiz de Montellano and Stearns[130] reported that pentamethylbenzyl alcohol is a metabolite of hexamethyl Dewar benzene, although these authors ascribed the formation of this product to rearrangement of the radical intermediate in α-hydroxylation (Scheme 32). Also, an additional metabolite, proposed to be the primary alcohol, was noted. In our studies (Scheme 33) (Macdonald and Gorycki,

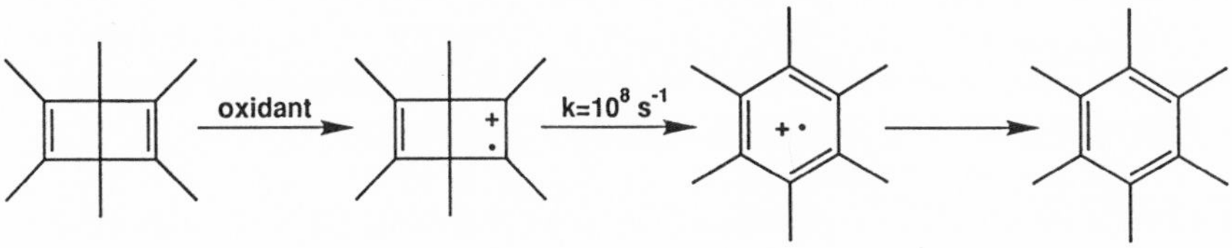

Scheme 31.

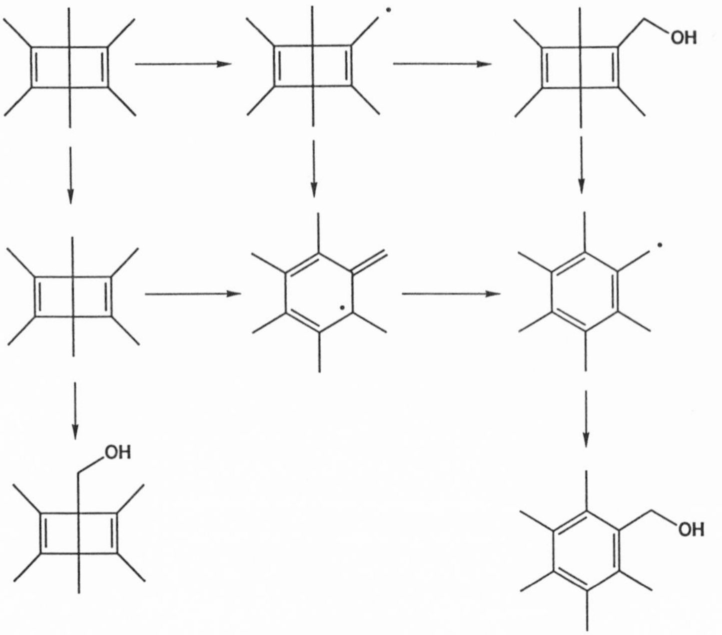

Scheme 32.

submitted for publication), only the allylic alcohol was observed; no pentamethyl-benzyl alcohol was found.

Although the differences between these studies are difficult to rationalize at present, both studies did not find the epoxide as a metabolite. Further studies of tetrasubstituted alkenes have confirmed that epoxides are not significant products. Although several ambiguities need to be resolved between the two studies, the lifetime must be exceedingly short ($<< 15 \times 10^{-9}$ s) of any putative alkene radical cation intermediate in either epoxidation or α-hydroxylation. However, related radical intermediates with fleeting lifetimes have been demonstrated in alkane hydroxylation through the use of the "radical clock" methodology and unambiguous dismissal of radical cation intermediates in alkene oxidation cannot at present be advanced. Finally, the relationship between steric factors and electronic factors in redirecting alkene epoxidation to the α-hydroxylation pathway needs to be resolved.

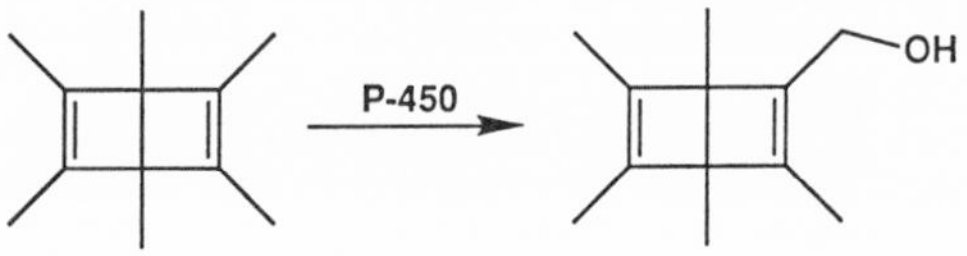

Scheme 33.

3.5. Oxidation of Alcohols

P-450s are known to catalyze the oxidations of many alcohols to the carbonyl oxidation state.[214] In addition, oxidation of aldehydes to carboxylic acids[215] and of *vic*-diol cleavage[45] are known. Several views have been presented regarding the oxidation of alcohols by P-450 enzymes. One possibility that has been considered is that "mobile" reactive oxygen species (such as OH• formed from O_2^{-} and H_2O_2 released in the normal mechanism, *vide supra*) are actually involved.[216] It is not easy to dismiss the possibility that such an entity is formed in the active site and is involved in some cases, but with more complex substrates observations regarding regio- and stereoselectivity the evidence favors a stepwise mechanism in which a high-valent iron complex is involved in abstraction and recombination reactions.[217]

The oxidation of alcohols can be formally described as a process involving abstraction of a carbinol hydrogen atom, oxygen rebound to yield a *gem*-diol, and spontaneous dehydration (Scheme 34). However, in the oxidation of (the alcohol) testosterone to (the ketone) androstenedione, [18]O from molecular oxygen is not incorporated into the product (Scheme 35).[218,219]

Surprisingly, when the α-alcohol epitestosterone is used as substrate (instead of the β-alcohol testosterone), then 84% theoretical enrichment from $^{18}O_2$ was observed.[219] Therefore a *gem*-diol intermediate does not appear likely for testosterone. A mechanism involving sequential abstraction of a hydrogen atom from C-17,

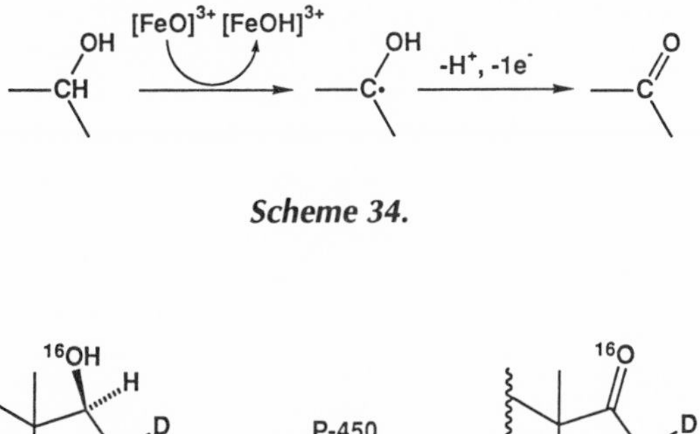

Scheme 34.

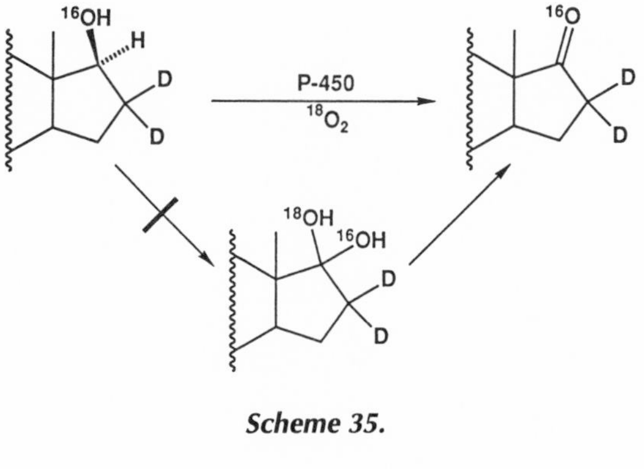

Scheme 35.

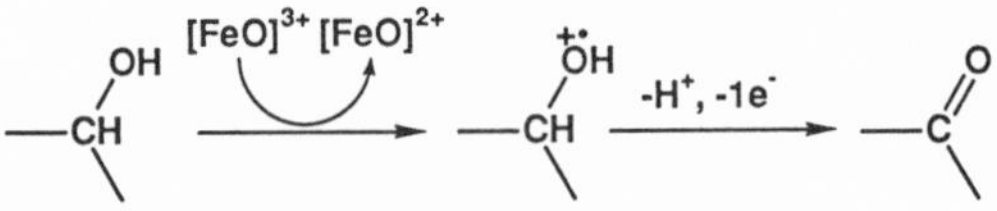

Scheme 36.

deprotonation, and loss of an electron has been proposed as a mechanism to account for these results.[219] Kinetic deuterium isotope effects of 2.4–3.4 have been offered as evidence for the proposed mechanism. However, noncompetitive kinetic isotope effect measurements were not reported; the size of the effect is not in itself sufficient evidence for the mechanism.

A viable alternative is that the initial reaction in the P-450-catalyzed oxidation of alcohols is the abstraction of an electron from the oxygen atom, followed by deprotonation, recombination, and dehydration (or deprotonation and formal hydrogen abstraction, Scheme 36). The $E_{1/2}$ values of alcohols (~ 2 V) are higher than those of nitrogens and sulfur atoms, but they are still low enough to be accessible for the high-valent iron complex. Indeed, alcohols have been known for many years to be susceptible to oxidation by catalase Compound I and peroxidases.[27] Biomimetic models also oxidize alcohols, as exemplified by the formation of formaldehyde by an iron(III) complex in CH_3OH.[220]

Acetaminophen is oxidized to both an iminoquinone and a catechol.[221] The forces involved in partitioning between the two products in different P-450 enzymes are not known, but one hypothesis is that abstraction of an electron from the phenolic oxygen atom leads to the catechol (after rehybridization and oxygen rebound to the ring), and that iminoquinone formation is initiated by initial abstraction (of a hydrogen atom ?) from the amide moiety (Scheme 37).[221] To date, only negative evidence has been obtained that an epoxide, gem-diol, or hydroxamic acid is formed in the oxidation of acetaminophen. It is highly likely that initial oxidation of the phenolic oxygen by electron transfer is involved in at least some of the oxidation reactions, and iminoquinone formation can also be catalyzed by peroxidases such as horseradish peroxidase[222,223] which appear to have only electron transfer capability.[72]

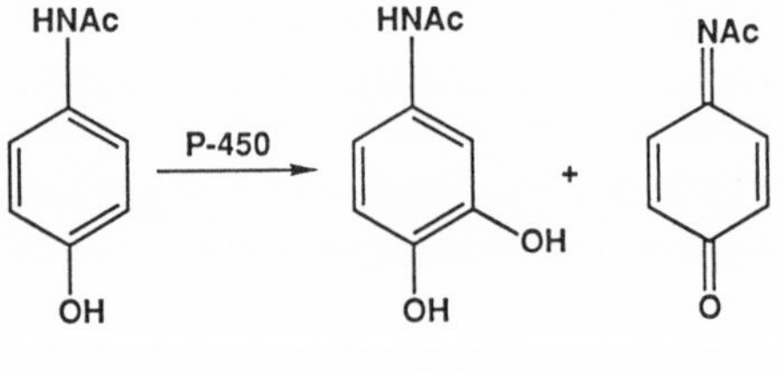

Scheme 37.

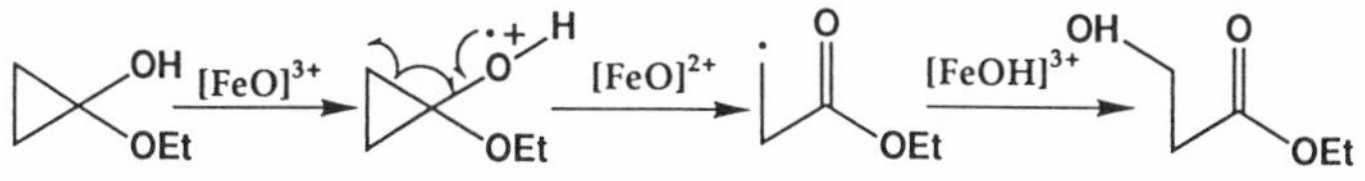

$$\text{OH} \quad \xrightarrow{[FeO]^{3+}} \quad \xrightarrow{[FeO]^{2+}} \quad \xrightarrow{[FeOH]^{3+}}$$

Scheme 38.

Another example of an alcohol oxidation apparently proceeding via electron transfer is provided by the formation of ethyl 3-hydroxypropionate from the monoethyl ether of cyclopropanone hydrate (Scheme 38).[128] The product, whose identity was confirmed using combined capillary gas chromatography/mass spectrometry and its electron impact fragmentation spectrum, was formed in the presence of rat P-450 2B1 and the usual system including NADPH-P-450 reductase, O_2, and NADPH. Further, the same reaction occurred when the starting material was mixed with oxidized mangano tetraphenylporphin.[128] In this system, there is little opportunity for oxidation at sites other than the three methylenes (and the methyl), and the observed product is best explained by a sequence involving electron abstraction, rearrangement, and radical recombination to the putative methylene radical.

4. SUBSTRATE REDUCTIONS

All of the P-450-catalyzed reactions discussed thus far have been oxidations. However, P-450 can also catalyze reductions as well, cycling between the Fe^{2+} and Fe^{3+} forms.[214] The preponderance of oxidation reactions is not surprising because, in the sequence of oxygen activation steps (see Scheme 2), ferrous P-450 binds O_2 rapidly and tightly. Reduction of a substrate, therefore, must involve (competitive) binding of the substrate in a manner analogous to molecular oxygen and with a redox potential sufficient to enable facile reductase-mediated reduction. Since the oxygen tension in cells of the liver and other tissues is rather low in some cases,[224] reductive reactions do proceed under some conditions. All of the reductions must be one-electron processes (Fe^{2+}/Fe^{3+} couple). Sometimes these apparently occur in sequence to yield what appears to be an overall two- (or even a four)-electron reduction; the pathways for the import of electrons into the heme and out are probably different and the substrate could stay attached while electrons flow in to it.

Nitro compounds may be reduced all the way to amines (Scheme 39). The reduction may stop at certain intermediates, and these intermediates may also serve as starting points.[225,226] In the case of aromatic nitro compounds, there is evidence that nitro anion radicals can leave the protein and be reoxidized by molecular oxygen, yielding superoxide anion.[227] Azo compounds are reduced to hydrazines.[228,229] It should be pointed out that these reactions are not unique to P-450s, and are catalyzed by several classes of redox-active enzymes.[229]

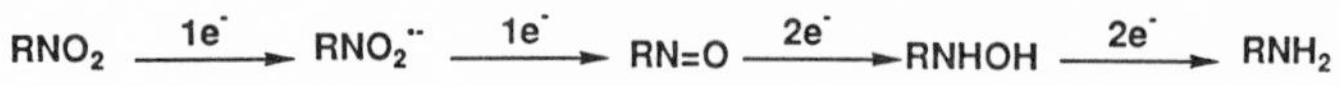

$$RNO_2 \xrightarrow{1e^-} RNO_2^{\cdot-} \xrightarrow{1e^-} RN=O \xrightarrow{2e^-} RNHOH \xrightarrow{2e^-} RNH_2$$

Scheme 39.

As in the case of hydroxylamines, tertiary amine oxides can be reduced to amines enzymatically.[230,231] The most feasible mechanism is shown in Scheme 4.[56] The reaction is acid catalyzed, and, as in the case of oxidations, involves radical cation intermediates. Normally, oxygen rebound to the carbon radical yields the carbinolamine (see Schemes 5, 6, 8, 13, 14, 16); here the radical is produced with the Fe^{3+} species accompanying it; further reduction would appear to require input of more electrons. It is not known if ferric P-450s are readily reduced by the aminium radicals (or by $R_2\overset{\cdot\cdot}{N}\overset{\cdot}{C}CHR$); while ferric P-450 will catalyze a very slow N-dealkylation of N,N-dimethylaniline N-oxide,[47] apparently the effect of electron donors on this reaction has never been investigated in a systematic manner. This reaction would regenerate the ferrous form of P-450 and yield the dealkylated product.

Arene oxides have been reported to be reduced to the parent (polycyclic) hydrocarbons by P-450s.[232–234] Little mechanistic information is available, and one can only surmise that the events may involve reversal of the sequence of epoxidation. However, metalloporphyrins tend to be poor Lewis acids toward epoxides,[235] and it is unclear what the fate of the oxygen atom is in such a reaction.

Reductive dehalogenation reactions catalyzed by P-450 have been studied extensively, primarily because of interest in these compounds as anesthetics, pesticides, and potentially toxic industrial solvents.[236] Halothane is defluorinated (Scheme 40); the bulk of evidence indicates that the products of oxidative transformation (i.e., acyl halides) are probably more relevant to toxic reactions.

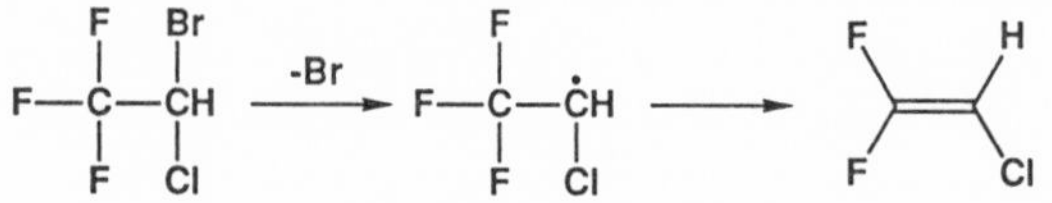

Scheme 40.

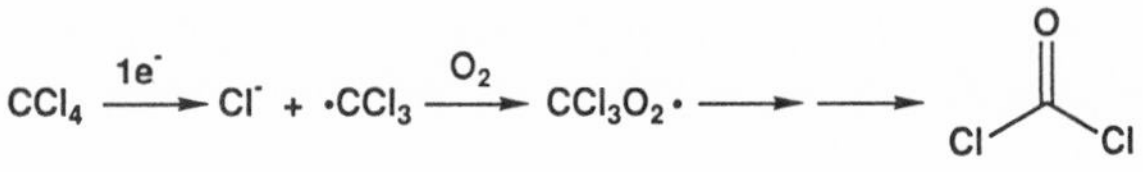

Scheme 41.

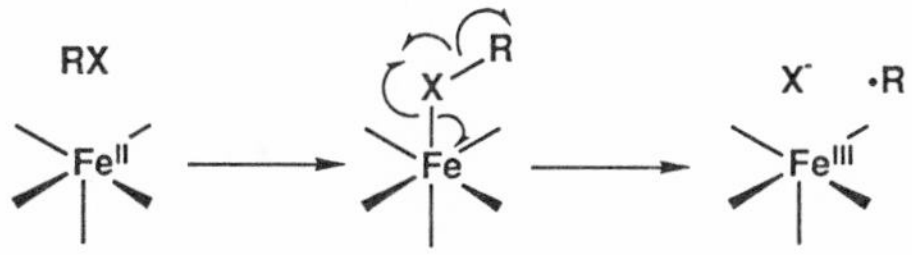

Scheme 39.

The metabolism of carbon tetrachloride has also been of interest, because the only conceivable oxidative reaction involves chloroso formation and is probably unlikely. The reaction appears to involve one-electron reduction of CCl_4 to yield chloride ion and the trichloromethyl radical, which can then react with oxygen (Scheme 41).[237] The oxygenated radical can break down to generate reactive phosgene and also an electrophilic chlorine atom. This appears to be the sequence most relevant to CCl_4 metabolism. Reduction of haloalkanes appears to be restricted to those bearing more than one (*gem*) constituents.[237] The mechanism has been considered by Castro and is postulated to involve inner-sphere reduction[238] (Scheme 42). Under some conditions further reduction of CCl_4 to a putative dichlorocarbene has been observed;[239–241] a tight ligand yielding an unusual heme spectrum is seen, but there is not much evidence that this is a predominant reaction in the overall biotransformation of haloalkanes such as CCl_4 and halothane.[238]

5. CONCLUSION

Much of the chemistry of P-450 catalysis occurs *via* sequential electron transfer processes. Both the reductive activation of molecular oxygen to the enzymatic active oxygen species and the oxidative transformations of substrates by this active species occur through single-electron transfer mechanisms. Because many of the initial intermediates produced through P-450-mediated substrate reduction generate radicals that can be characterized through classical techniques, P-450-catalyzed single-electron reductive transfer has long been recognized as a mechanistic mode for the enzyme. The unequivocal demonstration of substrate oxidation through single-electron transfer has been more challenging. Often the differentiation between sequential single-electron transfers and "concerted" two-electron transfer has been simply a function of the experimental capacity to "clock" the electron transfer processes. Recently a battery of techniques has confirmed the capacity of P-450 to oxidize by sequential single-electron transfer a diverse structural family of substrates containing sites with low oxidation potentials ($< \sim 1.5$ V *vs.* SCE). Future studies will be required to elucidate the extent of this mechanism in substrates with higher oxidation potentials.

NOTE ADDED IN PROOF

The reader is referred to recent studies on the role of base catalysis in alkylamine dealkylation[242] and the mechanism of amine oxygenation.[243]

REFERENCES

1. Guengerich, F. P. *J. Biol. Chem.* **1991**, *266*, 10019.
2. Porter, T. D.; Coon, M. J. *J. Biol. Chem.* **1991**, *266*, 13469.
3. Nebert, D. W.; Nelson, D. R.; Coon, M. J.; Estabrook, R. W.; Feyereisen, R.; Fujii-Kuriyama, Y.; Gonzalez, F. J.; Guengerich, F. P.; Gunsalus, I. C.; Johnson, E. F.; Loper, J. C.; Sato, R.; Waterman, M. R.; Waxman, D. J. *DNA Cell Biol.* **1991**, *10*, 1.
4. Hecker, M.; Ullrich, V. *J. Biol. Chem.* **1989**, *264*, 141.
5. Gopalakrishnan, S.; Harris, T. M.; Stone, M. P. *Biochemistry* **1990**, *29*, 10438.
6. Fisher, M. T.; Sligar, S. G. *J. Am. Chem. Soc.* **1985**, *107*, 5018.
7. Guengerich, F. P. *Biochemistry* **1983**, *22*, 2811.
8. Guengerich, F. P.; Ballou, D. P.; Coon, M. J. *Biochem. Biophys. Res. Commun.* **1976**, *70*, 951.
9. Oprian, D. D.; Gorsky, L. D.; Coon, M. J. *J. Biol. Chem.* **1983**, *258*, 8684.
10. Bonfils, C.; Debey, P.; Maurel, P. *Biochem. Biophys. Res. Commun.* **1979**, *88*, 1301.
11. Mansuy, D.; Bartoli, J. F.; Momenteau, M. *Tetrahedron Lett.* **1982**, *23*, 2781.
12. Blake, R. C., II; Coon, M. J. *J. Biol. Chem.* **1981**, *256*, 5755.
13. Blake, R. C., II; Coon, M. J. *J. Biol. Chem.* **1989**, *264*, 3694.
14. Larroque, C.; Lange, R.; Maurin, L.; Bienvenue, A.; van Lier, J. E. *Arch. Biochem. Biophys.* **1990**, *282*, 198.
15. Groves, J. T.; Kruper, W. J., Jr.; Haushalter, R. C. *J. Am. Chem. Soc.* **1980**, *102*, 6375.
16. Mansuy, D.; Battioni, P. *Frontiers of Biotransformation* (Ruckpaul, K.; Rein, H., Eds.). Taylor and Francis, London, 1989, Vol. 1, pp. 66–98.
17. McMurry, T. J.; Groves, J. T. *Cytochrome P-450* (Ortiz de Montellano, P. R., Ed.). Plenum Press, New York, 1986, pp. 1–28.
18. Strobel, H. W.; Coon, M. J. *J. Biol. Chem.* **1971**, *246*, 7826.
19. Sugimoto, H.; Tung, H. C.; Sawyer, D. T. *J. Am. Chem. Soc.* **1988**, *110*, 2465.
20. Goldblum, A.; Loew, G. H. *J. Am. Chem. Soc.* **1985**, *107*, 4265.
21. Ullrich, V. *Angew. Chem. Int. Ed. Eng.* **1972**, *11*, 701.
22. Gustafsson, J. Å.; Rondahl, L.; Bergman, J. *Biochemistry* **1979**, *18*, 865.
23. Kadlubar, F. F.; Morton, K. C.; Ziegler, D. M. *Biochem. Biophys. Res. Commun.* **1973**, *54*, 1255.
24. Lichtenberger, F.; Nastainczyk, W.; Ullrich, V. *Biochem. Biophys. Res. Commun.* **1976**, *70*, 939.
25. Groves, J. T.; McClusky, G. A. *J. Am. Chem. Soc.* **1976**, *98*, 859.
26. George, P. *J. Biol. Chem.* **1953**, *201*, 413.
27. Marnett, L. J.; Weller, P.; Battista, J. R. *Cytochrome P-450* (Ortiz de Montellano, P. R., Ed.). Plenum Press, New York, 1986, pp. 29–76.
28. McMahon, R. E.; Culp, H. W.; Occolowitz, J. C. *J. Am. Chem. Soc.* **1969**, *91*, 3389.
29. Ashley, P. L.; Griffin, B. W. *Mol. Pharmacol.* **1981**, *19*, 146.
30. Groves, J. T.; Nemo, T. E.; Myers, R. S. *J. Am. Chem. Soc.* **1979**, *101*, 1032.
31. Groves, J. T.; Watanabe, Y.; McMurry, T. J. *J. Am. Chem. Soc.* **1983**, *105*, 4489.
32. Takeuchi, K. J.; Busch, D. H. *J. Am. Chem. Soc.* **1981**, *103*, 2421.
33. Murugesan, N.; Ehrenfeld, G. M.; Hecht, S. M. *J. Biol. Chem.* **1982**, *257*, 8600.
34. Traylor, T. G.; Miksztal, A. R. *J. Am. Chem. Soc.* **1989**, *111*, 7443.
35. Lee, R. W.; Nakagaki, P. C.; Bruice, T. C. *J. Am. Chem. Soc.* **1989**, *111*, 1368.
36. Blackburn, N. J.; Pettingill, T. M.; Seagraves, K. S.; Shigeta, R. T. *J. Biol. Chem.* **1990**, *265*, 15383.

37. Kimura, E.; Machida, R. *J. Chem. Soc., Chem. Commun.* **1984**, 499.
38. Higuchi, T.; Uzu, S.; Hirobe, M. *J. Am. Chem. Soc.* **1990**, *112*, 7051.
39. Stassinopoulos, A.; Caradonna, J. P. *J. Am. Chem. Soc.* **1990**, *112*, 7071.
40. Lee, W. A.; Calderwood, T. S.; Bruice, T. C. *Proc. Natl. Acad. Sci. USA* **1985**, *82*, 4301.
41. Groves, J. T.; Gilbert, J. A. *Inorg. Chem.* **1986**, *25*, 123.
42. Hayashi, Y.; Yamazaki, I. *J. Biol. Chem.* **1979**, *254*, 9101.
43. Dolphin, D.; Felton, R. H. *Acc. Chem. Res.* **1974**, *7*, 26.
44. Erman, J. E.; Vitello, L. B.; Mauro, J. M.; Kraut, J. *Biochemistry* **1989**, *28*, 7992.
45. Ortiz de Montellano, P. R. *Cytochrome P-450* (Ortiz de Montellano, P. R., Ed.). Plenum Press, New York, 1986, pp. 217–271.
46. Ortiz de Montellano, P. R. *Trends Pharmacol. Sci.* **1989**, *10*, 354.
47. Burka, L. T.; Guengerich, F. P.; Willard, R. J.; Macdonald, T. L. *J. Am. Chem. Soc.* **1985**, *107*, 2549.
48. Macdonald, T. L.; Gutheim, W. G.; Martin, R. B.; Guengerich, F. P. *Biochemistry* **1989**, *28*, 2071.
49. Guengerich, F. P.; Böcker, R. H. *J. Biol. Chem.* **1988**, *263*, 8168.
50. Guengerich, F. P. *Chem. Res. Toxicol.* **1990**, *3*, 21.
51. Guengerich, F. P.; Peterson, L. A.; Böcker, R. H. *J. Biol. Chem.* **1988**, *263*, 8176.
52. Mueller, G. C.; Miller, J. A. *J. Biol. Chem.* **1953**, *202*, 579.
53. Axelrod, J. *J. Biol. Chem.* **1955**, *214*, 753.
54. Oae, S.; Kitao, T.; Kawamura, S. *Tetrahedron* **1963**, *19*, 1783.
55. Craig, J. C.; Dwyer, F. P.; Glazer, A. N.; Horning, E. C. *J. Am. Chem. Soc.* **1961**, *83*, 1871.
56. Ferris, J. P.; Gerwe, R. D.; Gapski, G. R. *J. Org. Chem.* **1968**, *33*, 3493.
57. Kurebayashi, H. *Arch. Biochem. Biophys.* **1989**, *270*, 320.
58. Parli, C. J.; Wang, N.; McMahon, R. E. *Biochem. Biophys. Res. Commun.* **1971**, *43*, 1204.
59. McMahon, R. E.; Culp, H. W.; Craig, J. C.; Ekwuribe, N. *J. Med. Chem.* **1979**, *22*, 1100.
60. Brodie, B. B.; Gillette, J. R.; LaDu, B. N. *Ann. Rev. Biochem.* **1958**, *27*, 427.
61. Shea, J. P.; Valentine, G. L.; Nelson, S. D. *Biochem. Biophys. Res. Commun.* **1982**, *109*, 231.
62. Kedderis, G. L.; Dwyer, L. A.; Rickert, D. E.; Hollenberg, P. F. *Mol. Pharmacol.* **1983**, *758*, 760.
63. Walsh, C. *Enzymatic Reaction Mechanisms.* W. H. Freeman, San Francisco, 1979.
64. Ortiz de Montellano, P. R.; Beilan, H. S.; Kunze, K. L. *J. Biol. Chem.* **1981**, *256*, 6708.
65. Augusto, O.; Beilan, H. S.; Ortiz de Montellano, P. R. *J. Biol. Chem.* **1982**, *257*, 11288.
66. Guengerich, F. P.; Martin, M. V.; Beaune, P. H.; Kremers, P.; Wolff, T.; Waxman, D. J. *J. Biol. Chem.* **1986**, *261*, 5051.
67. Böcker, R. H.; Guengerich, F. P. *J. Med. Chem.* **1986**, *29*, 1596.
68. Guengerich, F. P.; Brian, W. R.; Iwasaki, M.; Sari, M-A.; Bäärnhielm, C.; Berntsson, P. *J. Med. Chem.* **1991**, *34*, 1838.
69. Lee, J. S.; Jacobsen, N. E.; Ortiz de Montellano, P. R. *Biochemistry* **1988**, *27*, 7703.
70. Born, J. L.; Hadley, W. M. *Chem. Res. Toxicol.* **1989**, *2*, 57.
71. Kennedy, C. H.; Mason, R. P. *J. Biol. Chem.* **1990**, *265*, 11425.
72. Ortiz de Montellano, P. R. *Acc. Chem. Res.* **1987**, *20*, 289.
73. Bäärnhielm, C.; Hansson, G. *Biochem. Pharmacol.* **1986**, *35*, 1419.
74. Maeda, Y.; Ingold, K. U. *J. Am. Chem. Soc.* **1980**, *102*, 328.
75. Silverman, R. B.; Hiebert, C. K. *Biochemistry* **1988**, *27*, 8448.
76. Hanzlik, R. P.; Kishore, V.; Tullman, R. *J. Med. Chem.* **1979**, *22*, 759.
77. Macdonald, T. L.; Zirvi, K.; Burka, L. T.; Peyman, P.; Guengerich, F. P. *J. Am. Chem. Soc.* **1982**, *104*, 2050.
78. Hanzlik, R. P.; Tullman, R. H. *J. Am. Chem. Soc.* **1982**, *104*, 2048.
79. Silverman, R. B.; Zieske, P. A. *Biochemistry* **1986**, *25*, 341.
80. Bondon, A.; Macdonald, T. L.; Harris, T. M.; Guengerich, F. P. *J. Biol. Chem.* **1989**, *264*, 1988.
81. Wiseman, J. S.; Nichols, J. S.; Kolpak, M. X. *J. Biol. Chem.* **1982**, *257*, 6328.
82. Guengerich, F. P. *Biochem. Biophys. Res. Commun.* **1986**, *138*, 193.

83. Osawa, Y.; Pohl, L. R. *Chem. Res. Toxicol.* **1989**, *2*, 131.
84. Kadlubar, F. F.; Hammons, G. J. *Mammalian Cytochromes P-450* (Guengerich, F. P., Ed.). CRC Press, Boca Raton, FL, 1987; Vol. 2, pp. 81–130.
85. Ziegler, D. M. *Drug Metab. Rev.* **1988**, *19*, 1.
86. Ziegler, D. M.; Pettit, F. H. *Biochem. Biophys. Res. Commun.* **1964**, *15*, 188.
87. Spiteller, M.; Spiteller, G. *Biological Oxidation of Nitrogen* (Gorrod, J. W., Ed.). Elsevier, North Holland, 1978, pp. 109–112.
88. Guengerich, F. P. *J. Med. Chem.* **1984**, *27*, 1101.
89. Cashman, J. R. *Mol. Pharmacol.* **1989**, *36*, 497.
90. Cope, A. C.; LeBel, N. A. *J. Am. Chem. Soc.* **1960**, *82*, 4656.
91. Poulsen, L. L.; Kadlubar, F. F.; Ziegler, D. M. *Arch. Biochem. Biophys.* **1974**, *164*, 774.
92. Guengerich, F. P.; Macdonald, T. L. *Acc. Chem. Res.* **1984**, *17*, 9.
93. Guengerich, F. P.; Müller-Enoch, D.; Blair, I. A. *Mol. Pharmacol.* **1986**, *30*, 287.
94. Prough, R. A.; Brown, M. I.; Dannan, G. A.; Guengerich, F. P. *Cancer Res.* **1984**, *44*, 543.
95. Bowry, V. W.; Ingold, K. U. *J. Am. Chem. Soc.* **1991**, *113*, 5699.
96. Hammerich, O.; Parker, V. D. *Advances in Physical Organic Chemistry* (Gold, V.; Bethell, D., Eds.). Academic Press, London, 1984, pp. 55–148.
97. Bordwell, F. G.; Cheng, J. P.; Bausch, M. J. *J. Am. Chem. Soc.* **1988**, *110*, 2867.
98. Dinnocenzo, J. P.; Banach, T. E. *J. Am. Chem. Soc.* **1989**, *111*, 8646.
99. Nelsen, S. F.; Ippoliti, J. T. *J. Am. Chem. Soc.* **1986**, *108*, 4879.
100. Williams, D. E.; Reed, R. L.; Kedzierski, B.; Guengerich, F. P.; Buhler, D. C. *Drug Metab. Disp.* **1989**, *17*, 387.
101. Miranda, C. L.; Reed, R. L.; Guengerich, F. P.; Buhler, D. R. *Carcinogenesis* **1991**, *12*, 515.
102. Mattocks, A. R.; Bird, I. *Toxicol. Lett.* **1983**, *16*, 1.
103. Baba, T.; Yamada, H.; Oguri, K.; Yoshimura, H. *Xenobiotica* **1988**, *18*, 475.
104. Mansuy, D.; Battioni, P.; Chottard, J. C.; Riche, C.; Chiaroni, A. *J. Am. Chem. Soc.* **1983**, *105*, 455.
105. Burstyn, J. N.; Iskandar, M.; Brady, J. F.; Fukuto, J. M.; Cho, A. K. *Chem. Res. Toxicol.* **1991**, *4*, 70.
106. Hammons, G. J.; Guengerich, F. P.; Weis, C. C.; Beland, F. A.; Kadlubar, F. F. *Cancer Res.* **1985**, *45*, 3578.
107. Thompson, J. A.; Holtzman, J. L. *Drug Metab. Disp.* **1974**, *2*, 577.
108. Nelson, S. D.; Pohl, L. R.; Trager, W. F. *J. Med. Chem.* **1975**, *18*, 1062.
109. Shea, J. P.; Nelson, S. D.; Ford, G. P. *J. Am. Chem. Soc.* **1983**, *105*, 5451.
110. Hall, L. R.; Hanzlik, R. P. *J. Biol. Chem.* **1990**, *265*, 12349.
111. Miwa, G. T.; Walsh, J. S.; Kedderis, G. L.; Hollenberg, P. F. *J. Biol. Chem.* **1983**, *258*, 14445.
112. Ottoboni, S.; Carlson, T. J.; Trager, W. F.; Castagnoli, K.; Castagnoli, N., Jr. *Chem. Res. Toxicol.* **1990**, *3*, 423.
113. Guengerich, F. P.; Macdonald, T. L. *FASEB J.* **1990**, *4*, 2453.
114. Wei, M. M.; Stewart, R. *J. Am. Chem. Soc.* **1966**, *88*, 1974.
115. Groves, J. T.; McClusky, G. A.; White, R. E.; Coon, M. J. *Biochem. Biophys. Res. Commun.* **1978**, *81*, 154.
116. Hjelmeland, L. M.; Aronow, L.; Trudell, J. R. *Biochem. Biophys. Res. Commun.* **1977**, *76*, 541.
117. Miwa, G. T.; Walsh, J. S.; Lu, A. Y. H. *J. Biol. Chem.* **1984**, *259*, 3000.
118. Harada, N.; Miwa, G. T.; Walsh, J. S.; Lu, A. Y. H. *J. Biol. Chem.* **1984**, *259*, 3005.
119. Shono, T.; Toda, T.; Oshino, N. *J. Am. Chem. Soc.* **1982**, *104*, 2639.
120. Hull, L. A.; Davis, G. T.; Rosenblatt, D. H.; Williams, H. K. R.; Weglein, R. C. *J. Am. Chem. Soc.* **1967**, *89*, 1163.
121. Rosenblatt, D. H.; Davis, G. T.; Hull, L. A.; Forberg, G. D. *J. Org. Chem.* **1968**, *33*, 1649.
122. Lindsay Smith, J. R.; Mead, L. A. V. *J. Chem. Soc. Perkin Trans. II* **1973**, 206.
123. Sinha, A.; Bruice, T. C. *J. Am. Chem. Soc.* **1984**, *106*, 7291.

124. Powell, M. F.; Bruice, T. C. *Oxidases and Related Redox Systems*. Alan R. Liss, New York, 1988, pp. 369–385.
125. Lewis, F. D.; Ho, T. I. *J. Am. Chem. Soc.* **1980**, *102*, 1751.
126. Guengerich, F. P. *J. Biol. Chem.* **1987**, *262*, 8459.
127. Funck-Brentano, C.; Kroemer, H. K.; Pavlou, H.; Woosley, R. L.; Roden, D. M. *Br. J. clin. Pharmacol.* **1989**, *27*, 435.
128. Guengerich, F. P.; Willard, R. J.; Shea, J. P.; Richards, L. E.; Macdonald, T. L. *J. Am. Chem. Soc.* **1984**, *106*, 6446.
129. Guengerich, F. P. *J. Biol. Chem.* **1989**, *264*, 17198.
130. Ortiz de Montellano, P. R.; Stearns, R. A. *J. Am. Chem. Soc.* **1987**, *109*, 3415.
131. Northrop, D. B. *Methods Enzymol.* **1982**, *87*, 607.
132. Jones, J. P.; Korzekwa, K. R.; Rettie, A. E.; Trager, W. F. *J. Am. Chem. Soc.* **1986**, *108*, 7074.
133. Ortiz de Montellano, P. R.; Choe, Y. S.; DePillis, G.; Catalano, C. E. *J. Biol. Chem.* **1987**, *262*, 11641.
134. Ottoboni, S.; Caldera, P.; Trevor, A.; Castagnoli, N., Jr. *J. Biol. Chem.* **1989**, *264*, 13684.
135. Silverman, R. B. *Biochem. Soc. Trans.* **1991**, *19*, 201.
136. Yelekci, K.; Lu, X.; Silverman, R. B. *J. Am. Chem. Soc.* **1989**, *111*, 1138.
137 Vazquez, M. L.; Silverman, R. B. *Biochemistry* **1985**, *24*, 6538.
138. Manring, L. E.; Peters, K. S. *J. Am. Chem. Soc.* **1983**, *105*, 5708.
139. Nicholas, A. M. deP.; Arnold, D. R. *Canad. J. Chem.* **1982**, *60*, 2165.
140. Padgette, S. R.; Wimalasena, K.; Herman, H. H.; Sirimanne, S. R.; May, S. W. *Biochemistry* **1985**, *24*, 5826.
141. Katopodis, A. G.; Wimalasena, K.; Lee, J.; May, S. W. *J. Am. Chem. Soc.* **1984**, *106*, 7928.
142. Guengerich, F. P. *Crit. Rev. Biochem. Mol. Biol.* **1990**, *25*, 97.
143. Poulos, T. P. *Pharmaceut. Res.* **1988**, *5*, 67.
144. Galliani, G.; Nali, M.; Rindone, B.; Tollari, S.; Rocchetti, M.; Salmona, M. *Xenobiotica* **1986**, *16*, 511.
145. Galliani, G.; Rindone, B.; Dagnino, G.; Salmona, M. *Eur. J. Drug Metab. Pharmacokin.* **1984**, *9*, 289.
146. Lindsay Smith, J. R.; Audeh, C. A. *J. Chem. Soc. (B)* **1971**, 1741.
147. Davis, G. T.; Demek, M. M.; Rosenblatt, D. H. *J. Am. Chem. Soc.* **1972**, *94*, 3321.
148. Lindsay Smith, J. R.; Mortimer, D. N. *J. Chem. Soc., Chem. Commun.* **1985**, 64.
149. Watanabe, Y.; Iyanagi, T.; Oae, S. *Tetrahedron Lett.* **1982**, *23*, 533.
150. watanabe, Y.; Oae, S.; Iyanagi, T. *Bull. Chem. Soc. Jpn.* **1982**, *55*, 188.
151. Watanabe, Y.; Numata, T.; Iyanagi, T.; Oae, S. *Bull. Chem. Soc. Jpn.* **1981**, *54*, 1163.
152. Watanabe, Y.; Iyanagi, T.; Oae, S. *Tetrahedron Lett.* **1980**, *21*, 3685.
153. Gorsky, L. D.; Koop, D. R.; Coon, M. J. *J. Biol. Chem.* **1984**, *259*, 6812.
154. Sakurada, J.; Sekiguchi, R.; Sato, K.; Hosoya, T. *Biochemistry* **1990**, *29*, 4093.
155. Thanabal, V.; La Mar, G. N.; de Ropp, J. S. *Biochemistry* **1988**, *27*, 5400.
156. Oae, S.; Asada, K.; Yoshimura, T. *Tetrahedron Lett.* **1983**, *24*, 1265.
157. Oae, S.; Mikami, A.; Matsuura, T.; Ogawa-Asada, K.; Watanabe, Y.; Fujimori, K.; Iyanagi, T. *Biochem. Biophys. Res. Commun.* **1985**, *131*, 567.
158. Marcus, R. A.; Sutin, N. *Biochim. Biophys. Acta* **1985**, *811*, 265.
159. Moore, G. R.; Pettigrew, G. W.; Rogers, N. K. *Proc. Natl. Acad. Sci. USA* **1986**, *83*, 4998.
160. McLendon, G. *Acc. Chem. Res.* **1988**, *21*, 160.
161. Murdoch, J. *J. Am. Chem. Soc.* **1983**, *105*, 2159.
162. Eberson, L. *Acta Chem. Scand.* **1990**, *44*, 733.
163. Eberson, L. *Prog. Phys. Org. Chem.* **1981**, *18*, 79.
164. Meyer, T. E.; Przysiecki, C. T.; Watkins, J. A.; Bhattacharyya, A.; Simondsen, R. P.; Cusanovich, M. A.; Tollin, G. *Proc. Natl. Acad. Sci. USA* **1983**, *80*, 6740.
165. Eberson, L. *Acta Chem. Scand.* **1982**, *36*, 533.

166. Miller, J. R.; Calcaterra, L. T.; Closs, G. L. *J. Am. Chem. Soc.* **1984**, *106*, 3047.
167. Ballardini, R.; Varani, G.; Indelli, M. T.; Scandola, T.; Balzani, V. *J. Am. Chem. Soc.* **1978**, *100*, 7219.
168. Bock, C. R.; Connor, J. A.; Gutierrez, A. R.; Meyer, T. J.; Whitten, G.; Sullivan, B. P.; Nagle, J. K. *J. Am. Chem. Soc.* **1979**, *101*, 1455.
169. Chung, A.; Weiss, R.; Warshel, A.; Takano, T. *J. Phys. Chem.* **1983**, *87*, 1683.
170. McClendon, G.; Miller, J. R. *J. Am. Chem. Soc.* **1985**, *107*, 7811.
171. Cheung, E.; Taylor, K.; Kornblatt, J. A.; English, A. M.; McLendon, G.; Miller, J. R. *Proc. Natl. Acad. Sci. USA* **1986**, *83*, 1330.
172. Peterson-Kennedy, S. E.; McGourty, J. L.; Hoffman, B. M. *J. Am. Chem. Soc.* **1984**, *106*, 5010.
173. McLendon, G.; Guarr, T.; McGuire, M.; Simolo, K.; Stranch, S.; Taylor, K. *Coor. Chem. Rev.* **1985**, *64*, 113.
174. Peterson-Kennedy, S. E.; McGourty, J. L.; Ho, P. S.; Sutoris, C. J.; Liang, N.; Zemal, H.; Blough, N. V.; Margoliash, E.; Hoffman, B. M. *Coor. Chem. Rev.* **1985**, *64*, 125.
175. Bortolini, O.; Meunier, B. *J. Chem. Soc., Chem. Commun.* **1983**, 1364.
176. Eberson, L. *Adv. Free Radical Biol. Med.* **1985**, *1*, 19.
177. Poulos, T. L.; Finzel, B. C.; Howard, A. J. *J. Mol. Biol.* **1987**, *195*, 687.
178. Gilson, M. K.; Honig, B. H. *Biopolymers* **1986**, *25*, 2097.
179. Gilson, M. K.; Honig, B. H. *Nature* **1987**, *330*, 84.
180. Warshel, A.; Russell, S. T. *Q. Rev. Biophys.* **1984**, *17*, 283.
181. Warshel, A. *Nature* **1987**, *330*, 15.
182. Sternberg, M. J. E.; Hayes, F. R. F.; Russell, A. J.; Thomas, P. G.; Gersht, A. R. *Nature* **1987**, *330*, 86.
183. Dawson, J. H.; Sono, M. *Chem. Rev.* **1987**, *87*, 1255.
184. Dawson, J. H. *Science* **1988**, *240*, 433.
185. Imai, M.; Shimada, H.; Watanabe, Y.; Matsushima-Hibiya, Y.; Makino, R.; Koga, H.; Horiuchi, T.; Ishimura, Y. *Proc. Natl. Acad. Sci. USA* **1989**, *86*, 7823.
186. Neal, R. A.; Halpert, J. *Ann. Rev. Pharmacol. Toxicol.* **1982**, *22*, 321.
187. Smyser, B. P.; Levi, P. E.; Hodgson, E. *Biochem. Pharmacol.* **1986**, *35*, 1719.
188. Chin, D. H.; La Mar, G. N.; Balch, A. L. *J. Am. Chem. Soc.* **1980**, *102*, 5945.
189. Shapiro, S.; Arunachalam, T.; Caspi, E. *J. Am. Chem. Soc.* **1983**, *105*, 1642.
190. White, R. E.; Miller, J. P.; Favreau, L. V.; Bhattacharyya, A. *J. Am. Chem. Soc.* **1986**, *108*, 6024.
191. Gelb, M. H.; Heimbrook, D. C.; Mälkönen, P.; Sligar, S. G. *Biochemistry* **1982**, *21*, 370.
192. Groves, J. T.; Subramanian, D. V. *J. Am. Chem. Soc.* **1984**, *106*, 2177.
193. McClanahan, R. H.; Huitric, A. C.; Pearson, P. G.; Desper, J. C.; Nelson, S. D. *J. Am. Chem. Soc.* **1988**, *110*, 1979.
194. Wimalasena, K.; May, S. W. *J. Am. Chem. Soc.* **1987**, *109*, 4036.
195. Fitzpatrick, P. F.; Villafranca, J. J. *J. Am. Chem. Soc.* **1985**, *107*, 5022.
196. Stearns, R. A.; Ortiz de Montellano, P. R. *J. Am. Chem. Soc.* **1985**, *107*, 4081.
197. Lashmet Johnson, P. R.; Ziegler, D. M. *J. Biochem. Toxicol.* **1986**, *1*, 15.
198. Rettie, A. E.; Rettenmeier, A. W.; Howald, W. N.; Baillie, T. A. *Science* **1987**, *235*, 890.
199. Nagata, K.; Liberato, D. J.; Gillette, J. R.; Sasame, H. A. *Drug Metab. Dispos.* **1986**, *14*, 559.
200. Wang, R. W.; Kari, P. H.; Lu, A. Y. H.; Thomas, P. E.; Guengerich, F. P.; Vyas, K. P. *Arch. Biochem. Biophys.* **1991**, *290*, 355.
201. Guengerich, F. P.; Kim, D-H. *Chem. Res. Toxicol.* **1991**, *4*, 413.
202. Chadwick, R. W.; Chuang, L. T.; Williams, K. *Pest. Biochem. Physiol.* **1975**, *5*, 575.
203. Hata, S.; Nishino, T.; Katsuki, H.; Aoyama, Y.; Yoshida, Y. *Biochem. Biophys. Res. Commun.* **1983**, *116*, 162.
204. Kaminsky, L. S.; Fasco, M. J.; Guengerich, F. P. *J. Biol. Chem.* **1980**, *255*, 85.
205. Fujiwara, Y.; Okayasu, T.; Ishibashi, T.; Imai, Y. *Biochem. Biophys. Res. Commun.* **1983**, *110*, 36.
206. Korzekawa, K. R.; Jones, J. P.; Gillette, J. R. *J. Am. Chem. Soc.* **1990**, *112*, 7042.

207. Burka, L. T.; Thorsen, A.; Guengerich, F. P. *J. Am. Chem. Soc.* **1980**, *102*, 7615.
208. Fontana, A.; Dalzoppo, D.; Grandi, C.; Zambonin, M. *Methods Enzymol.* **1983**, *91*, 311.
209. Rogan, E. G.; Cavalieri, E. L.; Tibbels, S. R.; Cremonesi, P.; Warner, C. D.; Nagel, D. L.; Tomer, K. B.; Cerny, R. L.; Gross, M. L. *J. Am. Chem. Soc.* **1988**, *110*, 4023.
210. Cavalieri, E. L.; Rogan, E. G.; Devanesan, P. D.; Cremonesi, P.; Cerny, R. L.; Gross, M. L.; Bodell, W. J. *Biochemistry* **1990**, *29*, 4820.
211. Rogan, E. G.; RamaKrishna, N. V. S.; Higginbotham, S.; Cavalieri, E. L.; Jeong, H.; Jankowiak, R.; Small, G. J. *Chem. Res. Toxicol.* **1990**, *3*, 441.
212. Lehr, R. E.; Kumar, S.; Levin, W.; Wood, A. W.; Chang, R. L.; Conney, A. H.; Yagi, H.; Sayer, J. M.; Jerina, D. M. *Polycyclic Hydrocarbons and Carcinogenesis*. American Chemical Society, Washington, D.C., 1985, pp. 63–84.
213. Burka, L. T.; Plucinski, T. M.; Macdonald, T. L. *Proc. Natl. Acad. Sci. USA* **1983**, *80*, 6680.
214. Wislocki, P. G.; Miwa, G. T.; Lu, A. Y. H. *Enzymatic Basis of Detoxication* (Jakoby, W. B., Ed.). Academic Press, New York, 1980, Vol. 1, pp. 135–182.
215. Watanabe, K.; Narimatsu, S.; Yamamoto, I.; Yoshimura, H. *J. Biol. Chem.* **1991**, *266*, 2709.
216. Ingelman-Sundberg, M.; Hagbjörk, A. L. *Xenobiotica* **1982**, *12*, 673.
217. Gorsky, L. D.; Coon, M. J. *Drug Metab. Dispos.* **1985**, *13*, 169.
218. Cheng, K. C.; Schenkman, J. B. *J. Biol. Chem.* **1983**, *258*, 11738.
219. Wood, A. W.; Swinney, D. C.; Thomas, P. E.; Ryan, D. E.; Hall, P. F.; Levin, W.; Garland, W. A. *J. Biol. Chem.* **1988**, *263*, 17322.
220. Bell, S. E. J.; Cooke, P. R.; Inchley, P.; Leanord, D. R.; Smith, J. R. L.; Robbins, A. *J. Chem. Soc. Perkin. Trans. II* **1991**, 549.
221. Harvison, P. J.; Guengerich, F. P.; Rashed, M. S.; Nelson, S. D. *Chem. Res. Toxicol.* **1988**, *1*, 47.
222. Nelson, S. D.; Dahlin, D. C.; Rauckman, E. J.; Rosen, G. M. *Mol. Pharmacol.* **1981**, *20*, 195.
223. Potter, D. W.; Hinson, J. A. *J. Biol. Chem.* **1987**, *262*, 966.
224. Wu, Y-R.; Kauffman, F. C.; Qu, W.; Ganey, P.; Thurman, R. G. *Mol. Pharmacol.* **1990**, *38*, 128.
225. Kato, R.; Oshima, T.; Takanaka, A. *Mol. Pharmacol.* **1969**, *5*, 487.
226. Sum, C. Y.; Cho, A. K. *Drug Metab. Disp.* **1976**, *4*, 436.
227. Mason, R. P.; Holtzman, J. L. *Biochemistry* **1975**, *14*, 1626.
228. Hernandez, P. H.; Mazel, P.; Gillette, J. R. *Biochem. Pharmacol.* **1967**, *16*, 1877.
229. Fujita, S.; Peisach, J. *J. Biol. Chem.* **1978**, *253*, 4512.
230. Sugiura, M.; Iwasaki, K.; Kato, R. *Mol. Pharmacol.* **1976**, *12*, 322.
231. Iwasaki, K.; Noguchi, H.; Kato, R.; Imai, Y.; Sato, R. *Biochem. Biophys. Res. Commun.* **1977**, *77*, 1143.
232. Booth, J.; Hewer, A.; Keysell, G. R.; Sims, P. *Xenobiotica* **1975**, *5*, 197.
233. Yamazoe, Y.; Sugiura, M.; Kamataki, T.; Kato, R. *FEBS Lett.* **1978**, *88*, 337.
234. Kato, R.; Iwasaki, K.; Shiraga, T.; Noguchi, H. *Biochem. Biophys. Res. Commun.* **1976**, *70*, 681.
235. Liebler, D. C.; Guengerich, F. P. *Biochemistry* **1983**, *22*, 5482.
236. Van Dyke, R. A.; Gandolfi, A. J. *Drug Metab. Disp.* **1976**, *4*, 40.
237. Mico, B. A.; Pohl, L. R. *Arch. Biochem. Biophys.* **1983**, *225*, 596.
238. Castro, C. E.; Wade, R. S.; Belser, N. O. *Biochemistry* **1985**, *24*, 204.
239. Ahr, H. J.; King, L. J.; Nastainczyk, W.; Ullrich, V. *Biochem. Pharmacol.* **1982**, *31*, 383.
240. Wolf, C. R.; Mansuy, D.; Nastainczyk, W.; Deutschmann, G.; Ullrich, V. *Mol. Pharmacol.* **1977**, *13*, 698.
241. Mansuy, D.; Fontecave, M. *Biochem. Pharmacol.* **1983**, *32*, 1871.
242. Okazaki, O.; Guengerich, F. P. *J. Biol. Chem.* **1993**, *268*, 1546.
243. Seto, Y.; Guengerich, F. P. *J. Biol. Chem.* **1993**, *268*, 9986.

INDEX

Advances in
Electron Transfer Chemistry

Edited by **Patrick S. Mariano,** *Department of Chemistry and Biochemistry, University of Maryland, College Park*

Coverage in this series will focus on chemical and biochemical aspects of electron transfer chemistry. Recognition over the past decade that a wide variety of chemical processes operate by single electron transfer mechanisms has stimulated numerous efforts in this area. These range from (1) theoretical and experimental investigations of the rates of electron transfer in donor-acceptor systems, (2) studies of photo electron transfer reactions, (3) exploratory efforts probing electron transfer mechanisms for traditional nucleophilic substitution and addition processes, and (4) investigations of electron transfer mechanisms which operate in biochemical processes.

Advances in Electron Transfer Chemistry will cover topics in the recently developed and important areas. The coverage will span the broad areas of organic, physical, inorganic, and biological chemistry. Each of the contributions will be written on a level to make them understandable for graduate students and workers in the chemical and biochemical sciences, and will emphasize recent work of the contributing authors.

Volume 1, 1991, 197 pp. $90.25
ISBN 1-55938-167-1

J A I   P R E S S

JAI PRESS

Volume 2, 1992, 286 pp. $90.25
ISBN 1-55938-168-X

JAI PRESS INC.

55 Old Post Road # 2 - P.O. Box 1678
Greenwich, Connecticut 06836-1678
Tel: (203) 661-7602   Fax:(203) 661-0792

Advances in Supramolecular Chemistry

Edited by **George W. Gokel,** *Department of Chemistry, University of Miami*

Volume 1, 1990, 197 pp. $90.25
ISBN 1-55938-181-7

Volume 2, 1992, 195 pp. $90.25
ISBN 1-55938-329-1

Volume 3, 1993, 219 pp. $90.25
ISBN 1-55938-546-4

Advances in Cycloaddition

Edited by **Dennis P. Curran,** *Department of Chemistry, University of Pittsburgh*

REVIEW: "This volume is highly recommended to all those who want to stay abreast of developments in the mechanisms and synthetic applications of 1,3-dipolar cycloaddition reactions. The writers have realized a good balance between the summary of achievements and the reporting of gaps in understanding or remaining synthetic challenges. The articles are well written, they are amply illustrated with equations or schemes."

- Journal of the American Chemical Society

Volume 1, 1988, 208 pp.　　　　　　　　　　　　　$90.25
ISBN 0-89232-861-4

Volume 2, 1990, 220 pp.　　　　　　　　　　　　　$90.25
ISBN 0-89232-951-3

Volume 3, 1993, 210 pp. $90.25
ISBN 1-55938-319-4

CONTENTS: π-**Facial Diastereoselection in Diels-Alder Cycloadditions and Related Reactions: Understanding Planar Interactions and Establishing Synthetic Potential,** *A. G. Fallis and Yee-Fung Lu, Ottawa Carleton Chemistry Institute, University of Ottawa.* **Substituent and Structural Effects in the Ozonolysis of Cyclic Vinylogous Esters,** *W. H. Bunnelle, University of Missouri-Columbia.* **N-Metalated Azomethine Ylides,** *S. Kanemasa, Kyushu University and Otohiko Tsuge, Kumamoto Institute of Technology.* **Azomethine Ylide Cycloadditions via 1,2- Prototropy and Metallo-Dipole Formation from Imines,** *R. Grigg and V. Sridharan.*

JAI PRESS INC.

55 Old Post Road # 2 - P.O. Box 1678
Greenwich, Connecticut 06836-1678
Tel: (203) 661-7602   Fax:(203) 661-0792

JAI PRESS

Im
Blc